Security Informatics and Law Enforcement

Series editor
Babak Akhgar
CENTRIC (Centre of Excellence in Terrorism, Resilience,
Intelligence and Organised Crime Research),
Sheffield Hallam University
Sheffield, UK

Editor's Note:

The primary objective of this book series is to explore contemporary issues related to law enforcement agencies, security services and industries dealing with security related challenges (e.g., government organizations, financial sector insurance companies and internet service providers) from an engineering and computer science perspective. Each book in the series provides a handbook style practical guide to one of the following security challenges:

Cyber Crime - Focuses on new and evolving forms of crimes. Books describe the current status of cybercrime and cyber terrorism developments, security requirements and practices.

Big Data Analytics, Situational Awareness and OSINT- Provides unique insight for computer scientists as well as practitioners in security and policing domains on big data possibilities and challenges for the security domain, current and best practices as well as recommendations.

Serious Games – Provides an introduction into the use of serious games for training in the security domain, including advise for designers/programmers, trainers and strategic decision makers.

Social Media in Crisis Management – explores how social media enables citizens to empower themselves during a crisis, from terrorism, public disorder, and natural disasters

Law enforcement, Counterterrorism, and Anti-Trafficking – Presents tools from those designing the computing and engineering techniques, architecture or policies related to applications confronting radicalisation, terrorism, and trafficking.

The books pertain to engineers working in law enforcement and researchers who are researching on capabilities of LEAs, though the series is truly multidisciplinary – each book will have hard core computer science, application of ICT in security and security / policing domain chapters. The books strike a balance between theory and practice.

More information about this series at
http://www.springer.com/series/15902

Babak Akhgar
Dimitrios Kavallieros • Evangelos Sdongos
Editors

Technology Development for Security Practitioners

 Springer

Editors
Babak Akhgar
CENTRIC (Centre of Excellence
in Terrorism, Resilience, Intelligence
and Organised Crime Research)
Sheffield Hallam University
Sheffield, UK

Dimitrios Kavallieros
Center for Security Studies-KEMEA
Athens, Greece

Evangelos Sdongos
Crisis Management and Security Unit
Institute of Communication
and Computer
Athens, Greece

ISSN 2523-8507   ISSN 2523-8515   (electronic)
Security Informatics and Law Enforcement
ISBN 978-3-030-69462-3   ISBN 978-3-030-69460-9   (eBook)
https://doi.org/10.1007/978-3-030-69460-9

This Springer imprint is published by the registered company Springer Nature
Switzerland AG
The registered company address is: Gewerbestrasse 11, 6330 Cham, Switzerland

PREFACE

MSE2019 was initiated by the Mediterranean Network of Security Practitioners (MEDEA), a network supported by the EC Horizon 2020 program of DG HOME in the General Matters (GM) line of topics under the coordination of the Center for Security Studies (KEMEA) of the Greek Ministry of Citizen Protection. The event was organized with the support and cooperation of 36 ongoing European R&D projects, funded by the European Commission. Furthermore, MSE2019 runs under the auspices of the Commission's Directorate-General for Home Affairs (DG HOME), the Community of Users (CoU) for Research in the Field of Security, the European Association of Research and Technology Organisations (EARTO), and the European Network of Technology Services of Law Enforcement Agencies (ENLETS).

The Mediterranean Security Event (MSE2019) has been a first attempt for strengthening the interaction and exchange of knowledge between the various communities of the EU security stakeholders including practitioners, academia, researchers, and industry representatives.

The scientific excellence and the progress made by the European research in various topics of homeland security presented in context of MSE2019 is captured in this volume. This volume presents knowledge gained and knowledge applied within security domains from various applied research projects.

The Mediterranean Security Event aims to become the largest security forum, collaboratively organized by the community of security R&D stakeholders in the European Union. The success of MSE2019 engages the R&D community in the field of security to strengthen their synergy with the practitioners and policy makers and keep the momentum going.

Athens, Greece   Georgios Eftychidis

CONTENTS

PART I

Cyber Crime, Cyber Terrorism and Cyber Security

CHAPTER 1

ASGARD: A Novel Approach for Collaboration in Security Research Projects

Juan Arraiza, Esther Novo, Seán Gaines,
Aitor García Pablos, and Haizea Erostarbe

1.1 Introduction

Research on security at European level is a field that took greater importance until it became one of the key thematic areas of the Seventh Framework Programme (FP7) and then one of the seven societal challenges of the Horizon 2020 Programme (FP8).

In 2014, a study conducted from Czech security research funding programmes concluded, among other things, that "RTOs tend to propose projects with results which are achievable, yet without regard to their usefulness in practice; they rarely initiate voluntary involvement of the end user (except single cases); and their proposals tend to limit their ambition to results comprehendible by the RTO without much regard to the

J. Arraiza (✉) · E. Novo · S. Gaines · A. G. Pablos · H. Erostarbe
Vicomtech, Donostia, San Sebastian, Spain
e-mail: jarraiza@vicomtech.org; enovo@vicomtech.org; sgaines@vicomtech.org;
agarciap@vicomtech.org; herostarbe@vicomtech.org

B. Akhgar et al. (eds.), *Technology Development for Security Practitioners*, Security Informatics and Law Enforcement,
https://doi.org/10.1007/978-3-030-69460-9_1

specifics, limits and boundaries of its future use in law enforcement practices" [1].

Also, in 2014 the European Commission, Directorate-General Enterprise and Industry commissioned a study on the final evaluation of Security Research under the FP7 [2]. The research methodology included desk research, statistical analysis of data from the CORDA database, surveys of participants and end-users, stakeholder interviews, a series of case studies and a stakeholder workshop. The last two conclusions and recommendations of this study were to "further buttress the role of end-users in all phases of Security Research Actions" on one hand and to "do more to maximise the benefits derived from the FP7 Security Research Programme and to reduce the tendency for insights and tools produced within projects to be left behind as partners move on to new projects or other priorities" on the other hand [2, p. 80]. In addition to those two key aspects, another important issue was identified as a major problem, which was that the law enforcement agencies' (LEA) expectations were not being appropriately met.

The LEAs and other security practitioners participating in FP7 security projects were normally treated as "customers", who were there to define their requirements at the beginning of the project and to evaluate the results during the final trials or demonstrations. In many occasions, they were not even members of the Consortia, but instead, they were participating as members of the projects' advisory boards. Most of the LEAs were not familiar with Research and Development, and by the end of the projects, they were expecting fully operational solutions that they could use right away, even when the research projects were targeting technology readiness levels (TRL) of "5" Technology validated in relevant environment, "6" Technology demonstrated in relevant environment, or "7" System demonstrated in relevant environment.[1] The vast majority of FP7 and then H2020 projects were never targeting TRL levels beyond "7" ("8" System complete and qualified or "9" Actual system proven in operational environment). Showing that there is still the need to improve how security project results deliver results that meet end-user needs, the main theme of the 2019 Security Research Event has been "Building Bridges:

---

[1] Technology Readiness Levels (Horizon 2020 work programme) – https://ec.europa.eu/research/participants/data/ref/h2020/wp/2014_2015/annexes/h2020-wp1415-annex-g-trl_en.pdf

Promoting Market Uptake by Reinforcing Synergies Between Security Research and Other Funding Instruments" [3].

Back in the second half of 2013, a group of project coordinators and other principals of FP7 security research projects[2] met and decided they wanted to change some of the fundamental things on how this type of projects were being implemented. After some meetings and discussions, they agreed that they were going to collaborate in the preparation of a new project proposal designed to tackle precisely the same two key aspects that were soon afterwards included as conclusions and recommendations of the aforementioned final evaluation of security research under FP7.

To strengthen the role of end-users in security research, the new project proposal was going to adopt and adapt to the specific needs of the field of security research several of the open-source model principles and practices, which were in use in other domains such as ICT, but not yet in the field of European security research. Among these principles and practices are decentralization, open collaboration, peer-review production and iterative and incremental full-development life cycles.

To reduce the tendency for insights and tools produced within projects to be left behind as partners move on to new projects or other priorities, the new project proposal decided that the main goal of the project was going to be "to support LEA Technological Autonomy by building a sustainable, long-lasting community formed by LEAs, Researchers and Industry that will create (at little or no cost to LEAs), maintain and evolve a best of class tool set for the extraction, fusion, exchange and analysis of Big Data including cyber-offenses data for forensic investigation". The idea therefore was to create a sustainable and long-lasting "restricted" community composed by mutually trusted actors which was going to adapt and adopt open-source model principles and practices to conduct in a more efficient way European security research projects.

To appropriately manage the expectations of the LEA partners, the project's work plan was structured so that they were able to evaluate intermediate results several times before the end of the project, so that their feedback after each of the short full-development cycles could be used to re-adjust the scope and work plan of the subsequent ones.

In summary, based on their experience, the leaders of the new proposal designed the project under the hypothesis that a Consortium of LEA, Researchers and Industry, closely collaborating in a security research

[2] I.e., SAVASA, VALCRI, EPOOLICE, RECOBIA, VOXPOL, CAPER and VIRTUOSO.

project, following open-source model principles and practices, including iterative and incremental full development life cycles, and with the goal of building a sustainable and long-lasting restricted community was, on one hand, going to strengthen the role of end users in security research, it was on the other hand going to help market uptake of security research project results, and that it was also going to help setting better LEA expectations.

This chapter presents the H2020 ASGARD project, which is the project that was presented and that won the FCT-01-2015 topic, scoring 14.5 out of 15, that was funded with 12M Euro, and that started in September 2016 and is scheduled to end at the end of February 2020. This project has been identified by the European Commission as a success story that has built best practices which should be followed by future research projects [4, p. 24].

The rest of the chapter is structured as follows. In the "Related Work" section a brief state-of-the-art study and a brief description on how many, if not most, of the security research projects were executed at the time when the ASGARD project was being defined is presented. In the "Methods" section, the ASGARD project is presented as a case study. In the "Results" section, the findings of the case study are presented. Finally, in the "Discussion and Future Research" section, an analysis and explanation of the results of the research conducted, some of its limitations and some ideas for future research are presented.

1.2   RELATED WORK

During Framework Programmes 6 (2002–2006), the participation of European law enforcement agencies in security research projects was small or rare that in general it could be considered as insignificant. The Framework Programme 7 (2007–2013) was a significant step forward in the recognition of the importance of this research field, and security became one of the key thematic areas of the programme. However, at the end of the Framework Programme 7, only a few European LEAs were participating in research projects. Out of the 320 projects listed in FP7 Projects section of the Horizon Dashboard[3] under the "SECURITY" thematic priority, the authors have only been able to identify 64 projects including at least one law enforcement agency or other relevant security

---

[3] https://webgate.ec.europa.eu/dashboard/sense/app/eaf1621c-67cc-4972-a07b-dddba31815c1

practitioners that are public legal entities. Based on these results, 80% of the projects under the "SECURITY" thematic priority were not including any relevant public security practitioners.

A 2011 study from the Harvard's Executive Session on Policing and Public Safety calls for "a shift in ownership of police science from the universities to police agencies" [5, p. 1]. This study states that such ownership would facilitate the implementation of evidence-based practices and policies in policing and would change the fundamental relationship between research and practice.

A 2014 study for the LIBE Committee of the European Parliament also revealed that technological tools and services cannot be developed without a thorough legal, social and political assessment, in order to determine their impact and effects, and it anticipated that funded security research in the future was mainly going to be put at the service of industry rather than society [6].

The role of LEAs participating in FP7 projects was mainly focused on requirements gathering phase and on participating on evaluations or demonstrations conducted at the end of the projects. In most cases of the 20% of the projects that included relevant public security practitioners, case apart from a few exceptions, the collaboration between research technology organisations, industry and LEAs was superficial, and the LEAs were not integrated into the project teams at the same level as the rest of their partners. The FP7 final evaluation report states that "End-users are thought to have constituted a significant minority of the organisations participating directly in FP7 Security Research projects, and are known to have also been engaged with the programme through other routes such as project advisory boards and dissemination events" [2, p. 118].

A trust relationship is one of the basic elements of any efficient collaboration. Most likely due to the nature of their work, LEAs had the tendency to follow the security through obscurity paradigm, and therefore the exchange of information between them and their partners in security research projects was limited. This issue was even more exacerbated by the intrinsic characteristics of European research projects, which include heterogeneous partners from multiple countries.

In addition, legitimate interests but different from different types of partners were not managed in the most appropriate way possible. It was not rare that LEAs hoped that the research projects would provide them with operational solutions, if not during the execution of the project, at least at the end of the projects. The report on Final Evaluation of Security

Research under the Seventh Framework Programme for Research established the following conclusions about the forms of end-users involvement in projects: "the development of outputs that don't always correspond to end-users' needs and requirements, and so cannot be immediately deployed in operational environments" and "the only way for projects to deliver outputs that are fit for purpose, immediately deployable at operational level, and that can contribute to solving real-life needs, is to actively involve them throughout the entire process of the preparation, management and review/evaluation of the Security Research programme" [2, p. 124–125].

During the definition of the proposals for new projects, many LEAs were asking for solutions that they needed at that time, not considering the times and deadlines that the research and development cycle entails. A proposal takes months to be prepared, then the evaluation process takes a few more months, then, if your proposal has been successful, the grant preparation also takes a few extra months. Once the grant has been signed and the project starts, the duration can easily be of 3 or more years. And the end results of the project are normally at technology readiness levels of 5–7; these are prototypes tested and/or demonstrated in laboratories or in operational environments but not final products and services that can be commercialized straight away. The whole process could normally last 4–6 years, and by the time the projects were finished, the results were in most cases still not ready to be used by end-users in their real cases. The (in many cases unrealistic) expectations of LEAs were not being met, and their level of frustration and dissatisfaction started to grow.

On the other hand, many industrial partners were using the research projects as part of their overall technological research and development pipeline process, being the technology itself their main interest, and not the pursue of developing new products for the law enforcement agencies. The technological results of the security research projects could serve as foundations for developing products and services in other (more profitable) markets. Besides, the fragmentation of the security market and the existence of a few global providers, often non-European, discouraged SMEs' investments oriented towards crossing the innovation "valley of death".

And with regard to research and technology organisations, their main interest was in many cases focused on producing scientific results that go beyond the state of the art and publishing those results in scientific journals and conferences, as those are the things that boost the careers of successful researchers.

Therefore, in the twilight of the FP7 programme, there was a clear misalignment between the interests, and in consequently the efforts and focus, of the different types of partners. In the context of decentralised, distributed, multidisciplinary and collaborative actions such as the European security research projects, this misalignment was as lethal as a torpedo in the waterline of the project.

In the twilight of the FP7 programme, a group of project coordinators and other principals of several security projects met and agreed that they wanted to tackle those known problems. They agreed to jointly prepare a proposal for a new project which should aim at building trust among LEAs, research technology organisations and industry, should also build upon best practices and lessons learnt from those previous projects and should do its best to deliver results which could be valuable to LEAs as soon as possible.

1.3 METHODS

The main paradigm followed has been constructivism, which proclaims that reality is not discovered, but that it is constructed [7]. In this way, it is not intended to measure or control the real world but to know about it and to rebuild it in the most reliable way possible. The authors understand that knowledge is socially constructed by the participants in the research process and the research itself is not alien to the values of the researcher. The authors, as understood in the constructivist paradigm, believe that there is no single and (pre)determined reality but constructions that respond to the individual perception of each participant in the phenomenon. Therefore, the different interpretations of the phenomenon studied by a representative sample of the individuals participating in it have been studied.

The questions that this chapter will try to answer are: "How was the ASGARD project designed, how has it been executed, and why is it considered a success story?"

To answer these questions, the method chosen was a case study. According to the following definition for a case study, "a case study is an empirical inquiry that investigates a contemporary phenomenon within its real-life context, especially when the boundaries between phenomenon and context are not clearly evident" [8, p. 13]. This research method was chosen because the authors deliberately believed that contextual conditions could be highly pertinent to the phenomenon of study and, therefore, they wanted to cover them.

To build trust among LEAs, research technology organisations and industry, to build upon best practices and lessons learnt from previous related projects, and to deliver results which were of value to LEAs as soon as possible, the project was designed following open-source model principles.

Framework Programme security research projects are decentralised, distributed and collaborative endeavours, by nature. Also, the members of the Consortia are heterogeneous. Under these conditions, the open-source model principles of open collaboration, peer production, decentralised and iterative and incremental full-development cycles fit nicely. These principles, and the associated best practices, methods and techniques, were already common practice in other domains (such as information and communications technology – ICT) by the time the ASGARD project was designed, but they were not being applied in European security research projects.

At operational level, the management structure of the project gives task and work package leaders a great deal of flexibility to organise and coordinate the work within their work packages. Project coordination focused mainly on inter-work packages dependencies and issues.

The project's work plan was structured with a 3-month duration ramp-up phase, six full-development cycles of 6 months duration each and a final ramp-down phase of 3 months. This structure provides the flexibility needed as it allows adjusting the scope and the detailed plan of each of the stages as/if needed. Figure 1.1 below describes this agile approach. This flexibility in the design of the work plan allows keeping track of the evolution of the expectations from the relevant stakeholders (both internal and external to the Consortium) and making changes to the plan to try to meet them as much as possible whilst respecting the terms of the contractual agreement with the European Commission (EC).

But not only the expectations of relevant stakeholders are to be considered when thinking about changes to the plan of the subsequent project periods, it is also the continuous improvement of the processes of the project that matters. For this, it is important to conduct regular self-reflection of the project processes to identify, reduce and eliminate suboptimal processes and to strengthen those that are considered best practices.

In ASGARD, this was achieved by jointly conducting a lesson-learned exercise by all the project partners at the end of each of the project stages. This exercise aims at identifying what went right, what when wrong, which are the best practices that should be further implemented and improved,

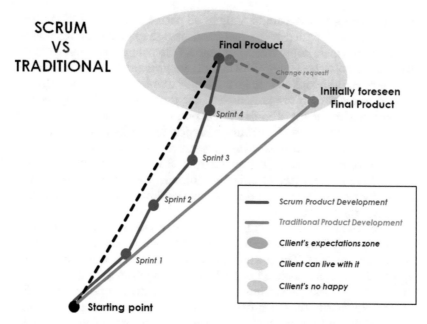

Fig. 1.1 How an agile approach allows adjusting the work plan to monitor and meet as much as possible the expectations of the stakeholders (source: "Scrum VS Traditional", Jorge Abad, http://www.lecciones-aprendidas.info/2016/07/Scrum-vs-traditional.html)

and which are the things that should be changed or avoided because they did not deliver the results that were desired.

Therefore, throughout the project, the Consortium can agree to adjust the scope and the detailed work plan for the subsequent periods based on the feedback that is collected and jointly assessed on regular basis (in the case of the ASGARD project, every 6 months).

The ramp-up phase served mainly to build the team, promptly launch requirements gathering and system specification and architecture design work streams, so that early drafts of all these deliverables could be produced to feed the first full-development cycle starting in month four of the project.

The first full-development cycle served to build the process, integrate several background technologies and allow conducting the first "hackathon" event on month nine of the project, at which these background technologies were evaluated jointly by all partners. The ASGARD "hack-athons" were designed to be hands-on workshops for experimentation. These events team computer programmers, domain experts and users collaborate intensively. The ASGARD tools are presented and made available to set of multidisciplinary teams. Then a number of scenarios that are related to the project are presented to the teams so that they try to build a solution with the list of tools and datasets available to address those scenarios. The "hackathons" encourage participants to form ad hoc multidisciplinary teams, brainstorm ideas, implement and present a demo from which a winner is picked by popular vote. But most importantly, "hackathons" aid in team (trust) building, efficient peer collaboration and re-setting or adjusting project priorities and deadlines based on the feedback collected during the events.

At the end of each of the full-development cycles, self-assessment audits are conducted. These audits are based on anonymous feedback assessment by project partners. The purpose of the audits is team development and appraisal of project goals. They facilitate communication and team development within the consortium by providing feedback on partner performance. Furthermore, the process enables the partners to participate in goal setting and ensure they are married to the project goals, provide motivation to the consortium by demonstrating the participative nature of the project management process, provide clarity on the definition of project goals and most importantly make the communication and coordination processes of the project more effective and agile.

The audits are simple 360-degree feedback process in order to foster open and frank discussion on the progress of the project and performance of partners and project principals. The expected benefit of the processes is to formalize periods of reflection after distinct phases of the project life cycle so that inefficiencies and conflict can be identified and the appropriate measures adopted. The purpose of this review style is not intended to drive or maximize performance in the project but rather to ensure the expectations of partners and other stakeholders are reasonable and achievable to meet project goals. The audit's questionnaire is structured around the goals of the project period and the core values and ambitions of the project purpose. As a standing agenda item at project meetings, the governance body of the project assesses the results of these audits and takes appropriate action. As a consequence, a plan of action that addresses corrective actions, that re-inforces the strengths shown in the previous project period, and that works to remove the barriers identified to achieve the project goals is defined and implemented.

Face-to-face meetings, evaluation forms, and other means have been used to collect the feedback from other relevant project stakeholders such as European Commission officers or members of the project's Stakeholder Advisory Group (SAG).

1.4   RESULTS

Up until now, six detailed project level (inter-work packages) work plans have been produced during the project, one for the ramp-up phase and one per each of the full-development cycles. These detailed work plans are discussed and agreed at the beginning of each of the periods, and they focus on the period at hand, maintaining the rest of the project periods at high level only. As described in Sect. 1.3, these detailed work plans were produced considering the overall project plan, the feedback collected and jointly discussed from the self-assessment audits and the outcome of the lesson-learned exercises conducted at the end of each of the stages. As an example, note how item #1 of the lesson-learned exercise conducted at the end of the fifth hackathon (Fig. 1.2) was included as part of the detailed plan of action for the subsequent period (Fig. 1.3) and even explicitly added to the plan as task #1.

As described in Sect. 1.3, the main tool to collect feedback from the members of the Consortium, and to measure their level of satisfaction and motivation, were the self-assessment audits. The audits consist on a set of 13 questions. All questions allow respondents to make comments, whilst questions 1–8 include in addition five-level Likert scale questions for specific statements. This is the meaning of each of the five level Likert questions:

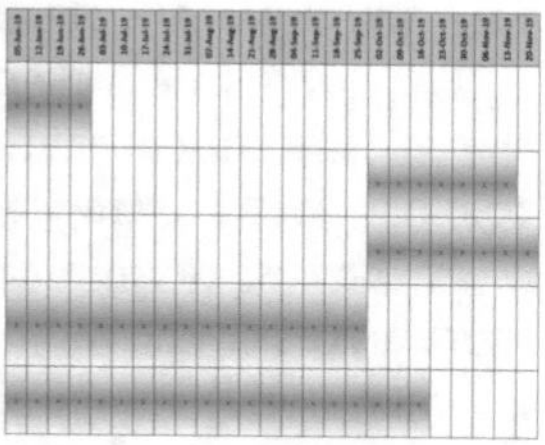

Fig. 1.2   Sample extract of one of the lesson-learned exercises

Fig. 1.3   Sample extract of one of the detailed work plans

1. Insufficient
2. Barely sufficient
3. Sufficient
4. Good
5. Very good

Figure 1.4 presents an anonymized extract of the self-assessment summary corresponding to month 39 of the project. This extract excludes the comments, and it includes the average score obtained in questions 1–8 of the previous self-assessment audits.

For example, in the case of question 1, it can be observed that there were 32 responses, scoring an average of 4.63. The minimum score obtained was 4 and the maximum 5, being the variance quite low (0.242). Also, in comparison to the previous audit (month M33), this question obtained a better score, 0.26 points higher. When looking to all the average scores obtained for question 1 in all previous audits, it can be observed that the minimum score obtained was precisely on the previous audit (month M33) and the maximum score was 4.68 in month M21; therefore, the variance obtained throughout all periods was also low (0.017).

For question 4, one of the key roles has got an average score across all the self-assessment audits in the range of 3.61–4.10, whilst another key role has got an average score in the range of 4.42–4.73.

In the case of question 5, one of the work packages has got an average score across all the self-assessment audits in the range of 3.23–3.77, whilst another work package has got average scores in the range of 4.27–4.50. The average score for all work packages and all audits is 3.93.

ASGARD project >> M39 Self-Assessment Audit

Responses: 33 out of 33 partners

QUESTION	AVERAGE	# responses	Min	Max	VAR	M6	M15	M21	M27	M33	M39	Diff M39<>M33
Q1 Coordination	4,63	32	4	5	0,242	4,44	4,45	4,68	4,39	4,36	4,63	0,26
Q2 Communication	4,24	33	3	5	0,314	4,27	4,23	4,39	4,05	4,03	4,24	0,21
Q3 Conflict	4,47	19	4	5	0,263	4,30	4,33	4,47	4,33	4,50	4,47	-0,03
Project Coordination	4,73	33	4	5	0,20	4,42	4,58	4,74	4,56	4,69	4,73	0,04
Technical Management	3,61	33	1	5	0,81	3,71	3,94	4,10	3,75	3,71	3,61	-0,11
Administrative Management	4,50	32	3	5	0,39	4,41	4,45	4,74	4,44	4,54	4,50	-0,04
Quality management	4,00	31	3	5	0,47	4,27	4,15	4,38	4,00	4,08	4,00	-0,08
Q5 Work packages (comments)		4				#¡DIV/0!						
WPxx	4,17	24	3	5	0,49	3,64	4,00	4,41	4,16	4,05	4,17	0,11
WPxx	3,87	23	2	5	0,57	3,52	4,07	4,27	3,92	3,83	3,87	0,04
WPxx	3,67	21	3	5	0,43	3,50	3,71	3,50	3,82	3,76	3,67	-0,10
WPxx	3,95	21	3	5	0,35	4,20	4,00	4,18	3,60	3,97	3,95	-0,02
WPxx	3,85	20	3	5	0,34	4,45	4,13	4,36	3,78	3,87	3,85	-0,02
WPxx	3,92	12	2	5	0,81	4,25	4,07	4,27	4,00	4,08	3,92	-0,17
WPxx	3,40	15	2	5	0,69	3,65	3,77	3,64	3,23	3,40	3,40	0,00
WPxx	3,57	14	2	5	0,88	3,50	3,92	4,11	3,64	3,71	3,57	-0,14
WPxx	4,23	13	3	5	0,36	3,72	4,07	4,10	3,92	4,17	4,23	0,06
WPxx	4,32	19	3	5	0,34	4,42	4,35	4,50	4,27	4,42	4,32	-0,10
WPxx	3,95	19	3	5	0,50	4,14	3,85	4,20	4,00	3,93	3,95	0,02
WPxx	3,75	20	2	5	0,51	3,50	3,65	3,56	3,62	4,00	3,75	-0,25
Q6 Partners		0										
ACME	4,65	26	4	5	0,24	N/A	N/A	4,80	4,61	4,58	4,65	0,07
ACME	4,00	1	4	4	#¡DIV/0!	4,10	3,00	3,57	3,25	4,00	4,00	0,00
ACME	3,67	12	2	5	0,79	4,07	4,00	3,97	3,36	3,71	3,67	-0,04
ACME	4,00	2	4	4	0,00	3,86	4,20	3,50	3,86	3,80	4,00	0,20
ACME	4,00	1	4	4	#¡DIV/0!	3,50	3,00	3,75	4,00	#¡DIV/0!	4,00	#¡DIV/0!
ACME	4,25	4	3	5	0,92	3,94	4,20	4,30	3,86	4,25	4,25	0,00
ACME	4,29	14	3	5	0,37	4,08	4,07	4,18	3,82	4,13	4,29	0,16
ACME	3,86	7	2	5	0,81	3,47	3,90	4,17	3,88	4,33	3,86	-0,48
ACME	4,53	15	4	5	0,27	4,31	4,07	4,27	4,18	4,73	4,53	-0,19
ACME	3,88	8	3	5	0,41	4,07	3,70	3,75	3,40	3,75	3,88	0,13
ACME	4,27	15	3	5	0,35	4,36	4,06	4,08	4,00	4,10	4,27	0,17
ACME	4,07	15	3	5	0,64	3,70	3,80	3,91	4,07	3,92	4,07	0,14
ACME	4,00	5	4	4	0,00	4,09	4,21	4,31	3,91	4,17	4,00	-0,17
ACME	4,64	14	3	5	0,40	3,85	4,44	4,74	4,43	4,56	4,64	0,08
ACME	3,50	2	3	4	0,50	4,15	3,33	3,61	3,00	3,00	3,50	0,50
ACME	5,00	1	5	5	#¡DIV/0!	3,58	2,00	2,75	3,50	5,00	5,00	0,00
ACME	#¡DIV/0!	0	0	0	#¡DIV/0!	3,25	3,00	3,00	3,50	4,00	#¡DIV/0!	#¡DIV/0!
ACME	#¡DIV/0!	0	0	0	#¡DIV/0!	4,31	4,00	3,80	4,00	4,00	#¡DIV/0!	#¡DIV/0!
ACME	4,50	4	3	5	1,00	4,00	4,43	4,11	3,83	4,50	4,50	0,00
ACME	4,29	7	4	5	0,24	4,07	4,00	4,33	4,00	4,00	4,29	0,29
ACME	4,00	1	4	4	#¡DIV/0!	3,58	3,00	3,75	3,75	4,33	4,00	-0,33
ACME	3,81	16	2	5	0,70	3,85	3,90	3,90	3,77	3,86	3,81	-0,04
ACME	4,26	23	2	5	0,66	4,68	4,30	4,43	4,31	4,50	4,26	-0,24
ACME	4,67	6	4	5	0,27	4,06	4,50	4,11	4,50	4,67	4,67	0,17
ACME	4,00	1	4	4	#¡DIV/0!	3,83		3,50	3,00	4,00	4,00	0,00
ACME	4,29	7	3	5	0,57	3,63	4,25	3,83	3,67	4,33	4,29	-0,05
ACME	3,50	2	2	5	4,50	3,75	4,00	3,40	3,00	3,67	3,50	-0,17
ACME	4,54	13	4	5	0,27	4,54	4,45	4,46	4,44	4,36	4,54	0,17
ACME	4,00	4	3	5	0,67	3,20	3,00	3,50	3,20	3,25	4,00	0,75
ACME	3,75	4	2	5	2,25	4,00	3,80	3,60	4,25	3,00	3,75	0,75
ACME	4,29	7	3	5	0,57	4,42	4,00	4,30	4,00	4,25	4,29	0,04
ACME	4,50	2	4	5	0,50	3,83	4,00	3,50	3,75	4,75	4,50	-0,25
ACME	4,38	8	4	5	0,27	3,78	3,73	3,75	4,25	4,25	4,38	0,13
Q7 Progress Period	3,93	30	3	5	0,34	4,00	4,03	4,15	3,95	3,70	3,93	0,24
Q8 Progress against objectives - project	4,03	32	3	5	0,16028226	3,94	4,10	4,17	3,73	3,80	4,03	0,23
Q9 Weaknesses in the project		26										
Q10 Strengths		30										
Q11 Potential problems		19										
Q12 Changes to suggest		17										
Q13 Potential Information to be disseminated		19										

Fig. 1.4 Extract from the month 39 self-assessment audit summary

In the case of question 6, one of the partners has got an average score across all the self-assessment audits in the range of 3.00 and 4.00 (an average across all audits of 3.23), whilst another partner got average scores in the range of 4.58–4.80 (an average across all audits of 4.63). The average score for all partners and all audits is 3.94.

Note that for certain questions, the number of responses obtained could be none or low; thus, the score/result obtained should be interpreted accordingly. The detailed results for the rest of the questions can be found in Fig. 1.4.

An analysis of the results obtained after each of the audits allows identifying potential issues, weak areas and conflicts. The comments provided by the respondents, after being anonymized, are shared with the whole group, allowing also joint discussions around conflicts, issues, risks and barriers, but also around suggested improvements. In addition, each partner, key management roles and work package leaders can also see what the rest of the partners comment and think about their participation in the project.

Except for the second period review meeting, the feedback from the EC was consistently positive throughout the project. During the second period review meeting, there was a complaint about not having provided a draft of the technical report prior to the review meeting. Apart from this, communication with the project officer(s) from the EC was fluid and in general terms provided positive feedback on the progress made by the project. It is also important to note that in multiple occasions this fluid communication helped finding the most appropriate way forward to tackle specific issues or unexpected situations. An example of this is when the project coordinator requested clarification to the EC on how to proceed with the protection of the European Union classified information (EUCI) of the project. The project includes 16 deliverables that have been classified, and several of them required frequent access by most or all the partners. It was very important to find a prompt and efficient implementation of the EUCI handling guidelines, and this was successfully achieved thanks to the support provided by the EC.

The feedback from the SAG members was mainly obtained via evaluation forms that the SAG members attending the project events filled in. The satisfaction level was high among them, and there were also a few constructive criticisms (which SAG members were explicitly requested to provide as part of the continuous improvement process established in the project). It is also worth mentioning that multiple SAG members requested access to the ASGARD results and/or ad hoc demonstrations, which is a clear symptom of the interest that the project arose among them.

1.5   Discussion and Future Research

In this chapter we have presented the results of an exploratory study that lays the ground for future studies which could determine if what has been observed might be explained by an emerging theory.

The results presented in Sect. 1.4 show that the management structure of the project has been successful. The two different levels of operational management, one at project (inter work package) level and the other at work package level, have provided sufficient flexibility to the needs and characteristics of each of the work packages whilst offering the space to discuss and jointly agree on the solutions to put in place to tackle the issues and dependencies that affect multiple or all work packages. It is also worth mentioning that not all work packages have received the same type and level of coordination, and the satisfaction of the affected partners reflect that as well.

The feedback about the tools delivered in each of the full-development cycles was used to re-adjust the development plans in the technical work packages. But it was also not rare to identify new things not initially foreseen which were necessary or convenient. In many of these occasions, the Consortium discussed and agreed to add a new task or a new piece of work to the original plan. A couple of illustrative examples are the decision to design and proof-test a privacy engine on one hand and the decision to design and implement a tool maturity evaluation model on the other hand. This level of flexibility is, in opinion of the authors, another very important success factor of the project, as it provides space for creativity and innovation within the project boundaries.

The self-assessment audits are a very valuable project management tool. In line with a continuous improvement spirit, this tool provides the process and the framework for individual and joint self-reflection, allowing prompt identification and managing of negative trends, issues and conflicts. In addition, the tool also allows providing valuable feedback and constructive criticism.

The experience of the ASGARD project seems to indicate that applying open-source model principles and practices, including iterative and incremental full-development life cycles, and with the goal of building a sustainable and long-lasting restricted community, does strengthen the role of end-users in security research, it helps market uptake of security research project results, and it helps setting better LEA expectations.

However, at the time of writing this chapter, November 2019, the ASGARD project is still ongoing, so no final conclusions should be established. Based on the results obtained so far and considering the feedback got both from internal Consortium partners as well as from the external stakeholders of the project, it can be assumed that the initial hypothesis seems to be correct. In any case, though the results of this study could probably be transferable to similar projects, generalizing them should be avoided, as one case cannot represent all similar cases or situations. Further causal or explanatory research would be needed to confirm or deny the conclusions of this study and to test the cause-and-effect relationship between the methods followed in the ASGARD project and the positive impacts that have been identified on it.

In addition, there is still much to learn about which is the most appropriate role of the end-users in security research projects, the role that adds the greater value to them and to the rest of stakeholders that participate on such research projects (i.e. research technology organisations and industry). For example, how does the internal organisational structure of the end-users and the role of their project team members affect in their contribution to the benefits gathering of the research project? Or which methodologies, processes and techniques are the most appropriate to implement in security research projects to maximize the value added by the participation of end-users?

And there is also much to learn about which are the factors that help maximizing the market uptake of the results from security research projects. For example, which is the best combination of type of research actions or instruments that promote market uptake and that minimizes the risk of reducing the tendency for insights and tools produced within projects to be left behind?

Acknowledgements

 Results incorporated in this chapter received funding from the European Union's Horizon 2020 research and innovation programme under grant agreement No 700381

References

1. Moravec, L. (2014). Research market gap in law enforcement technology: Lessons from Czech security research funding programmes. *Central European Journal of Public Policy, 8*(2), 28–49.
2. *Final evaluation of security research under the seventh framework programme for research, technological development and demonstration.* https://ec.europa.eu/home-affairs/sites/homeaffairs/files/e-library/documents/policies/security/reference-documents/docs/fp7_security_research_final_report_en.pdf. Accessed 10 Dec 2019.
3. *Security Research Event 2019* – https://www.sre2019.eu/. Accessed 12 Dec 2019.
4. *Horizon 2020 – Work Programme 2018–2020, Secure societies – Protecting freedom and security of Europe and its citizens (European Commission Decision C(2019)4575 of 2 July 2019)*, p. 24. https://ec.europa.eu/research/participants/data/ref/h2020/wp/2018-2020/main/h2020-wp1820-security_en.pdf. Accessed 10 Dec 2019.
5. Weisburd, D., & Neyroud, P. (2011). *Police science: Toward a new paradigm.* Washington DC: National Institute of Justice.
6. Bigo, D., Jeandesboz, J., Martin-Maze, M., & Ragazzi, F. (2014). *Review of security measures in the 7th research framework programme FP7 2007–2013.* Brussels: Committee on Civil Liberties, Justice, and Home Affairs.
7. Burr, V. (2003). *Social constructionism* (2nd ed.). Hove: Routledge.
8. Yin, R. K. (2014). *Case study research design and methods* (5th ed., p. 282). Thousand Oaks: Sage.

CHAPTER 2

SoK: Blockchain Solutions for Forensics

Thomas K. Dasaklis, Fran Casino,
and Constantinos Patsakis

2.1 INTRODUCTION

The undergoing digitization of information-intensive processes has a radical impact on our daily lives. Digitization affects almost all aspects of our lives from how we work to how we interact and communicate with each other. As a result, a myriad of devices is involved in almost every possible aspect of our daily lives. For instance, a smartphone may contain data with different levels of sensitivity (text messages, emails, financial transactions, etc.) which provide background information on its owner and his/her social connections. Digitization, however, comes at a cost. Financial frauds, intellectual property infringements, industrial espionage and digital terrorist networks are just a few among the various faces of

T. K. Dasaklis (✉)
Department of Informatics, University of Piraeus, Pireas, Greece
e-mail: dasaklis@unipi.gr

F. Casino · C. Patsakis
Department of Informatics, University of Piraeus, Pireas, Greece

Athena Research Center, Athens, Greece
e-mail: francasino@unipi.gr; kpatsak@unipi.gr

© The Author(s), under exclusive license to Springer Nature
Switzerland AG 2021
B. Akhgar et al. (eds.), *Technology Development for Security*
Practitioners, Security Informatics and Law Enforcement,
https://doi.org/10.1007/978-3-030-69460-9_2

21

cybercriminal behaviour. In the event of such deviant behaviour, digital evidence may be the only evidence of a case. Hence, digital evidence forms an integral part of the overall criminal investigation process.

Although the digital forensics community has established reliable scientific methodologies and common standards in its workflows, it still faces many challenges due to the volatile and malleable nature of the evidence and the continuous advances in technology that introduce new attack vectors. Moreover, most of the criminal activity on the Internet is transnational, generating cross-jurisdiction problems of cooperation and information exchange that can be alleviated by common standards and protocols. Several challenges have been identified in the literature regarding the development of robust digital forensics approaches. In [19], the authors identify four broad categories of challenges in digital forensics: (a) technical challenges, (b) legal systems and/or law enforcement challenges, (c) personnel-related challenges, and (d) operational challenges. Another major issue that directly impacts digital forensics is the ever-increasing volume of potential evidence generated along with the growing number of devices used [34]. Some domain-specific challenges are worth mentioning. For example, in the case of cloud forensics, identifying useful network events and recording the minimum representative attributes for each event remain a significant challenge [30]. Lack of international collaboration and legislative frameworks in cross-nation data access/exchange and the increased number of mobile devices accessing the cloud are also significant challenges in cloud forensics [37]. Arguably the most critical problem in digital forensics is the validity and trustworthiness of the evidence itself (safeguarding the chain of custody for the data related to a case), particularly when multiple stakeholders are involved in the overall forensics process.

2.1.1   *Blockchain as a Game Changer in Digital Forensics*

Blockchain, a novel disruptive technology, has emerged the past few years, enabling the development of a wide range of applications [8]. In principle, a blockchain can be considered a distributed append-only data structure which stores states efficiently and in a transparent way. While the initial concept of Nakamoto was to store transactions of bitcoins in a way that prevents double-spending [28], the created structure has many appealing properties. Setting aside the different "flavours" that blockchains have, they offer auditability, robustness and security. Blockchains also provide immutability to a large extent [31], posing significant challenges to the implementation of the right to be forgotten principle, as defined in the EU General Data Protection Regulation Directive (GDPR) [32].

Based on the above, it is apparent that the blockchain properties constitute a very promising baseline for forensics [1]. More precisely, during a forensic investigation, all the involved investigators would like to store their findings in an immutable way so that they cannot be altered and be brought to a court of law. Similarly, blockchains provide transparency and auditability, which is a requirement for the chain of custody of the corresponding evidence. In this regard, in the past few years, several researchers have investigated these opportunities and proposed blockchain-based solutions for forensics. While the field is rather recent, we argue that in the coming years blockchain solutions for managing forensics will become a default. This survey performs an in-depth analysis of the needs and gaps of the field and the different approaches in the literature. Therefore, we set the landscape and facilitate the design of the new solutions.

2.1.2   Goal and Plan of the Chapter

In this chapter, we analyse the current state of blockchain-based forensic methods applied in different fields. First, we provide a comprehensive classification and the main features of the state-of-the-art solutions, which are retrieved using a sound bibliographic analysis approach. Next, we analyse how blockchain's features can enhance digital forensics. Finally, we discuss the limitations and the main challenges that are at the intersection of both fields.

The rest of the chapter is organized as follows: Sect. 2.2 describes the research methodology used and the main quantitative literature findings. Next, Sect. 2.3 provides a topic classification of the blockchain-based digital forensics methods and a qualitative analysis of their features. Thereafter, Sect. 2.4 provides a discussion of the main limitations of blockchain technologies and the challenges to be faced by next-generation digital forensics solutions. Finally, the chapter offers some final remarks in Sect. 2.5.

2.2   METHODOLOGY

To survey the available blockchain-based forensics approaches, we have used a sound methodological framework. In particular, we performed a systematic search during November 2019 without time-frame restrictions.

Table 2.1 Year-based and source-type classification of the available literature

Publication type	Publication year	
	2018	2019
Journal articles	2	8
Serials	–	4
Conference proceedings	4	6

We used the Scopus scientific database as our primary source for identifying relevant literature. We used a predefined set of keywords for searching within the titles of all the available Scopus papers (the terms used included the words "blockchain" and "forensics"). To locate additional studies, we used the so-called snowball effect (additional literature was retrieved based on references of key articles found in the initial phase of our search). We excluded some papers based on certain exclusion criteria (relevant to document type, language and subject area). In total, 24 articles were selected for analysis. For the thematic content analysis of the selected literature, we used a qualitative analysis software (MAXQDA11). Finally, we adopted various qualitative analysis methods (i.e. narrative synthesis and thematic analysis) for the classification and synthesis of the extracted data. We present the results of our analysis in Sect. 2.3.

There exist some bibliographic analysis results worth mentioning. As seen in Table 2.1, the available blockchain-enabled forensics literature spans only 2 years (2018 and 2019). Therefore, it is not until very recently that the scientific community has focused on blockchain technology as a viable solution for establishing robust forensics mechanisms. Regarding the type of publications, there seems to be an even allocation between conference proceeding papers and articles published in international journals.

2.3   CLASSIFICATION OF THE AVAILABLE BLOCKCHAIN-BASED FORENSICS LITERATURE

In this section, we thoroughly analyse the literature and provide a topic-based classification (see Fig. 2.1). Next, we identify the main features and solutions proposed by each method in Table 2.2 and discuss them in the following paragraphs.

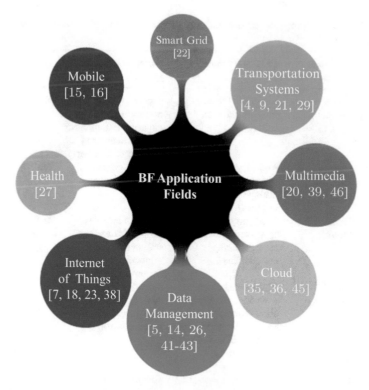

Fig. 2.1 Mind map abstraction the blockchain forensics research topics. The size of each topic has been weighted according to the number of contributions

2.3.1 Cloud Forensics

Cloud security threats remain a significant challenge nowadays. Cloud forensics, an umbrella term covering issues of cloud computing and digital forensics, may assist in investigating cloud environments and quickly respond to and report cloud security incidents [36]. Cloud forensics call for multiparty collaboration due to the multitude of stakeholders engaged. To this end, blockchain technology may enhance the collection of digital evidence in cloud environments and further improve different stakeholder coordination [45]. Another critical aspect of cloud forensics may refer to logs management. In particular, secure preservation and investigation of the various logs are essential elements of cloud forensics. However, due to

Table 2.2   Description of the features of the available blockchain-based forensic schemes

Refs.	Application domain	Problem addressed	Blockchain-enabled forensic features	Implementation
[35]	Cloud forensics	Integrity of logs, multi-stakeholder collusion	Secure logging as-a-service for cloud environment, integrity, confidentiality and immutability of logs	Yes
[36]	Cloud forensics	Multi-location storage of forensic evidence, multiple stakeholders engaged	Data encryption, distributed storage	No
[45]	Cloud forensics	Multiparty cooperation, trustworthiness of records among stakeholders	Chain of custody, proof of existence, privacy and anti-tampering preservation for process records	Yes
[33]	Cloud forensics	Multi-stakeholder collusion, security and access control, multiparty cooperation, trustworthiness of records among stakeholders	Integrity, chain of custody, data encryption, secure access control	Yes
[13]	Cloud/network forensics	Multiparty cooperation, trustworthiness of records among stakeholders, SDN log recording	Chain of custody, proof of existence, privacy and anti-tampering preservation for process records	Yes
[5]	Data management forensics	Validity of the digital evidence	Weighted digital evidence, digital evidence inventory, categorization according to each evidence relevance, assignment of confidence rating	No
[6]	Data management forensics	Integrity and validity of electronic evidence and ownership	Chain of custody, tracking of the stakeholders involved, credibility of the data provided	Yes

(*continued*)

Table 2.2 (continued)

Refs.	Application domain	Problem addressed	Blockchain-enabled forensic features	Implementation
[14]	Data management forensics	Integrity and validity of electronic evidence	Chain of custody, tracking of the stakeholders involved, credibility of the data provided	No
[26]	Data management forensics	Integrity and authenticity of digital evidence, authenticity and legality of processes and procedures used to gather and transfer the evidence	Chain of custody, safeguarding the integrity and tamper-resistance of digital forensics	Yes
[41]	Data management forensics	Tampering with evidence, data privacy issues, sensitive information leakages	Lightweight, scalable secure digital evidence framework, multi-signature schemes for evidence submission and retrieval	Yes
[42]	Data management forensics	Proof of existence of digital evidence	Tamper-proof chronology by means of OpenTimestamps	No
[43]	Data management forensics	Trust and security issues as derived by current centralized data management schemes	Electronic evidence preservation, different evidence access rights, data security protection, information integrity guarantees, traceability	No
[27]	Healthcare forensics	Different access levels (to both health data and devices), health data privacy	Log audit trails for integrity and provenance guarantees, health data privacy, fine-grained access	Yes
[7]	IoT forensics	Collection and preservation of evidence regarding alleged malicious behaviour in IoT networks	Private forensic data/metadata evidence collection, integrity, authentication, and non-repudiation of the data collected	No

(continued)

Table 2.2 (continued)

Refs.	Application domain	Problem addressed	Blockchain-enabled forensic features	Implementation
[18]	IoT forensics	Multiple IoT stakeholders, multiparty access to digital evidence	Integrity, confidentiality, anonymity, authenticity and non-repudiation	No
[23]	IoT forensics	Traceability, integrity and provenance of the evidence is limited due to the resource-constraint nature of IoT devices	Integrity, authenticity, non-repudiation, identity privacy, end-to-end forensic life cycle	No
[38]	IoT forensics	Heterogeneity and distribution characteristics of the IoT environment	Chain of custody for all the IoT-related forensics processes, security and data integrity, multiparty verification of the IoT-related forensics processes	Partial
[24]	IoT forensics	Traceability, integrity and provenance of the evidence	Security and data integrity, multiparty verification of the IoT-related forensics processes and evidences	Partial
[15]	Mobile forensics	Current limitations of static-based and dynamic-based code analysis tools (code obfuscation, encryption, malware in different families with various features)	Consortium blockchain framework to store and classify android malware, classification of different malware families	Partial
[16]	Mobile forensics	Tracking and recording of a very wide range of existing malicious programs, current limitations of static-based and dynamic-based code analysis tools	Enhanced malware detection features based on the usage of both private and consortium blockchain	No

(continued)

Table 2.2 (continued)

Refs.	Application domain	Problem addressed	Blockchain-enabled forensic features	Implementation
[20]	Multimedia forensics	Integrity and legal authenticity of video data produced as evidence in legal proceedings, privacy concerns of video data gathered by CCTV installations	Trustworthy evidence protection in distributed network environment, video data integrity (link to the primary video stream and its creation)	Yes
[39]	Multimedia forensics	Civilians/journalists who need to protect their identity while ensuring that the evidence they collect are forensically sound	Integrity and spatiotemporal properties of digital evidence	Yes
[46]	Multimedia forensics	Photo-faking, photo owners have limited control over their photos after uploading them on the Internet due to lack of copyright protection mechanisms	Customized access control rules, photo-tracing, creation of copyright-protected photos (resolving copyright dispute problems)	Yes
[22]	Smart grid forensics	Smart grid security, intrusion detection	Ensure the integrity of smart energy transaction platforms, keeping log information for effectively investigate cybercrimes and predict system failures	No
[4]	Transportation forensics	Contradictory use of personal data, privacy, multiple stakeholders involved	Integrity, veracity, authenticity, non-repudiation and identity privacy of vehicle-related data voluntarily and spontaneous release of data for forensic purposes	No

(*continued*)

Table 2.2 (continued)

Refs.	Application domain	Problem addressed	Blockchain-enabled forensic features	Implementation
[9]	Transportation forensics	Transportation data is overwritten shortly, no available system for integrating data from the various stakeholders involved (data from other vehicles, road conditions, manufacturers, and maintenance centres), only third-party solutions exist for vehicular forensics (such as surveillance cameras and eyewitnesses)	Lightweight privacy-aware blockchain framework to manage the collected vehicle-related data (maintenance information/history, car diagnosis reports)	No
[21]	Transportation forensics	Data privacy concerns (GPS sensitive info) due to third-party usage	Legal authority may run forensic analysis without unnecessary violation of the user anonymity and privacy	No
[29]	Transportation forensics	Unauthorized changes in vehicle hardware profiles, multiple stakeholders involved	Logs of all hardware profile changes are kept on blockchain, provision of customized access (only authenticated changes are allowed)	Yes

the inherent uncertainties of cloud environment, several difficulties exist concerning the collection of authentic logs from a cloud environment while preserving integrity and confidentiality. Blockchain technology may be used as a logging-as-a-service tool for securely storing and processing logs while coping with issues of multi-stakeholder collusion and the integrity and confidentiality of logs [13, 35].

2.3.2   Data Management Forensics

The works classified in data management include these proposing novel models for data processing and chain of custody preservation methodologies. The use of permissioned blockchains [14, 26, 43] is stated as a measure to enhance scalability, as well as the use of lightweight consensus mechanisms [41]. Advanced evidence collection and feature classification [5], as well as the relevance of the timeline of events [6, 42], are other features discussed by authors. However, the main drawback of the proposed solutions is that they only offer architectural designs and they do not provide full exploitation of blockchain, with only a few of them offering practical implementations [6, 26, 41].

2.3.3   Healthcare Forensics

With the prevalence of new regulatory frameworks brought forward (like the EU GDPR directive), healthcare organizations have started taking necessary steps towards protecting themselves against costly breaches of patients' sensitive information and further safeguarding their reputation. To this end, forensics may be a valuable ally for addressing litigation risks when it comes to data breaches and unauthorized access to medical data from both outside attacks or internal misuse [10]. Access control management is an essential feature of patient data protection. In [27], the authors propose a blockchain-enabled authorization framework for managing both the Internet of Medical Things (IoMT) devices and healthcare stakeholders. The proposed framework provides fine-grained access to patient health data and preserves the chain of custody of all logs by offering audit trails for integrity and provenance guarantees.

2.3.4   IoT Forensics

IoT forensics includes the study of IoT devices, their systems and interrelations between different parts of their ecosystems. In this regard, the result of our literature review showed that there is a relevant interest in IoT forensics in the blockchain. We observed that evidence collected from IoT devices and interactions between the different actors (e.g. through privacy-preserving identity management techniques) are the most relevant features studied in the literature [7, 18, 38]. Moreover, proper identity management and privacy preservation is also a mandatory requirement in such context [23]. Nevertheless, current solutions are not mature enough, since authors only proposed architectures and flows, except for [23], which only provided transaction performance tests.

2.3.5   Mobile Forensics

Mobile forensics includes the analysis of digital and physical evidence provided by smartphone devices and similar ones (i.e. these sharing similar architectural bodies and underlying operating systems, such as tablets or other handheld devices). Nevertheless, the identified blockchain-based forensic research mainly focuses on applications and malware detection. More concretely, authors propose the use of consortium blockchains and focus on malware detection and statistical analysis based on each application feature [15, 16]. Therefore, more work needs to be provided in this field, with special regard to hardware inclusion and holistic systems definition, as well as usable implementations.

2.3.6   Multimedia Forensics

Multimedia forensics employs various scientific techniques for examining a multimedia file (audio, video and/or image) concerning its (a) integrity (establish the linkage between a multimedia output and its source identification) and (b) authenticity (check for the veracity of the multimedia output). For example, in [20], a blockchain-based approach is proposed for cataloguing CCTV video evidence. The authors provide a functional implementation of the blockchain-based system that manages high volumes of CCTV evidence. In [39] the authors present E-Witness, a system that uses blockchain technology for safeguarding the integrity and spatiotemporal characteristics of digital evidence captured by smartphones. To verify the integrity and spatiotemporal claims of the evidence, the proposed system uses hashes of pictures/videos along with location certificates stored in the blockchain. A blockchain-based photo forensics scheme is presented in [46]. The proposed Ethereum-based scheme resolves photos' veracity issues like photo-faking, photo-tracing and copyright dispute problems.

2.3.7   Smart Grid Forensics

Smart grids offer significant improvements in terms of resources utilization in current electricity supply networks. Smart grids embrace digital communications technologies, smart metering, intelligent appliances and energy-efficient resources for better matching energy supply and demand. Like other cyberphysical systems, however, smart grids are vulnerable to cyberattacks, and intrusion detection might prove extremely important. In [22] the theoretical underpinnings of blockchain technology and its importance in smart grids forensics are discussed. The authors highlight how blockchain can enhance features such as energy optimization, system performance, managerial tasks and security of smart grids. Finally, the authors discuss the opportunities/open issues in the topic.

2.3.8   Intelligent Transportation Systems Forensics

Intelligent transportation systems (ITS) embrace a range of technological novelties like advanced sensing and control and IoT applications for improving safety, efficiency and services provision of both vehicles and road transport networks. However, the increased automation of ITS (e.g. self-driving cars) and the adoption of new data privacy frameworks (like the GDPR) call for the development of sound forensic mechanisms to analyse traffic accidents and protecting users' sensitive data. In [4], a blockchain-enabled system is proposed for managing users' requests (car navigation) and relevant data that fully complies with data privacy and protection legal frameworks. In [9], the authors propose a blockchain-based forensics system that enables the trustless, traceable and privacy-aware post-accident analysis with minimal requirements in storage and processing. A blockchain framework is proposed in [21] for managing sensitive navigation data (GPS position) within a fixed geographic zone while ensuring user anonymity. Cybersecurity threats may also prove critical in the context of current ITS. In [29] a blockchain-based framework is proposed for keeping logs of all hardware profile changes in a vehicle. Based on the inherent characteristics of blockchain technology, the proposed framework only allows authenticated changes, subject to user, time, geospatial and contextual constraints, as defined by automotive manufacturers.

2.4   DISCUSSION

In what follows, we describe the main limitations of blockchain technology and some strategies to overcome them. Moreover, we provide a detailed analysis of actual and future challenges of digital forensics and discuss possible countermeasures.

2.4.1   Limitations in Blockchain

The suitability of blockchain is a topic that has been extensively discussed in the literature [8, 11]. In this regard, the challenges to be faced by different blockchain technologies vary depending on their type and application scenario. For example, public blockchains face limitations such as scalability, performance and cost issues. In this regard, public blockchains are nowadays mainly used for cryptocurrencies and to commit small pieces of data (i.e. hashes) for verifiability purposes [44]. In the case of private blockchains, the performance and scalability challenges are overcome due to the use of more efficient consensus mechanisms and a reduced number of participants.

Moreover, the cost of memory, compared with public blockchain is negligible, yet off-chain data storage is a recommended strategy for most applications. Nevertheless, both public and private blockchains require the use of proper data management and architectural designs to provide security and privacy guarantees [25]. In this regard, the use of secure identity management systems [3], the proper analysis of the specific blockchain systems to be used [17] and a careful implementation development of smart contracts [2, 40] are mandatory.

2.4.2   Challenges in Blockchain Digital Forensics

We classified next-generation digital forensics' most relevant challenges in the following six domains:

Tokenization of Artefacts from Digital Evidence
Digital forensics imply the analysis of the digital evidence and the extraction of the corresponding knowledge regarding the events of a crime under investigation. However, this analysis is not performed by a single entity. For instance, an image of hard disk may contain different evidence that must be analysed by different people who will look into different parts. One person may study the log files, while another may investigate the file system and a third one might be needed to analyse a specific binary that requires reversing. Therefore, a single evidence is expected to be divided in an arbitrary amount of artefacts, each of which might have to be studied individually and from another person. Breaking down things and storing them in blockchains is not straightforward, and several existing solutions could be adopted (e.g. the use of tokens); however, the bulk

of them considers that the elements that something is decomposed to is predetermined. Despite the fact that some solutions for assigning tokens in blockchain for arbitrary decomposition of an object have been proposed in the supply chain field [12], storing tokenized artefacts in the blockchain during the course of a digital investigation remains a challenge.

Efficient Management of Data Volume in the Chain of Custody
One of the main concerns in digital forensics is the volume of data, since evidence may include thousands of multimedia files or log files per case. In this regard, although data storage of raw documents has to be provided for all cases, it should be based on off-chain technologies (e.g. IPFS, Storj). In this case, only hashes should be used in the blockchain (i.e. or meta-hashes if data are processed as blocks, to ease auditability).

Parse Forensic Sound Procedures in Blockchain Systems
Standard and sound forensic flows have to be provided, even when using blockchain as a platform to provide verifiability and chain of custody tamper-proof guarantees. Therefore, proper standardized flows and smart contracts that map the adequate functions have to be provided to enable final court validation as well as certification by digital forensic laboratories and law enforcement agencies.

Enable an Understandable Forensic Outcome/Reports
The use of blockchain provides a myriad of benefits, such as the efficient and verifiable provision of data flows. Still, the knowledge retrieving and report creation parts belong to a different stage. In this regard, even if automated, the reports and outcomes generated should be understandable in court. Therefore, even if blockchain facilitates this task, research efforts have to be done in this direction, providing a link between forensic sound procedures and their proper explanation.

Interoperability and Cross-Border Jurisdictions The use of international standardized flows and proper data management and sharing agreements will enhance the fight of cybercrime. Nowadays, international collaborations already exist in the scope of the European Union.[1] Nevertheless,

---

[1] https://ec.europa.eu/home-affairs/what-we-do/policies/cybercrime/e-evidence_en

further development of blockchain-based solutions[2] will serve as a ground truth platform for standardized solutions, enabling international interoperability.

Timeline of Events and Chronology
The relevance of data acquisition and timeline of events in digital forensics is key to identify patterns and relate similar cases, since the knowledge generated by forensic investigations has to be used in the future to prevent or minimize them. Therefore, the proper reporting and evidence collection procedures have to be done respecting the timeline of events.

Blockchain can provide proof of existence due to its immutability, which, combined with the use of block timestamps and hashes, can guarantee that evidence was collected at a specific moment and they have not been modified.

2.5   CONCLUSIONS

Digitization comes with a myriad of novel opportunities and services. Nevertheless, this heterogeneous landscape is also becoming a profitable playground for malicious users, which are continuously increasing the dynamism and complexity of cybercriminal activities. In this regard, digital forensics needs to be rapidly updated to deal with a set of multidisciplinary challenges, ranging from the advances in information and communication technologies, to jurisdictional and interoperability restrictions. To this end, we believe that digital forensics can be benefited from the widespread adoption of blockchain technology and its inherent characteristics.

In this chapter, we presented a literature review of the current blockchain-based forensic solutions and classified them according to their features as well as their application field. Thereafter, we identified the benefits and limitations of blockchain-based forensics and outlined the main challenges to be overcome in the future, providing a fertile ground for research.

Future work will focus on developing a blockchain-based forensic framework which enables the collection of heterogeneous digital evidence as well as forensic procedures in a standardized manner. To this end, we will study the tokenization of digital forensic evidences to provide a common layer of abstraction for different categories of cybercrime.

[2] https://locard.eu/

Acknowledgement

 This work was supported by the European Commission under the Horizon 2020 Programme (H2020), as part of the project LOCARD (https://locard. eu) (Grant Agreement no. 832735).

BIBLIOGRAPHY

1. Al-Khateeb, H., Epiphaniou, G., & Daly, H. (2019). Blockchain for modern digital forensics: The chain-of-custody as a distributed ledger. In *Advanced Sciences and Technologies for Security Applications* (pp. 149–168).
2. Atzei, N., Bartoletti, M., & Cimoli, T. (2017). A survey of attacks on ethereum smart contracts (sok). In *International Conference on Principles of Security and Trust* (pp. 164–186). Berlin: Springer.
3. Bernal Bernabe, J., Canovas, J. L., Hernandez-Ramos, J. L., Torres Moreno, R., & Skarmeta, A. (2019). Privacy-preserving solutions for blockchain: Review and challenges. *IEEE Access, 7*, 164908–164940.
4. Billard, D., & Bartolomei, B. (2019). Digital forensics and privacy-by-design: Example in a Blockchain-based dynamic navigation system. In *Annual Privacy Forum* (pp. 151–160). Cham: Springer.
5. Billard, D. (2018). Weighted forensics evidence using blockchain. In *Proceedings of the 2018 International Conference on Computing and Data Engineering* (pp. 57–61). New York: ICCDE 2018, ACM. https://doi. org/10.1145/3219788.3219792.
6. Bonomi, S., Casini, M., & Ciccotelli, C. (2020). B-coc: A blockchain-based chain of custody for evidences management in digital forensics. In *OpenAccess Series in Informatics, 71.*
7. Brotsis, S., Kolokotronis, N., Limniotis, K., Shiaeles, S., Kavallieros, D., Bellini, E., & Pavue, C. (2019). Blockchain solutions for forensic evidence preservation in iot environments. In *2019 IEEE Conference on Network Softwarization (NetSoft)* (pp. 110–114). IEEE.
8. Casino, F., Dasaklis, T. K., & Patsakis, C. (2018). A systematic literature review of blockchain-based applications: Current status, classification and open issues. *Telematics and Informatics, 36*, 55–81.
9. Cebe, M., Erdin, E., Akkaya, K., Aksu, H., & Uluagac, S. (2018). Block4Forensic: An integrated lightweight blockchain framework for forensics applications of connected vehicles. *IEEE Communications Magazine, 56*(10), 50–57.
10. Chernyshev, M., Zeadally, S., & Baig, Z. (2019). Healthcare data breaches: Implications for digital forensic readiness. *Journal of Medical Systems, 43*(1), 1–12.

11. Chowdhury, M. J. M., Ferdous, M. S., Biswas, K., Chowdhury, N., Kayes, A. S. M., Alazab, M., & Watters, P. (2019). A comparative analysis of distributed ledger technology platforms. *IEEE Access, 7,* 167930–167943.
12. Dasaklis, T., et al. (2019). A framework for supply chain traceability based on blockchain tokens. In *International Conference on Business Process Management.* Cham: Springer.
13. Duy, P., Do Hoang, H., Thu Hien, D., Ba Khanh, N., & Pham, V.H. (2019). Sdnlog-foren: Ensuring the integrity and tamper resistance of log files for sdn forensics using blockchain. In: *Proceedings - 2019 6th NAFOSTED Conference on Information and Computer Science, NICS 2019.* pp. 416–421.
14. Gopalan, S. H., Suba, S. A., Ashmithashree, C., Gayathri, A., & Jebin Andrews, V. (2019). Digital forensics using blockchain. *International Journal of Recent Technology and Engineering, 8*(2 Special Issue 11), 182–184.
15. Gu, J., Sun, B., Du, X., Wang, J., Zhuang, Y., & Wang, Z. (2018). Consortium blockchain-based malware detection in mobile devices. *IEEE Access, 6,* 12118–12128.
16. Homayoun, S., Dehghantanha, A., Parizi, R. M., & Choo, K. K. R. (2019). A blockchain- based framework for detecting malicious mobile applications in App stores. In *2019 IEEE Canadian Conference of Electrical and Computer Engineering (CCECE)* (pp. 1–4). IEEE.
17. Homoliak, I., Venugopalan, S., Hum, Q., Reijsbergen, D., Schumi, R., & Szalachowski, P. (2019). The security reference architecture for blockchains: Towards a standardized model for studying vulnerabilities, threats, and defenses. *arXiv.* https://doi.org/10.1109/COMST.2020.3033665.
18. Hossain, M. M., Hasan, R., & Zawoad, S. (2018). Probe-IoT: A public digital ledger based forensic investigation framework for IoT. In *INFOCOM Workshops* (pp. 1–2).
19. Karie, N. M., & Venter, H. S. (2015). Taxonomy of challenges for digital forensics. *Journal of Forensic Sciences, 60*(4), 885–893.
20. Kerr, M., Han, F. V., & Schyndel, R. (2018). A blockchain implementation for the cataloguing of cctv video evidence. In *2018 15th IEEE International Conference on Advanced Video and Signal Based Surveillance (AVSS)* (pp. 1–6).
21. Kevin, D., & David, B. (2019). HACIT2: A privacy preserving, region based and blockchain application for dynamic navigation and forensics in VANET. In *International Conference on Ad Hoc Networks* (pp. 225–236). Cham: Springer.
22. Kotsiuba, I., Velykzhanin, A., Biloborodov, O., Skarga-Bandurova, I., Biloborodova, T., Yanovich, Y., & Zhygulin, V. (2018). Blockchain evolution: From bitcoin to forensic in smart grids. In *2018 IEEE International Conference on Big Data (Big Data)* (pp. 3100–3106). IEEE.
23. Le, D. P., Meng, H., Su, L., Yeo, S. L., & Thing, V. (2018). Biff: A blockchain-based iot forensics framework with identity privacy. In *TENCON 2018–2018 IEEE Region 10 Conference* (pp. 2372–2377). IEEE.

24. Li, S., Qin, T., & Min, G. (2019). Blockchain-based digital forensics investigation framework in the internet of things and social systems. *IEEE Transactions on Computational Social Systems 6*(6), 1433–1441.
25. Li, X., Jiang, P., Chen, T., Luo, X., & Wen, Q. (2017). A survey on the security of blockchain systems. *Future Generation Computer Systems, 107*, 841–853.
26. Lone, A. H., & Mir, R. N. (2019). Forensic-chain: Blockchain based digital forensics chain of custody with PoC in Hyperledger composer. *Digital Investigation, 28*, 44–55.
27. Malamas, V., Dasaklis, T., Kotzanikolaou, P., Burmester, M., & Katsikas, S. (2019). A forensics-by-design management framework for medical devices based on blockchain. In 2019 IEEE World Congress on Services (SERVICES) (Vol. 2642, pp. 35–40). IEEE.
28. Nakamoto, S. (2008). *Bitcoin: A peer-to-peer electronic cash system.* Manubot.
29. Patsakis, C., Dellios, K., De Fuentes, J. M., Casino, F., & Solanas, A. (2019). External monitoring changes in vehicle hardware profiles: Enhancing automotive cyber-security. *Journal of Hardware and Systems Security, 3*(3), 289–303.
30. Pilli, E. S., Joshi, R. C., & Niyogi, R. (2010). Network forensic frameworks: Survey and re-search challenges. *Digital Investigation, 7*(1–2), 14–27.
31. Politou, E., Casino, F., Alepis, E., & Patsakis, C. (2019). Blockchain mutability: Challenges and proposed solutions. *IEEE Transactions on Emerging Topics in Computing*, 1–1. https://doi.org/10.1109/TETC.2019.2949510.
32. Politou, E., Alepis, E., & Patsakis, C. (2018). Forgetting personal data and revoking consent under the gdpr: Challenges and proposed solutions. *Journal of Cybersecurity, 4*(1), tyy001.
33. Pourvahab, M., & Ekbatanifard, G. (2019): Digital forensics architecture for evidence collection and provenance preservation in iaas cloud environment using sdn and blockchain technology. IEEE Access 7, 153349–153364.
34. Quick, D., & Choo, K. K. R. (2016). Big forensic data reduction: Digital forensic images and electronic evidence. *Cluster Computing, 19*(2), 723–740.
35. Rane, S., & Dixit, A. (2019). BlockSLaaS: Blockchain assisted secure logging-as-a-service for cloud forensics. In *International Conference on Security & Privacy* (pp. 77–88). Singapore: Springer.
36. Ricci, J., Baggili, I., & Breitinger, F. (2019). Blockchain-based distributed cloud storage digital forensics: Where's the beef? *IEEE Security and Privacy, 17*(1), 34–42.
37. Ruan, K., Carthy, J., Kechadi, T., & Baggili, I. (2013). Cloud forensics definitions and critical criteria for cloud forensic capability: An overview of survey results. *Digital Investigation, 10*(1), 34–43.
38. Ryu, J. H., Sharma, P. K., Jo, J. H., & Park, J. H. (2019). A blockchain-based decentralized efficient investigation framework for IoT digital forensics. *Journal of Supercomputing, 75*(8), 4372–4387.

39. Samanta, P., & Jain, S. (2018). E-Witness: Preserve and prove forensic soundness of digital evidence. In *Proceedings of the 24th Annual International Conference on Mobile Computing and Networking* (pp. 832–834). ACM.

40. Singh, A., Parizi, R. M., Zhang, Q., Choo, K. K. R., & Dehghantanha, A. (2020). Blockchain smart contracts formalization: Approaches and challenges to address vulnerabilities. *Computers & Security, 88,* 101654.

41. Tian, Z., Li, M., Qiu, M., Sun, Y., & Su, S. (2019). Block-DEF: A secure digital evidence framework using blockchain. *Information Sciences, 491,* 151–165.

42. Weilbach, W. T., & Motara, Y. M. (2019). Distributed ledger technology to support digital evidence integrity verification processes. In *International Information Security Conference* (pp. 1–15). Cham: Springer.

43. Xiong, Y., & Du, J. (2019). Electronic evidence preservation model based on blockchain. In *Proceedings of the 3rd International Conference on Cryptography, Security and Privacy* (pp. 1–5). ACM.

44. Yousaf, H., Kappos, G., & Meiklejohn, S. (2019). Tracing transactions across cryptocurrency ledgers. In *28th {USENIX} Security Symposium ({USENIX} Security 19)* (pp. 837–850).

45. Zhang, Y., Wu, S., Jin, B., & Du, J. (2017). A blockchain-based process provenance for cloud forensics. In *2017 3rd IEEE International Conference on Computer and Communications (ICCC)* (pp. 2470–2473). IEEE.

46. Zou, R., Lv, X., & Wang, B. (2019). Blockchain-based photo forensics with permissible trans-formations. *Computers and Security, 87,* 101567.

CHAPTER 3

Query Reformulation Based on Word Embeddings: A Comparative Study

Panos Panagiotou, George Kalpakis, Theodora Tsikrika, Stefanos Vrochidis, and Ioannis Kompatsiaris

3.1 Introduction

Given the abundance of online information, the discovery of content of interest by formulating and submitting queries to search engines and social media platforms is of paramount importance for practitioners in several fields, including experts involved in crime and terrorism-related investigations. Effective information retrieval requires though that the submitted query includes terms relevant to the vocabulary used in the sought content, so that the query and the available information are successfully matched. As this is a challenging task, automatic query reformulation, including term expansion, substitution, and reduction, can be

P. Panagiotou (✉) · G. Kalpakis · T. Tsikrika · S. Vrochidis · I. Kompatsiaris
Information Technologies Institute, Centre for Research and Technology Hellas (CERTH), Thermi-Thessaloniki, Greece
e-mail: panagiotou@iti.gr; kalpakis@iti.gr; theodora.tsikrika@iti.gr; stefanos@iti.gr; ikom@iti.gr

B. Akhgar et al. (eds.), *Technology Development for Security Practitioners*, Security Informatics and Law Enforcement, https://doi.org/10.1007/978-3-030-69460-9_3

41

employed so as to increase the likelihood of retrieving relevant documents higher in the rankings, even if they do not contain the terms in the original query.

The task of query reformulation usually requires the representation of the terms occurring in documents and the query in a way that effectively depicts their meaning and overall semantics; typically, vector representations are employed for this purpose. Into this direction, word embeddings have recently attracted much attention due to their effectiveness.

Word embeddings are real-valued vector representations of terms that are produced by neural network-based algorithms and that rely on the co-occurrence statistics of terms in a document corpus. The word embedding models are distinguished between global and local, based on the corpus used for their generation; the former entails the use of broad corpora covering a variety of topics, whereas the latter are based on more domain-specific corpora. The most popular word embedding algorithms are all neural network-based approaches and include Word2Vec [1], GloVe [2], and FastText [3].

In the particular case of terrorism-related material, the submission of effective queries to search engines and social media platforms is of vital significance for law enforcement and intelligence services, in terms of discovering and retrieving online content of interest for their ongoing investigations. To this end, query reformulation is an important tool that helps the investigators construct more effective queries, thus quickly reaching online content of interest that may not be discovered through the manual query formulation.

In this work, we compare the performance of global versus local embedding models when applied for query expansion using five datasets (four benchmark and one terrorism-related dataset). In particular, we apply two query expansion methods (i.e. CombSUM and Centroid) for identifying the most similar terms to each query term, using global and local word embeddings models, trained on our datasets. We focus on the GloVe algorithm, where co-occurrences are calculated by moving a sliding n-words window over each sentence in the corpus. We assess the effectiveness of 100- and 300-dimensional global and local word embedding models on the four benchmark datasets based on commonly used evaluation metrics in information retrieval, and we also perform a qualitative evaluation of the efficacy of the respective models on the terrorism-related dataset.

The remainder of this chapter is organised as follows: Sect. 3.2 discusses related work, and Sect. 3.3 describes word embedding approaches and the query expansion methods. Section 3.4 outlines the evaluation process, and Sect. 3.5 presents the experimental results. Finally, Sect. 3.6 summarises our conclusions.

3.2   RELATED WORK

The effectiveness of different query expansion methods using word embeddings on the retrieval task is discussed in [4] which reports that both the CombSUM and the Centroid methods (originally proposed in [5]) for combining the word embeddings of the query terms yield similar results. In addition, recent work has shown that a retrieval process employing query expansion based on local word embeddings can outperform a solution that uses global word embeddings [6]. As far as the appropriate dimensionality of an embedding model is concerned, it has been shown that, although there is a bias-variance trade-off in the dimensionality selection for word embeddings, the GloVe algorithm, as well as the skip-gram variation of Word2Vec (which uses a word to predict a target context), is robust to overfitting [7]. This means that although there exists an optimal dimensionality that is dependent on the training corpus, using a greater number of dimensions is not so harmful for the performance of the aforementioned embeddings, according to experiments in natural language processing tasks. In this work, we compare 100- and 300-dimensional embedding models.

3.3   METHODS

This section presents in more detail (i) word embeddings and their applicability in query expansion and (ii) the query expansion methods employed in this chapter.

3.3.1   *Word Embedding*

Word embeddings are real-valued vector representations of terms, produced by neural network-based algorithms that adopt the distributional hypothesis [8, 9], which states that words occurring in similar context tend to have similar meanings. Formally, in a word embeddings model, a term t in a vocabulary V is represented in a latent space of k dimensions by a dense vector $t \in R^{|k|}$. In the trained word embedding space, similar words converge to similar locations in the k-dimensional space.

The neural network-based algorithms can be applied on any available corpus of documents in order to learn the word embedding representations of the terms that exist in the given corpus. The most typical data sources for generating new terms include (i) large-scale external corpora that can be considered to reflect the overall term distribution in a given language, such as all Wikipedia articles in a given language [10], (ii) a document collection being searched in the current setting [11] that can be viewed as modelling term distribution in a particular domain, and (iii) documents relevant to the submitted query which are identified either interactively by the user or automatically by the system; in the former case, i.e. in the so-called relevance feedback cycle, the user proactively provides guidance in the form or relevant reference documents [12], while in the latter case, referred to as pseudo-relevance feedback, the top retrieved documents are assumed to be relevant [11].

If large-scale corpora, covering a sufficient number of diverse topics, are employed, the word embeddings generated on their basis are able to encode a broad context that enables their applicability in a variety of domains; we refer to such embedding models as global or universal. On the other hand, word embedding models learned on domain-specific corpora may be more beneficial in uncovering term relationships for terms with specific interpretations in those particular domains and contexts; such embedding models are referred to as local.

Given a user query and a trained (global or local) word embedding model, the goal of query expansion is to identify the top-r most relevant terms to (i) each individual query term or to (ii) the query as a whole, with r being a parameter to be defined by the user or the system. Those identified terms are then used for expanding the original query. The first option is the simplest in its implementation; the r most similar terms to each individual query term are identified using a similarity metric, such as cosine similarity, and they are added to the query. The second option requires more intricate techniques for queries that contain more than one term, but it is a more powerful solution.

3.3.2   Query Expansion

Irrespective of the algorithm deployed to produce the word embeddings, the linguistic or semantic similarity between two terms w_i, w_j is typically measured using the cosine similarity between their corresponding embedding vectors:

$$sim\left(\mathbf{w}_i,\mathbf{w}_j\right) = \cos\left(\mathbf{w}_i,\mathbf{w}_j\right) = \frac{\mathbf{w}_i^T \mathbf{w}_j}{\mathbf{w}_i \mathbf{w}_j}$$

Given a trained word embeddings model, the CombSUM and Centroid methods, presented in [4], are considered for the definition of the similarity of a term t (whose corresponding embedding is $\mathbf{t}$) to a query q consisting of M terms q_i where $i = 1,\ldots,$M (with corresponding embeddings $\mathbf{q}_i$).

CombSUM method The similarity score of each of the vocabulary terms to the query is calculated separately for every query term, and then a list L_{q_i} is produced for each query term q_i, containing the top n most similar terms. Subsequently, for each of the terms t that are included in L_{q_i}, the final similarity score is softmax normalised, so that it is in the form of probability:

$$p\left(t|q_i\right) = \frac{\exp\left(\cos\left(\mathbf{q}_i,\mathbf{t}\right)\right)}{\sum_{i' \in L_{q_i}} \exp\left(\cos\left(\mathbf{q}_i,\mathbf{t}'\right)\right)}, \text{while } p\left(t|q_i\right) = 0 \text{ for the terms } t \notin L_{q_i}.$$

Finally, the resulting term lists are fused so that the final similarity score between a query and a vocabulary term is defined as follows:

$$S_{CombSUM(t,q)} = \sum_{q_i \in q} p\left(t|q_i\right)$$

Centroid method The centroid method is based on the observation that the semantics of an expression can often be adequately represented by the sum of the vectors of its constituting terms. Consequently, a query q can be represented by a vector $\mathbf{Q}_{cent} = \sum_{q_i \in Q} \mathbf{q}_i$, and the similarity score between a vocabulary term and a query is defined as:

$$S_{cent(t,q)} = \exp\left(\cos\left(\mathbf{t},\mathbf{Q}_{cent}\right)\right)$$

where $\mathbf{t}$ denotes the L_2- normalised vector of a term t.

3.4   EVALUATION

This section describes the experiments performed in order to assess the performance of the global and local word embedding models in query expansion.

3.4.1   Experiments on Benchmark Datasets

The first set of our experiments was performed on benchmark datasets. In particular, we used the ClueWeb2009 Category B corpus[1] which has been extensively used by the TREC conference.[2] The corpus consists of 50,220,423 English-language Web pages which cover a wide range of subjects. These benchmark datasets are widely employed in order to assess the effectiveness of information retrieval and acquisition methods and thus allow us to determine the methods that are likely to provide the best results in an operational setting.

We used the topics of TREC 2009, 2010, 2011, and 2012 Web Tracks as queries in our evaluation experiments; each of these TREC tracks consists of a set of 50 topics (queries). Initially, we retrieved the top 1000 documents for each of the queries of a query set and then used the superset of those documents to train a local embedding model that corresponds to this query set. In fact, two GloVe models were produced by each query set, that differ in the dimensions of the embeddings. Specifically, we trained 100-dimensional and 300-dimensional embeddings. When applying the local models for the query expansion process in the retrieval experiments, we use those models that correspond to the query set where the specific query belongs.

The process of building each of the local GloVe models involved an initial step of extracting the main content of each retrieved Web page and removing its boilerplate using the python implementation of boilerpipe [13]. We have also experimented with the exact vocabulary for which the embeddings were built. More specifically, in an attempt to deal with the problem of misspelled terms, we considered embedding models where terms existing in only one Web document were not taken into account by the learning process. Indeed, those models completely outperformed models that included the complete set of words in the collection. In

---

[1] https://lemurproject.org/clueWeb09.php/
[2] https://trec.nist.gov/

addition, the exclusion of terms with a frequency of less than a threshold of five has led to improved retrieval performance in most of the cases. Therefore, we consider those local models built with this specific process for our further analysis.

As for the global embeddings, we used two of the GloVe embeddings trained on the union of Wikipedia 2014 and Gigaword 5 datasets, specifically the 100-dimensional and the 300-dimensional models.[3]

For each combination of query and embedding model, we performed retrieval using both the expansion methods presented above. In addition, we variated the number k of the expansion terms; k = 5, 10, 25, 50. For the retrieval process, we used the Indri search engine.[4] The initial query was combined with the expansion terms. Moreover, it was associated with a weight of 0.8, while the set of expansion terms was given a weight of 0.2. In total, we have conducted 32 experiments for each query, i.e. all the combinations of four embedding models (i.e. local and global GloVe-based models of 100 and 300 dimensions), two expansion methods (i.e. CombSUM and Centroid methods), and four different values for the parameter k.

We tested the performance of the retrieval processes using four commonly used evaluation metrics, namely, MAP (mean average precision), P @k (precision at k corresponds to the number of relevant results among the top k retrieved documents), nDCG@k (discounted cumulative gain), and ERR@k (expected reciprocal rank). For all the models with parameter k, we use k = 20. Both nDCG [14] and ERR [15] are designed for situations of non-binary notions of relevance, and ERR is an extension of the classical reciprocal rank.

3.4.2 Experiments on a Terrorism-Related Dataset

The second set of our experiments was performed on a terrorism-related dataset consisting of 329 Web pages containing text in English. This set was collected by domain experts and consists of Web pages referring to the religion of Islam and Islamism, to the Islamic State (ISIS), as well as pages containing news and references related to the region of Middle East (i.e. Israel, Palestine, Saudi Arabia, etc.). The content of the Web pages was downloaded and scraped. Similarly, with the experiments on the

---

[3] https://nlp.stanford.edu/projects/glove/
[4] http://boston.lti.cs.cmu.edu/Services/

benchmark datasets, we have employed the boilerpipe algorithm in order to remove content such as navigational elements, templates, advertisements, etc. The local embedding models were produced using the derived dataset based on the GloVe algorithm.

Given the small vocabulary size of this dataset (i.e. consisting of 7651 distinct terms), in order to produce the local GloVe models, we experimented with the *window size* and the number of epochs. After experimental tuning, we produced models with window sizes of 5 and 10, as well as epochs = 50 trained on 100 dimensions; we refer to the derived models as *local-wind5* and *local-wind10*, respectively. For our experiments, we used the 100-dimensional and the 300-dimensional global word embeddings employed at the experiments on the benchmark datasets; we refer to these models as *global100d* and *global300d*, respectively. In order to compare the efficacy of the retrieval process of the global versus the local word embedding models on the terrorism-related dataset, we extracted the top three terms generated by the two global and the two local word embedding variations for a number of terrorism-related search terms.

3.5   RESULTS

This section presents the evaluation results of the experiments on both the benchmark datasets and the terrorism-related dataset.

3.5.1   Benchmark Datasets

Following the experimental setting on the benchmark datasets, we computed the mean performance for each combination of an embeddings model with a query expansion process when applied to a query set, using the four evaluation metrics. We took this approach of analysis to better present and interpret the results.

Figures 3.1, 3.2, 3.3, and 3.4 present those mean performances, comparing the efficacy of the local models versus the global ones, for each query set and evaluation metric. Each plot depicts the results obtained with both the 100- and 300-dimensional models, in order to analyse the effect of the dimensionality of the models and the interdependence of the model's origin and dimensionality. In each plot, the eight different points corresponding to each embedding model (i.e. local or global) represent different combinations of the expansion method and the number of expansion terms used. Specifically, each point in the plots represents the

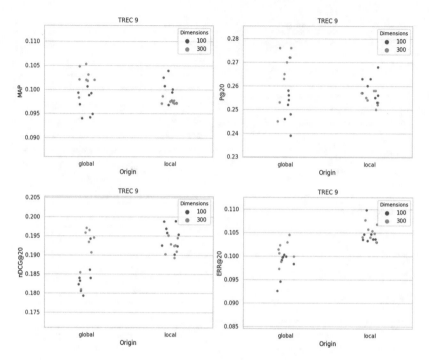

Fig. 3.1 The average performances of the local and global models for the query set of TREC 9, for the metrics MAP, P@20, nDCG@20, and ERR@20 on the retrieval task

average performance of experiments that use the same model, expansion method, and number of expansion terms. Blue points represent the 100-dimensional models and orange the 300-dimensional ones.

At a first level of analysis, the local models outperform the global ones when measured by the ERR@20 metric for the query sets of TREC 9 and 10, by MAP and P @20 for TREC 11, and by ERR@20 for TREC 12. On the other hand, the global models perform better than the local ones when measured by MAP for the queries of TREC 10. Overall, the results of Wilcoxon signed-rank test (nonparametric statistical hypothesis test) imply that both the origin of the model and its dimensions are important parameters in the retrieval process, since a modification in our choice for any of those parameters yields statistically significant change in the performance. As far as the dimensionality is concerned, the 300-dimensional models

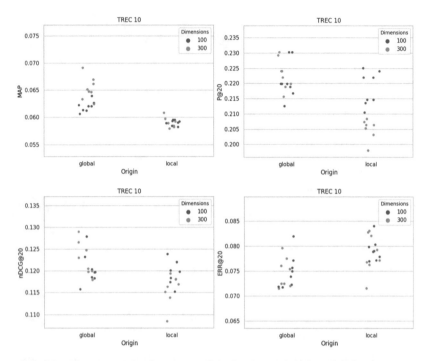

Fig. 3.2 The average performances of the local and global models for the query set of TREC 10, for the metrics MAP, P@20, nDCG@20, and ERR@20 on the retrieval task

outperform the 100-dimensional ones in MAP, P @20, and nDCG@20 for the queries of TREC 12.

At a second level, we observe that the optimal decision regarding the origin of an embedding model and its dimensionality are interdependent in many cases. On the one hand, there are cases where the comparison of the local models versus the global ones gives a specific outcome when considering only the 100-dimensional models, but a different one when considering only the 300- dimensional models. As an example, consider the MAP for the TREC 9 queries; the 100-dimensional local models are better than the 100-dimensional global models, but the opposite is observed for the 300 dimensions. Similarly, when observing from the dimensionality point of view, in many cases it is clear that the origin of the model also affects the outcome. For example, in TREC 9 and according

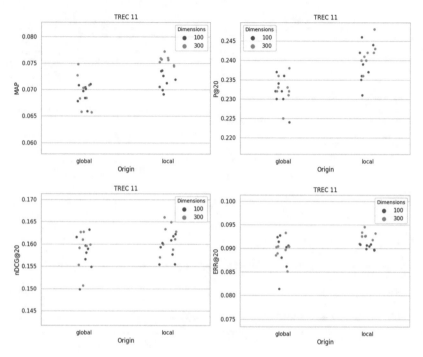

Fig. 3.3 The average performances of the local and global models for the query set of TREC 11, for the metrics MAP, P@20, nDCG@20, and ERR@20 on the retrieval task

to all metrics, the 300-dimensional global models outperform the 100-dimensional global ones, but among the local models, the 100-dimensionals are better, according to MAP, P @20, and nDCG@20.

As far as the expansion method is concerned, we paired experiments that share the same query, type of embeddings model, and number of expansion terms but differ on the expansion method. The Wilcoxon signed-rank test between those pairs has shown that choosing among the investigated expansion methods does not elicit a statistically significant change in the performance of the retrieval, for any of the evaluation metrics considered. This outcome is on par with the findings of Kuzi (2016), and it is important, especially when we consider the efficiency of the two expansion methods, since the centroid method is much more preferable than the CombSUM method in terms of execution time.

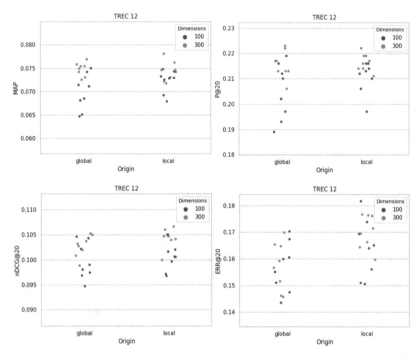

Fig. 3.4 The average performances of the local and global models for the query set of TREC 12, for the metrics MAP, P@20, nDCG@20 and ERR@20 on the retrieval task

3.5.2   Terrorism-Related Dataset

Table 3.1 presents the most relevant words to a number of search terms provided by domain experts based on their relevance to the terrorism domain, after employing the four embedding models used in this experimental setup.

The results illustrate the differences and complementarity between the local and global word embeddings on the presented search terms. While global word embeddings capture the overall context, local word embeddings provide interpretations relevant to the particular domain.

Consider, for instance, the term "karbala" that is relevant to "martyrdom" according to the local-wind10 model. This term refers to the Battle of Karbala that was fought in October 680 between the army of the second Umayyad caliph Yazid I and a small army led by Husayn ibn Ali, the

Table 3.1 Top 3 most similar terms to the search terms based on global and local word embeddings

Search term	local-wind5	local-wind10	global100d	global300d
Allah	Messenger	Exalted	God	God
	Blessings	Messenger	Almighty	Almighty
	Merciful	Blessings	Unto	Bless
Almighty	God	Exalted	Allah	Allah
	Blessings	Attributes	God	God
	Allah	Accept	Bless	Merciful
Apostates	Victory	Al-raqqah	Infidels	Infidels
	Software	Arabulus	Unbelievers	Unbelievers
	Fight	Alab	Traitors	Heretics
Believers	Thabit	Camp	Christians	Christians
	Rah	Thabit l	Adherents	Adherents
	Rabbi	Earned	Catholics	Nonbelievers
Jihad	Terrorism	Terrorism	Militant	Militant
	Tomorrow	Compared	Hamas	Hamas
	Intro	Converting	Islamic	Islamic
Martyrdom	Hell	Karbala	Martyr	Martyr
	Paradise	Thinking	Resurrection	Martyred
	Interview	Al-saleheen	Martyrs	Martyrs
Soldiers	Perish	Tools	Troops	Troops
	Fiqhnamaz	Rabab	Army	Army
	Libya	Peaceful	Policemen	Policemen
War	Crime	Crime	Conflict	Conflict
	Syria	Syria	Battle	Battle
	Crimes	Crimes	Civil	Civil

grandson of the Islamic prophet Muhammad; Husayn and his companions are widely regarded as martyrs by both Sunni and Shi'a Muslims. The Battle of Karbala has been promoted online as an example of religiously motivated martyrdom for revolutionary causes in the aftermath of the Arab Spring [16]. It is thus evident in this case that the local models produce related terms within the particular context of interest, while the global ones provide more universally related terms and in particular terms with the same root as the term "martyrdom".

Furthermore, the local models output "syria" as a term relevant to "war", while the global models have a preference over more general terms. The same also applies to the outputs for the search term "believers", such as "thabit" vs. "adherents"; the former is indeed related to the particular context of interest, while the latter is virtually a synonym to the search term "believers" and therefore could be considered in any context, and not only in this specific one.

Finally, there are cases, where the local models yield possibly unrelated terms; however, this may be attributed to the very small size of the domain-specific dataset on which those models were built.

3.6   Conclusions

In this work, we compared the performance of global versus local word embedding models for the task of query expansion based on four large-scale benchmark datasets and one domain-specific dataset related to terrorism. With regard to the benchmark datasets, our findings indicate that local models outperform global ones for the majority of the experiments run and the metrics employed. At the same time, it is evident that there is an interdependency among the origin of a model and its dimensionality.

Regarding the terrorism-related dataset, we found that the local models delivered relevant words to a number of terrorism-related search terms, despite the small size of the corpus. The domain could benefit from larger domain-specific corpora for building embedding models that can better capture the semantic relationships in a relevant vocabulary.

Acknowledgements This work was supported by the TENSOR (H2020-700024) and the CONNEXIONs (H2020-786731) projects, both funded by the European Commission.

References

1. Mikolov, T., Sutskever, I., Chen, K., Corrado, G. S., & Dean, J. (2013). Distributed representations of words and phrases and their compositionality. In *Advances in neural information processing systems* (pp. 3111–3119).

2. Pennington, J., Socher, R., & Manning, C. (2014). Glove: Global vectors for word representation. In *Proceedings of the 2014 conference on empirical methods in natural language processing (EMNLP)* (pp. 1532–1543).

3. Bojanowski, P., Grave, E., Joulin, A., & Mikolov, T. (2017). Enriching word vectors with subword information. *Transactions of the Association for Computational Linguistics, 5*, 135–146.

4. Kuzi, S., Shtok, A., & Kurland, O. (2016). Query expansion using word embeddings. In *Proceedings of the 25th ACM international on conference on information and knowledge management* (pp. 1929–1932). ACM.

5. Fox, E. A., & Shaw, J. A. (1994). Combination of multiple searches. *NIST special publication SP, 243*.

6. Diaz, F., Mitra, B., & Craswell, N. (2016). Query expansion with locally-trained word embeddings. *arXiv.* https://doi.org/10.18653/v1/P16-1035.
7. Yin, Z., & Shen, Y. (2018). On the dimensionality of word embedding. In *Advances in neural information processing systems* (pp. 887–898).
8. Harris, Z. S. (1954). Distributional structure. *Word, 10*(2–3), 146–162.
9. Deerwester, S., Dumais, S. T., Furnas, G. W., Landauer, T. K., & Harshman, R. (1990). Indexing by latent semantic analysis. *Journal of the American Society for Information Science, 41*(6), 391–407.
10. Balog, K., Weerkamp, W., & De Rijke, M. (2008). A few examples go a long way: Constructing query models from elaborate query formulations. In *Proceedings of the 31st annual international ACM SIGIR conference on Research and development in information retrieval* (pp. 371–378). ACM.
11. Xu, J., & Croft, W. B. (2017). Query expansion using local and global document analysis. In *Acm sigir forum* (Vol. 51, pp. 168–175). ACM.
12. Efthimiadis, E. N. (2000). Interactive query expansion: A user-based evaluation in a relevance feedback environment. *Journal of the American Society for Information Science, 51*(11), 989–1003.
13. Mikolov, T., Sutskever, I., Chen, K., Corrado, G., & Dean, J. (2013). Distributed representations of words and phrases and their compositionality. arXiv preprint arXiv:1310.4546.
14. Jarvelin, K., & Kekäläinen, J. (2002). Cumulated gain-based evaluation of ir techniques. *ACM Transactions on Information Systems (TOIS), 20*(4), 422–446.
15. Chapelle, O., Metlzer, D., Zhang, Y., & Grinspan, P. (2009). Expected reciprocal rank for graded relevance. In *Proceedings of the 18th ACM conference on Information and knowledge management* (pp. 621–630). ACM.
16. Karolak, M. (2016). Online aesthetics of martyrdom. *Political Islam and Global Media: The boundaries of religious identity*, 48.

CHAPTER 4

Evolving from Data to Knowledge Mining to Uncover Hidden Relationships

Konstantinos Demestichas, Konstantina Remoundou,
Ioannis Loumiotis, Evgenia Adamopoulou,
Wilmuth Mueller, Dirk Pallmer, Dirk Mühlenberg,
Rafal Kozik, Michael Choras, David Faure,
Roxana Horincar, Edward Benedict Brodie of Brodie,
Charlotte Jacobe de Naurois, Krishna Chandramouli,
and Alexandra Rosca

K. Demestichas (✉) · K. Remoundou · I. Loumiotis · E. Adamopoulou
Institute of Communication and Computer Systems, Athens, Greece
e-mail: cdemest@cn.ntua.gr; kremoundou@cn.ntua.gr; i_loumiotis@cn.ntua.gr;
eadam@cn.ntua.gr

W. Mueller · D. Pallmer · D. Mühlenberg
Fraunhofer Institute of Optronics, System Technologies and Image Exploitation,
Karlsruhe, Germany
e-mail: wilmuth.mueller@iosb.fraunhofer.de; dirk.pallmer@iosb.fraunhofer.de;
dirk.muehlenberg@iosb.fraunhofer.de

4.1   Introduction

Over the last decades, human activities have progressively moved from person-to-person to seamless interactions between the physical and the information technology (IT) worlds; crime has naturally followed the same path, with imagination as the sole limit. This, in addition to the diversity of the events that constitute a crime to be prevented, has forced the law enforcement agencies (LEAs) to research for evidence data from different data sources, such as video, audio, text/documents, social media and web data, telecom data, surveillance systems data, police databases, etc. making it a very difficult and time-consuming task to analyse them and conclude to end evidence results in order to fight terrorism and crime cases in time. So, the main challenge presented in this chapter is to assist the LEAs in their fight against crime, by resolving the problem of heterogeneity of the massive volumes of primary data collected, by fusing and analysing them in order to uncover hidden relationships among data items, compute trends for the evolution of security incidents, ultimately (and at a faster pace) reaching solid evidence that can be used in court, gaining also better awareness and understanding of current or past security-related situations.

To this end, novel approaches that will address the significant needs of LEAs in their fight against terrorism and organized crime, related to the prevention, investigation and prosecution of criminal offences, are

R. Kozik · M. Choras
ITTI SP. o.o., Poznań, Poland
e-mail: rafal.kozik@itti.com.pl; mchoras@itti.com.pl

D. Faure · R. Horincar · Edward Benedict Brodie of Brodie · C. J. de Naurois
Thales Research and Technology, Palaiseau, France
e-mail: david.faure@thalesgroup.com; roxana.horincar@thalesgroup.com;
edward-benedict.brodieofbrodie@thalesgroup.com;
charlotte.jacobedenaurois@thalesgroup.com

K. Chandramouli
Venaka Media, Kent, UK
e-mail: k.chandramouli@venaka.co.uk

A. Rosca
Siveco, Bucharest, Romania
e-mail: Alexandra.Rosca@siveco.ro

required. In the current work, the authors propose a system based on sophisticated knowledge representation, advanced semantic reasoning and augmented intelligence, well integrated in a common, modular platform with open interfaces. In order to produce court-proof evidence for the LEA's criminal investigation actions, the collection and unification of different evidence data sources is presented, as well as a common representation model for internal data representation. This unified data model is placed in the ontology that enables joint exploration and exploitation of the multiple diverse data sources, allowing the anticipation and prediction of the future trends (e.g. threats) and establishing the ground for reasoning and cognition (operational and situational awareness). According to the authors' knowledge, there are no any recent attempts to integrate all the above technologies to a simple system for the LEAs.

The chapter is structured as follows. In Sect. 4.2, the current approaches of big data collection and processing in order to fight crime and terrorism are presented. Next, in Sect. 4.3, the proposed system is explained, while in Sect. 4.4 the proposed methodology, the tools and the initial results are described in more detail. Finally, Sect. 4.5 concludes the chapter.

4.2 State of the Art

During the last years, there have been many attempts towards enhancing LEAs work with modern technologies. These attempts have been organized and funded by the European Commission, including projects such as RED-ALERT [1] which aims to create data mining and predictive analytics tools for complex event processing targeting mainly the social media and LASIE [2] which scope is to assist forensic analysis with multiple sources.

Another attempt towards enhancing LEA's work is TENSOR [3] which aims to provide a terrorism intelligence platform that will allow LEAs fast and reliable planning and prevention functionalities for the early detection of terrorist organized activities, radicalization and recruitment. VICTORIA [4] is another attempt that focuses on creating a platform that accelerates the video analysis for investigating criminal and terrorism activities. Other attempts in the security domain include ROBORDER [5] which aims to create a fully functional autonomous border surveillance system with unmanned mobile robots including aerial, water surface, underwater and ground vehicles which will incorporate multimodal sensors as part of an interoperable network.

Although these attempts aim to increase the efficiency of LEAs in their daily work, they lack of semantic and fusion techniques to assist the processing of information from multiple data sources. Our current approach proposes a novel system that will provide reliable threat assessment and prediction by semantic fusion and trends analysis, that will allow the identification of correlations and hidden relationships among data.

4.3   PROPOSED SYSTEM

The proposed system includes a set of tools that aim to facilitate LEAs in their daily work. These tools will allow the processing of data collected by different data sources and the identification of hidden patterns within the data that will allow LEAs officer to solve a case faster. Data processing will be based on an ontology model specified for this purpose. An ontology is a formal explicit specification of a shared conceptualization where, conceptualization is an abstract, simplified view of the world that describes the objects, concepts and other entities, existing in a domain along with their relationships. One of the most critical contributions of an ontology is its ability for providing the higher-level distinction of concepts which help understand systematically and consistently the lower-level details with domain concepts which is hard to obtain without ontological ways of thinking. The whole process of data processing includes five main steps:

- Data acquisition, which is the process of collecting primary data.
- Data preprocessing, which is the first step in making the input information understandable by the computer. It is different for each source of data. For example, for textual data, it is based on natural language processing techniques.
- Feature extraction is converting a set of input information into a set of numerical features.
- Feature representations. The proposed approach will rely on the use of bipartite graphs, more specifically a subset of the conceptual graphs to represent semantic information and knowledge.
- Semantic fusion and classification. Semantic information fusion typically covers two phases: (i) building the knowledge and (ii) pattern matching. The first phase incorporates the most appropriate knowledge into semantic information. Then, the second phase fuses relevant attributes and provides a semantic interpretation of the input data.

In the following sections, the reasoning, the semantic processing and the trend analysis will be presented in more detail.

4.4 Methodology and Tools

Knowledge representation [6] is the field of artificial intelligence that focuses on designing computer representations that capture information about the world that can be used to solve complex problems. It goes hand in hand with automated reasoning because one of the main purposes of explicitly representing knowledge is to be able to reason about that knowledge, to make inferences, assert new knowledge, etc. Virtually all knowledge representation languages have a reasoning or inference engine as part of the system. The advanced semantic reasoning services presented in this chapter enables a computable framework for systems to deal with knowledge in a formalized manner, allowing navigation through the different pieces of data and discovery of relations and correlations among them, thus broadening the spectrum of knowledge capabilities for the LEAs.

Reasoning is a procedure that allows the addition of rich semantics to data and helps the system to automatically gather and use deep-level new information. Specifically, by logical reasoning, the system is able to uncover derived facts that are not expressed in the knowledge base explicitly, as well as discover new knowledge of relations between different objects and items of data.

A reasoner is a piece of software that is capable of inferring logical consequences from stated facts in accordance with the ontology's axioms and of determining whether those axioms are complete and consistent. Reasoning with technologies like Resource Description Framework Schema (RDFS) and Web Ontology Language (OWL) allows adding rich semantics to data, and it helps the system to automatically gather and use deep-level new information, allowing also to derive facts that are not expressed in the knowledge base explicitly, as well as discover new knowledge of relations between different objects and items of data. In other words, reasoners are able to infer logical consequences from a set of asserted facts or axioms. In the last decade, due to growth of the Semantic Web field, some of the most popular reasoners were developed to fulfil this task on different domains and with the ability to cover use cases on different levels of complexity and expressivity. For the purposes of the proposed system, the ontology attributes have been categorized in the following types [7]:

- Reasoning characteristics: Basic features of ontology reasoners can be described by this category (e.g. methodology, sound, expressivity, incremental classification).
- Practical usability characteristics: This category describes the view angle of a developer using the OWL API support, availability and type of license.
- Performance indicators: This category addresses performance aspects such as classification performance and consistency checking performance, in particular for ontologies that are time critical or expect fast query, including reasoning and response time.

In order for a reasoner to infer new axioms from the ontology's asserted axioms, a set of rules should be provided to the reasoner. Rules are of the form of an implication between an antecedent (body) and consequent (head). The intended meaning can be read as: whenever the conditions specified in the antecedent hold, then the conditions specified in the consequent must also hold.

The antecedent is the precondition that has to be fulfilled that the rule will be applied; the consequent is the result of the rule that will be true in this case. An empty antecedent is treated as trivially true, so the consequent must also be satisfied by every interpretation; an empty consequent is treated as trivially false, so the antecedent must also not be satisfied by any interpretation [8].

For the reasoning process, the Pellet [9] reasoner has been employed, which can support the Semantic Web Rule Language (SWRL) rules. Furthermore, in order to allow the reasoner to infer the new axioms, the OWL Application Programming Interface (API) and the SWRL API have been also used.

4.4.1   Semantic Fusion Tools

The proposed system described in this chapter will benefit from the heterogeneous information collected by the different data sources, and it will combine the collected information in order to enable actions and decisions that would be more accurate than those that were produced by a single data source. However, the data from the different data sources might contain instances related to the same persons and/or events that cannot be easily identified. Data fusion [10] is the integration of multiple information and knowledge about the same object in order to obtain a more accurate description. The goal of data fusion is to improve data quality of that object, which can be achieved if the information is stored separately obtaining synergy, which can be defined as the representation of a whole is better than the representation of the individual components.

The scope of the fusion tools is to utilize the redundancy of the information collected by the different data sources in order to eliminate duplicate instances regarding the person and event identities and increasing the credibility of the collected information. Four different fusion tools will be available in the system: person fusion, event fusion, trajectory fusion and graph-based semantic information fusion. The person fusion tool will be responsible for finding different person instances in the knowledge graph that refer to the same person and fuse these instances by checking the similarities between the different person instances in the knowledge graph and calculating the probability of these different person instances to refer to the same physical person. It is based on a variation of the k-nearest neighbours algorithm [11], and it becomes apparent that one of the key decisions in the implementation of the algorithm is the distance metric that will be used for calculating the distance between different observations. Based on the data type of the features that will be either numeric or symbolic, the appropriate distance metrics will be used. These includes the Euclidean distance for numeric features [12] and the Jaro distance [13] for the symbolic/alphanumeric features.

On the other hand, regarding the event fusion tool, the definition of an event can be very diverse inside an ontology. An event is represented by many types, such as crime, incident, killing, robbery, riot, burglary, etc., using also the moment and the place where the event took place. Similarity functions are defined to compare the concept instances involved in the definition of an event, such as the ones used in the person fusion tool. Furthermore, if it is decided that two concept instances are to be fused, the fusion is done following the defined fusion operators.

In addition, trajectory fusion tool refers to a time-ordered sequence of geographical positions; trajectory captures the movement of an agent that corresponds to a moving object whose position can change over time, represented in the ontology by concepts such as person or vehicle. A trajectory concept has at least two different location items representing the origin and the final destination of the movement and several optional intermediate ones. Several possible fusion strategies can be defined in order to fuse two trajectories. A necessary condition for two trajectories to be fused is for them to have the same starting and ending points, defined by a place and time pair. Furthermore, more constraints can be added that refer to the intermediate positions, defined as place and time pairs. Finally, if it is decided that two trajectory instances are to be fused, the fusion is done following the defined fusion operators.

Lastly, graph-based semantic information fusion module provides a high-level semantic information fusion tool based on graph data and algorithms. The goal of this tool is to fuse any two pieces of semantic information that come from different data sources. In particular, it can also help deduplicate the ontology and increase its quality level. It relies on the use of bipartite graphs, more specifically a subset of the conceptual graphs [14, 15], to represent semantic information and knowledge. The result of the semantic fusion may be further used by other information processing tools implemented in the system. The application of conceptual graphs on trajectories is particularly motivated by the similarity of trajectory points, including both time and place. The implementation of the similarity functions may go from the simplest, that is based on an exact match, to more complex ones based on distance measurements.

In the example presented in Fig. 4.1, there are two trajectories, one done by a "Mr. Blue" and the other one by a "Mrs. Red". After fusing the common trajectory point A, the original two conceptual graphs depicting the two initial trajectories are connected into a single one. This highlights that both Mr. Blue and Mrs. Red have passed by the same place in the same time, and therefore their trajectories can be compared. In the case when it is decided that two different trajectories have some common trajectory points, the two trajectories may be fused into a single one or kept separate. Once the trajectory fusion step is done, several types of requests may be done on top of the fused information, serving various applications.

4.4.2   Trend Prediction

Following the fusion of the data, next step in the system is the trend identification and prediction through data clustering, classification and regression analysis.

Data classification can be used for use cases such as detection of a specific behaviour and sentiment analysis. This process is typically trying to identify a specific topic from a natural language source of a large volume of data. Text classification techniques help doing that as they divide the data classification process in three parts: model training, model verification and prediction. For model training and prediction of a behaviour,

Fig. 4.1 Example of two trajectories with a common trajectory point

clustering algorithms, decision trees and random forests algorithms are used, as well as regression analysis and neural models for identifying and predict hidden patterns and suspicious actions.

K-means clustering [16] is one of the most popular unsupervised algorithms for data clustering, which is used when we have unlabelled data without defined categories or groups. This is usually used when some expected behaviour is studied and when the number of groups under study is previously known. It is an iterative algorithm that assigns the data points to a specific – from the k known – cluster based on the distance from the arbitrary cluster centroid. During the first iteration, the centroids are randomly defined, and the data points are assigned to the cluster based on the least distance from the centroid. Once the data points are allocated, within the subsequent iterations, the centroids are realigned to the mean of the data points, and the data points are once again added to the clusters based on the least vicinity from the centroids. These steps are iterated to the point where the centroids do not change more than the set threshold.

Another algorithm, for behavioural identification is the decision or regression trees [16] that can be used for classification or regression predictive modelling problems. Regression or prediction trees use the tree to represent the recursive partition. Each of the terminal nodes, or leaves, of the tree represents a cell of the partition and has attached to it a simple model which applies in that cell only. To figure out which cell we are in, we start at the root node of the tree and ask a sequence of questions about the features. The interior nodes are labelled with questions, and the edges or branches between them are labelled by the answers. In the classic version, each question refers to only a single attribute, a single input variable (x) and a split point on that variable which has a yes or no answer. In order to make a prediction for a given observation, we typically use the mean of the training data in the region to which it belongs.

A greedy approach is used to divide the space called recursive binary splitting, a numerical procedure where all the values are lined up and different split points are tried and tested using a cost function. The split with the best cost is selected, and the procedure stops when a predefined stopping criterion is used, such as a minimum number of training instances assigned to each leaf node of the tree. If the count is less than some minimum, then the split is not accepted and the node is taken as a final leaf node. The stopping criterion is important as it strongly influences the performance of your tree.

Figure 4.2 illustrates an example of a decision tree applied to a dataset regarding financial data records (FDR) based on the purposes of the proposed system.

A method that is similar to decision trees is the random forest [17] which can be also used for both classification and regression problems. A decision tree gives the set of rules that are used in building models, which can be executed against a test dataset for the prediction. In decision trees, first step is to calculate the root node. Similarly, in Random Forest, each tree will predict a different target variable that we will sum with respect to a key. The key with the highest count, predicted by the maximum number of trees, is the final. There are several advantages to this method, such as making fast predictions and easy implementation. Also, there might be no capability to go all the way down the tree, if some of the data is missing, but a prediction still can be made by averaging all the leaves in the sub-tree.

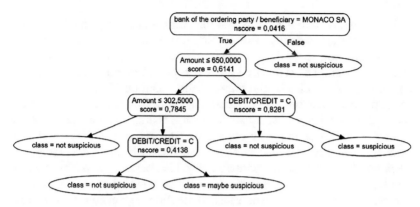

Fig. 4.2 Decision tree of the FDR dataset

The regression methods that are used for creating models which can identify and predict hidden patterns are statistical methods that examines the relationship between two or more variables of interest. The goal of the regression analysis is to predict the value of one or more target or response variables given the value of a vector of input or explanatory variables. In the simplest approach [17], this can be done by directly constructing an appropriate function y whose values for new inputs x constitute the predictions for the corresponding values of y, such as:

$$y = \beta_0 + \beta_1 * x_1 + \ldots + \beta_n * x_n$$

which is a linear regression model that combines a specific set of input values, the solution to which is the predicted output for that set of input values.

Learning a linear regression model means estimating the values of the coefficients used in the representation of the data. Below, there are some methods usually used for estimating the coefficients $\{\beta_0, \ldots \beta_n\}$ such as ordinary least squares [17], where we seek to minimize the sum of the squared residuals. This means that given a regression line through the data, the distance from each data point to the regression line is calculated, squared and summed. Another method is the gradient descent method [18], which is very useful in large datasets, works by starting with random values for each coefficient and the sum of the squared errors is calculated for each pair of input and output values. A learning rate (a) must be selected that determines the size of the improvement step to take on each iteration of the procedure as a scale factor, and the coefficients are updated in the direction towards minimizing the error. The process is repeated until a minimum sum squared error is achieved or no further improvement is possible. Finally, the regularization method [18] which seeks to both minimize the sum of the squared error of the model on the training data (using ordinary least squares) and also reduce the complexity of the model (like the number or absolute size of the sum of all coefficients in the model). This way also manages to avoid the problem of overfitting the data which can lead to model inaccuracy. Like before, the coefficients are chosen, such that they minimize the loss function through regularization procedures for linear regression such as the Ridge regression and the Lasso regression, whose main difference is the penalty of the high coefficients they use in order to minimize the loss function.

For non-linear problems, artificial neural networks and general regression neural networks are used. More specifically, ANNs [19] provide a general practical method for learning real-valued, discrete-valued and vector-valued functions from examples, and, thus, they can be used in any regression problem. They are based on simple units called perceptron that takes a vector of real-valued inputs x, calculates a linear combination of these inputs by using appropriate weights for each input and provide an output based on an activation function φ. Perceptron, as a simple unit, can only express linear decision surfaces; however, by using multiple perceptrons, we can build multilayer networks that can express a rich variety of non-linear decision surfaces. The most common type is the feedforward neural network where the perceptrons are fully connected and there is only one direction in the information flow in the network.

Usage of ANNs requires two phases: the training phase and the evaluation phase. In the training phase, the network tries to learn the weights of the neurons that better fit the desired output. For this purpose, the common backpropagation algorithm is employed. On the other hand, in the evaluation phase, the network is tested under unseen data that have been excluded from the training in order to evaluate its performance. Depending on how the data is split into the training set and the unknown dataset, the following techniques are commonly used: N-fold cross-validation, leave-one-out cross-validation and repeated random test-train splits [20].

Moreover, general regression neural networks (GRNN) [21] is a one-pass algorithm that provides estimates of variables and converges to the underlying (linear or nonlinear) regression surface. Assume that $f(x, y)$ represents the known joint probability density function of a vector random variable, x, and a scalar random variable, y. Let X be a particular measured value of the random variable x. The estimation of Y given x is:

$$\widehat{Y(x)} = \frac{\sum_{i=1}^{n} Y^i * expexp\left(-\frac{D_i^2}{2\sigma^2}\right)}{\sum_{i=1}^{n} expexp\left(-\frac{D_i^2}{2\sigma^2}\right)}$$

where $D_i^2 = \left(X - X^i\right)^T \left(X - X^i\right)$, and X is a specific value of the random variable x, n is the number of sample observations, X_i *and* Y_i are the sample values of the random variables x and y and σ is the smoothing parameter.

The selection of the smoothing parameter σ is an important issue in the creation of the GRNN network because it determines how closely the GRNN network matches to the prediction result with the training set data. A useful method of selecting σ is the holdout method. The main advantages of a GRNN network are their ability to learn fast and converge on the optimal regression surface as the number of samples increases. They are best used in cases where data is sparse, making it ideal in real-time scenarios, because the regression surface is directly defined throughout the space, even in the case of a single sample.

For the purposes of the proposed system, several machine learning approaches have been tested using a dataset regarding call data records (CDR). Specifically, in Table 4.1, the results of the GRNN compared to the other regression algorithms are presented regarding the estimated call duration.

Table 4.1 Results of regression models applied in CDR dataset

Model	Median absolute error (min)
Linear regression	6.03
Ridge regression	4.03
Lasso regression	4.42
Regression trees	5.89
Artificial neural networks (ANN) (4-5-2-1)	4.42
General regression neural networks (GRNN) ($\sigma = 0.09$)	3.63

4.5   Conclusions

In the current work, a novel approach based on state-of-the-art techniques in semantic processing and machine learning that aims to facilitate the work of the LEAs is proposed. Specifically, the proposed system employs semantic information fusion in order to infer assertions based on stated facts and rules provided by LEAs and fuses them in order to increase the reliability of the system. Then, machine learning techniques are applied in order to identify the trends and the patterns in the collected data and predict abnormal behaviour. The system is designed in order to provide evidence and back traceability that will allow the derived results to be used in court. The methodology and the initial results are presented in the current chapter, and it is expected that they will allow LEA's officers to solve crimes in less time.

Acknowledgement

 This work has been performed under the H2020 786629 project MAGNETO, which has received funding from the European Union's Horizon 2020 Programme. This chapter reflects only the authors' view, and the European Commission is not liable to any use that may be made of the information contained therein.

Bibliography

1. https://redalertproject.eu/
2. http://www.lasie-project.eu/
3. https://tensor-project.eu/
4. https://www.victoria-project.eu/
5. https://roborder.eu/
6. Markman, A. B. (2013). *Knowledge representation.* Cambridge: Psychology Press.
7. Abburu, S. (2012). A survey on ontology reasoners and comparison. *International Journal of Computer Applications, 57*(17), 33–39.
8. Lloyd, J. W. (1987). *Foundations of logic programming (second, extended edition). Springer series in symbolic computation.* New York: Springer Verlag.
9. Sirin, E., Parsia, B., Grau, B. C., Kalyanpur, A., & Katz, Y. (2007). Pellet: A practical OWL-DL reasoned. *Journal of Web Semantics: Science, Services and Agents on the World Wide Web, 5,* 51–53.

10. Dinca, L. M., & Hancke, G. P. (2017). The fall of one, the rise of many: A survey on multi-biometric fusion methods. *IEEE Access, 5,* 6247–6289. https://doi.org/10.1109/ACCESS.2017.2694050.
11. Hechenbichler, K., & Schliep, K. (2004). *Weighted k nearest neighbor techniques and ordinal classification.* Retrieved from http://epub.ub.unim-uenchen.de/
12. Chomboon, K., Chujai, P., Teerarassamee, P., Kerdprasop, K., & Kerdprasop, N. (2015). An empirical study of distance metrics for k nearest neighbor algorithm. In *Proceedings of 3rd International Conference on Industrial Application Engineering. Japan.*
13. Porter, E. H., & Winkler, W. E. (1997). *Advanced record linkage system.* U.S. Bureau of the Census, Research Report.
14. Sowa, J.-F. (1984). *Conceptual structures. Information processing in mind and machine.* Amsterdam: Addison-Wesley.
15. Chein, M., & Mugnier, M.-L. (2008). *Graph-based knowledge representation: Computational foundations of conceptual graphs.* London: Springer.
16. Deshpande, A. (2018). *Artificial intelligence for big data.* Birmingham: Packt Publishing.
17. Draper, N., & Smith, H. (1998). *Applied regression analysis.* New York: Wiley.
18. Bishop, C. (2011). *Pattern recognition and machine learning.* Cham: Springer.
19. Mitchell, T. (1997). *Machine learning* (1st ed.). Berlin: McGraw Hill.
20. Arlot, S., & Celisse, A. (2010). A survey of cross validation procedures for model selection. *Statistics Survey, 4,* 40–79.
21. Specht, D. (1991). A general regression neural network. *IEEE Transactions on Neural Networks, 2*(6), 568–576.

CHAPTER 5

Cyber-Trust: Meeting the Needs of Information Sharing Between ISPs and LEAs

Vasiliki Georgia Bilali, Dimitrios Kavallieros,
George Kokkinis, Pavlos Kolovos, Dimitrios Katsoulis,
Theodoros Anatolitis, Nikos Georgiou,
Nicholas Kolokotronis, Olga Gkotsopoulou, Clement Pavue,
Stefano Cuomo, Simone Naldini, Stravros Shiaeles,
and Gohar Sargsyan

5.1 Introduction

In the last decades, the Internet of things (IoT) has been reshaping the surrounding environment of our daily lives as well as the business models of small- and medium-size enterprises (SME) and large industries. The

V. G. Bilali (✉) · G. Kavallieros · P. Kolovos · D. Katsoulis · T. Anatolitis
N. Georgiou
Center for Security Studies, Athens, Greece
e-mail: g.bilali@kemea-research.gr; g.kokkinis@kemea-research.gr;
p.kolovos@kemea-research.gr; d.katsoulis@kemea-research.gr;
t.anatolitis@kemea-research.gr; n.georgiou@kemea-research.gr

B. Akhgar et al. (eds.), *Technology Development for Security*
Practitioners, Security Informatics and Law Enforcement,
https://doi.org/10.1007/978-3-030-69460-9_5

complex interconnection of large networked systems inevitably led to a considerable increase in the number of vulnerabilities related to IoT systems that are easily exploitable. The Cyber-Trust project aims at developing a platform that will reduce the potential cyber-threats towards IoT systems, secure evidentiary material for cyber-security incidents occurring in smart home networks, and facilitate the exchange of information (that might contain forensic evidence) on cyber-attacks between law enforcement agencies (LEAs) and Internet service providers (ISPs)[1]; the reduction of the time needed to electronically share such information in a secure way is a notable advantage. To achieve this goal, Cyber-Trust is

D. Kavallieros
Center for Security Studies, Athens, Greece

University of Peloponnese, Tripoli, Greece
e-mail: d.kavallieros@kemea-research.gr; d.kavallieros@uop.gr

N. Kolokotronis
University of Peloponnese, Tripoli, Greece
e-mail: nkolok@uop.gr

O. Gkotsopoulou
Vrije University Brussels, Ixelles, Belgium
e-mail: olga.gkotsopoulou@vub.be

C. Pavue
SCORECHAIN, Esch-sur-Alzette, Luxemburg
e-mail: clement.pavue@scorechain.com

S. Cuomo · S. Naldini
MATHEMA, Firenze, Italy
e-mail: stefano.cuomo@mathema.com; simone.naldini@mathema.com

S. Shiaeles
University of Portsmouth, Portsmouth, UK
e-mail: stavros.shiaeles@port.ac.uk

G. Sargsyan
CGI, Amsterdam, The Netherlands
e-mail: gohar.sargsyan@cgi.com

---

[1] In this chapter, by referring to Internet service providers, the authors refer to Internet access providers and telecommunications providers.

implementing a distributed ledger technology (DLT) – more specifically Blockchain – to store information relating to cyber-security incidents. Using Blockchain, Cyber-Trust will ensure the integrity and accountability of the information shared, along with fine-grained access control, enhanced privacy, scalability, and increased interoperability between the heterogeneous systems used by various actors.

Before choosing the most suitable type and architecture of the Blockchain to be implemented, among the other modules, Cyber-Trust gathered extensive end-user requirements that were translated to technical and architectural requirements. The requirements were gathered from (i) related work and state of the art of similar technologies (out of the self-products) and ongoing research activities; (ii) distribution of questionnaires; (iii) workshops with relevant stakeholders; and (iv) development and analysis of use cases.

One of the most significant questions in the Cyber-Trust project is if a Blockchain-based (i) communication between organizations and (ii) chain of custody is possible. The focus of the project is primarily in the communication through the Blockchain and the beginning of the chain of custody.

The use of a Blockchain in the particular context also gave birth to a number of questions from a legal point of view. Four major challenges had to be tackled during the conception and design phase. The first challenge emerges from the cross-border character of cyber-attacks whereas the law of evidence is primarily a domestic matter, creating implications for the admissibility of material that may contain evidence, to be stored and transferred through Blockchain, per concerned jurisdiction. To tackle this challenge, the different frameworks, including the international and European approaches in relation to evidence transfer and exchange, have been scrutinized, and an overall framework analysis has been provided. A second challenge was to make sure that the involved entities would have sufficient control over the information stored in and transferred through the Blockchain. This control is twofold: on the one hand, it refers to the control of the service provider over the information that it stores on the Blockchain – including the type of this information – and the information that it shares with the requesting authority and, on the other hand, the control of the information by the requesting authority after its receipt. Opting for a private permissioned solution and resorting to a combination of an on-chain/off-chain mechanism appeared to correspond better to the needs of such scheme. Another challenge focused on the selection of the Blockchain design/architecture that would ensure data integrity and confidentiality (security), by introducing different levels of access for the different entities. A fourth challenge was to provide sufficient safeguards for data subjects' rights, by foremost allowing the erasure of records, when no longer needed for the criminal proceedings, in accordance with the applicable national legislation, by making the system easily configurable and customizable.

Blockchain, thanks to some of their inherent features, may enable the cooperation between service providers and LEAs, by tracking the proceedings, including the handling of evidentiary material by all concerned entities as well as the communication between them. In the United States and the United Kingdom, Blockchain systems have been already piloted and/or used for the storage of electronic evidence, within the aim to amplify criminal proceedings and safeguard the validity of the methods deployed and the legality of the steps followed.

Cyber-Trust platform is focusing on two domains, namely, the smart home environment and the mobile/cellular devices. To demonstrate and validate the capabilities of the platform, a simulated environment is being built, which will emulate smart home environments and network traffic with an envisioned target of about 750 smart homes. The goal is to provide the service providers with a holistic mechanism to protect their services and seal their networks, by protecting all those devices where Cyber-Trust technologies have been deployed with the conduct of continuous vulnerability checks and the adoption of preventive or mitigation measures in case an attack has occurred or is very likely to happen. The service provider is also enabled in their communication with the LEA if information about the attack is requested by the latter for the prevention, detection, investigation, and mitigation of relevant criminal offences. To fully capture the wide range of functionalities, the platform aims at offering various types of cyber-security defences; 82 use cases were drawn [1].

The chapter is organized as follows. In Sect. 5.2, the related work is presented through selected products and research activities. The methodology of end-user requirement gathering is described in Sect. 5.3. The tools brought by Cyber-Trust (including the Blockchain) are presented in Sect. 5.4.

5.2 RELATED WORK

Cyber-Trust is aiming at developing an innovative platform that will gather information regarding cyber-threats and attacks and will detect and mitigate cyber-attacks, e.g. distributed denial of service (DDoS) against ecosystems in the domain of Internet of things. The domain of cyber-security is evolving in a daily basis, while the rapid growth of IoT does not allow to overlook the technologies already developed by companies and the ongoing research activities (mainly focused at the EU level).

5.2.1 Industry Solutions

Related industry solutions that are linked with Cyber-Trust framework and technologies are provided next; these include IBM Watson IoT platform, Motorola cyber-security services, F5 BIG-IP IoT intelligence platform, Intel IoT platform, Amazon web services IoT platform, and Ericsson IoT platform.

IBM Watson IoT platform [2–5]. It has been built with a security-by-design approach ensuring compliance with the ISO 27001 standard. Furthermore, Watson provides the following capabilities: (a) configuration and management of roles for users, applications, and gateways; (b) secure communications protocols, such as TLS v1.2; (c) high scalability and adaptability; (d) visualization analytics; (e) AI-driven analytics; (f) use of blockchain services to enable the validation of events generated by IoT devices; and (g) home appliance connectivity [6].

Motorola cyber-security services [7]. Motorola has four main services: (a) security patch installation to mitigate vulnerabilities by using pre-tested security updates as soon as they are available and validated; (b) remote services for security monitoring of security-generated events so as to implement countermeasures whenever necessary; (c) on-premise security operations centre (SOC) that delivers proactive security monitoring for unusual network activities; and (d) cyber-security risk assessment services relying on industry standards and frameworks to provide a comprehensive risk assessment and mitigation plan.

F5 BIG-IP IoT intelligence platform [8]. F5 Networks provide applications for IoT security, like IoT subscriber-aware firewall, protection against DDoS, SSL offload, protocol analysis, analytics, policy enforcement, and access control. The platform (a) can be integrated with third-party applications; (b) has the BIG-IP Local Traffic Manager (LTM) that handles network traffic offering services ranging from load balancing capabilities to complex traffic decisions depending on whether the applications are in a private data centre or the cloud; and (c) has BIG-IP domain name system (DNS) service that hyper-scales and secures the infrastructure during high query volumes and distributed denial of service (DDoS) attacks, ensuring that the applications are highly available across hybrid environments.

Intel IoT platform [9]. The platform is a family of products providing seamless and secure device connectivity. The three main characteristics of the platform are (a) security, delivery of trusted data with a tight integration of hardware- and software-based security that starts where data is most resilient to attacks; (b) scalability, achieve scalable computations from device to gateways and data centre solutions; and (c) manageability, get advanced data management and analytics from sensors to the data centre.

Amazon web services (AWS) IoT platform [10]. The set of services that are provided by Amazon enables high connectivity and management of IoT devices through the AWS cloud services. AWS offers three main services: (a) IoT services that includes *connected home*, i.e. appliances like voice-controlled lights, house-cleaning robots, machine learning-enabled security cameras, and Wi-Fi routers, and *industrial IoT*, i.e. a bundle of applications for predictive quality, maintenance, and remote monitoring of industrials equipment operations; (b) security, AWS IoT includes preventive security mechanisms, like encryption and access control to device data, while it also offers services to continuously monitor and audit security configurations; and (c) alerts, provides informative descriptions of security incidents via alerts so that the user can mitigate potential risks, like pushing a security fix to a device.

Ericsson IoT platform [11]. This platform-as-a-service (PaaS) accumulates sensor data from IP networks and focuses on the analytics and the management of the aggregated data. The PaaS includes a REST API, data storage functionalities, and OpenID access control for the data. The strength of the platform is the 'publish/subscribe mechanism' and the capability to query data streams (from local and external data sources) to perform analytical tasks.

5.2.2 Research Solutions

Research and innovation projects sharing many objectives with Cyber-Trust include SERIOT, SOFIE, REACT, ASTRID, SPEAR, SECUREIoT, CHARIOT, SEMIoTICS, and GHOST among others.

The *Secure and Safe Internet of Things (SERIOT)* project [12] aims to provide an open and reference framework for real-time monitoring of the traffic exchanged through heterogeneous IoT platforms within the IoT ecosystem. The goal is to recognize suspicious patterns, evaluate them, and finally decide on the detection of a security risk, privacy leak, and abnormal event detection, while offering parallel mitigation actions that are seamlessly exploited in the background.

The *Secure Open Federation for Internet Everywhere (SOFIE)* project [13, 14] creates a secure and open IoT federation architecture and framework. Distributed ledger technology (DLT) is employed, including blockchains and inter-ledger technologies, to enable actuation, auditability, smart contracts, and management of identities and encryption keys and to enable totally decentralized solutions with virtually unlimited scalability. The SOFIE project provides end-to-end security, key management, authorization, accountability, and auditability by utilizing DLTs where applicable.

The *Reactively Defending against Advanced Cybersecurity Threats (REACT)* project [15] aims to fight software exploitation and mitigate such advanced cyber-security threats in a timely fashion, based on four complementary actions: (a) probes actively, and in a transparent and ethical way, the network for identifying unknown vulnerabilities; (b) once aware of new vulnerabilities, automatically patches all vulnerable hosts of an organization, using software instrumentation, and secures them temporarily, until the official patch of the vulnerability is published; (c) detects exploited hosts and immediately isolates them from the rest of the network to limit malware propagation, and (d) analyses security incidents for forecasting future cyber-security threats. It also possesses a user interface with advanced visualizations to increase situational awareness for the entire life cycle of the product.

The *AddreSsing ThReats for virtualIseD services (ASTRID)* project [16] aims at shifting the detection and analysis logic outside the service graph by leveraging descriptive context models and their usage in ever smarter orchestration logic, hence shifting the responsibility for security, privacy, and trustworthiness from developers or end-users to service providers. This approach brings new opportunities for situational awareness in the growing domain of virtualized services, namely, unified access and encryption management, correlation of events and information among different services/applications, support for legal interception, and forensics investigation.

The *Secure and PrivatE smArt gRid (SPEAR)* project [17] aims at developing an integrated platform of methods, processes, tools, and supporting tools for (a) timely detection of evolved security attacks such as advanced persistent threats (APT), denial of service (DoS), and distributed DoS (DDoS) attacks using big data analytics, advanced visual-aided anomaly detection, and embedded smart node trust management; (b) developing an advanced forensic readiness framework, based on smart honeypot deployment, which will be able to collect attack traces and prepare the necessary legal evidence in court, ensuring at the same time user private information; and (c) implementing an anonymous smart grid channel for mitigating the lack of trust in exchanging sensitive information about cyber-attack incidents.

The SECUREIoT project [18] focuses on delivering predictive IoT security services that span multiple IoT platforms and networks of smart objects and are based on security building blocks at both the edge and the core of IoT systems. SECUREIoT will provide implementations of security data collection, security monitoring, and predictive security mechanisms to offer integrated services for risk assessment and compliance auditing against regulations and directives (e.g. GDPR, NIS, ePrivacy) and to support the IoT developers.

The *Cognitive Heterogeneous Architecture for Industrial IoT (CHARIOT)* project [19] will advance the state of the art by providing a design method and cognitive computing platform supporting a unified approach towards privacy, security, and safety (PSS) of IoT systems. It gives emphasis mainly in (a) a blockchain ledger in which IoT's physical, operational, and functional changes are both recorded and affirmed/approved; (b) IoT safety supervision engine for securing IoT data, devices, and functionality in new and existing industry-specific safety critical systems; and (c) a cognitive system accompanied by supervision, analytics, and prediction models.

The *Smart End-to-end Massive IoT Interoperability, Connectivity and Security (SEMIoTICS)* project's [20] main goal is to develop a pattern-driven framework, built upon existing IoT platforms, to enable and guarantee secure and dependable actuation and semi-autonomic behaviour in (industrial) IoT applications. The SEMIoTICS framework supports cross-layer intelligent dynamic adaptation, including heterogeneous smart objects, networks, and clouds. To address the complexity and scalability needs within horizontal and vertical domains, SEMIoTICS develops and integrates smart programmable networking and semantic interoperability mechanisms.

The *Safe-Guarding Home IoT Environments with Personalized Real-time Risk Control (GHOST)* project [21] aims at improving smart home security and privacy through the development of a user-friendly solution. The application will be based on technologies like DLT and techniques such as deep packet inspection. Furthermore, it will equip consumers with their own cyber-security inspection, discovery and decision toolset, and shift security focus paradigm from incoming data flows to the awareness and control of data going out. To this extent, the project has a threefold strategy: (a) implementation of extensively automated security; (b) exploitation of security-friendly behavioural patterns of the users; and (c) facilitation of the recovery process after security and/or privacy breach.

5.3 TOWARDS RESHAPING CYBER-CRIME INVESTIGATION PROCEDURES

As it was described in previous sections, given the volatile nature of cyber-attacks, one of the main aims of the Cyber-Trust platform is to reduce the time needed to exchange information, which might contain forensic evidence, regarding cyber-attacks between LEAs and Internet service providers. To effectively develop the tools targeting this specific challenge, a set of end-user requirement was gathered and analysed. Undoubtedly, it is a general truth that in most surveys the end-user's requirements may exceed the levels of feasible and create technical and legal hurdles. Also, they might harden the implementation phase of the project. For this reason, Cyber-Trust had selected the method of requirements prioritization (MoSCoW) before the creation of the questionnaires, in order to handle this risk.

5.3.1 *Platform User Requirements*

The requirements of the platform were derived from four sources [22], namely, (i) existing industry solutions and research activities-domain knowledge; (ii) analysis of Cyber-Trust use cases [1]; (iii) workshops with law enforcement agents; and (iv) questionnaires. It is important to highlight that the user requirements were gathered in an iterative manner so as to ensure that these are refined as the solution to be adopted becomes more mature. More specifically:

- 15 industry solutions and 16 research solutions were analysed throughout the project.
- A large number (82) use cases were considered.
- A single workshop with both LEA and ISP and multiple mini-workshops with LEA representatives were conducted.
- Distribution of questionnaires in two rounds – two (resp. three) questionnaires were designed for the first (resp. second) round of the process, leading to a total of 139 answered questionnaires (*see* Table 5.1), out of which 53 were completed by LEAs.

The end-user requirements were separated into functional and non-functional, based on the content of each requirement, and then they were prioritized according to the MoSCoW methodology.

As it is depicted in Fig. 5.1 the input from the aforementioned sources were the basis for the identification of the end-user requirements. After

Table 5.1 Total number of end-user requirements

Functional requirements				*Non-functional requirements*			
	1st round	*2nd round*	*Total*		*1st round*	*2nd round*	*Total*
Must	84	21	105	Must	46	9	55
Should	7	8	15	Should	5	6	11
Could	1	18	19	Could	0	0	0
Total	92	47	139	Total	51	15	66

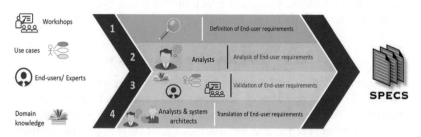

Fig. 5.1 Cyber-Trust platform specifications methodology

the analysis of the requirements, the end-users validated the requirements which were subsequently translated into a set of specifications for the development of Cyber-Trust's platform as well as the functionalities of its tools.

Through the needs deriving from the requirements, eleven (11) modules have been developed, which are depicted in Fig. 5.2. The platform's user interface (UI) has been developed to provide access to four types of users, namely, (i) platform's administrators; (ii) Internet service providers;

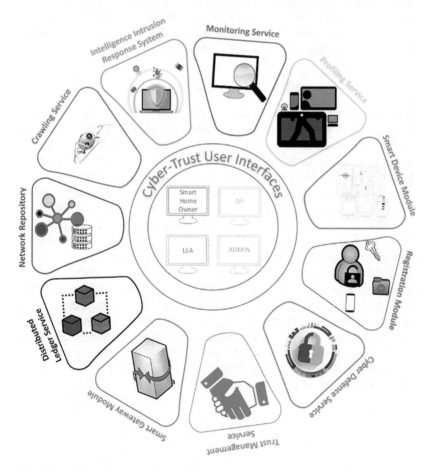

Fig. 5.2 Cyber-Trust modules

(iii) law enforcement agencies; and (iv) smart home (smart infrastructure in general) owners. The following sections are focused in the modules and the UI developed for the LEAs and are based on their needs, best practices, and applicable legal framework.

Table 5.2 presents briefly the Cyber-Trust modules without encompassing supplementary technological components and tools such as databases, open-source applications, etc.

Based on the framework of the project and the needs of the LEAs, the Blockchain will support the exchange of information related to forensic evidence and will also provide specific information through a tailor-made graphical user interface (GUI). Section 5.4 will elaborate and demonstrate both the Blockchain and the GUI of the LEAs.

5.3.2   LEAs Evidence Procedures in Cyber-Trust Platform

Cyber-Trust tools have been developed to enhance the communication between LEAs and ISPs in case of malicious activities while minimizing the time needed to review and exchange information; before their demonstration, two use cases will be provided based on DDoS and malware attacks against two different targets, smart home and the ISP's infrastructure. The two use cases are describing the steps and the actors involved, from the moment a cyber-attack has been detected.

The Cyber-Trust platform identifies abnormal behaviour on the smart home's network and/or on the device(s). Immediately and given the impact of the attack, it informs the owner of the smart home and of the device(s) about the security risk while initiating automated prevention and mitigation actions. In case a malicious activity is reported to the respective authorities, an investigation will be initiated to identify the cyber-criminals (e.g. malicious hacker) responsible for the malicious activity. Table 5.3 depicts the steps of an investigation; in the use cases presented below (DDOS and Malware attacks), it is assumed that the police has a Cyber-Crime Division (CCD) and a Forensic Science Division (FSD), acting as different divisions of the police.

Table 5.2 Cyber-Trust modules

Distributed ledger service (Blockchain) is a component of the Cyber-Trust platform related basically to integrity storage and enhanced sharing capabilities through the blockchain. It is divided to DLT service which is responsible for the core operational part and DLT admin module that is responsible for the component's administrative part. Some principal operations of the component are storage of data related to forensic evidence, validation of the transactions, consensus, etc.

Monitoring service is installed on the smart gateway, gathering data from the network and the devices with short-term goal of monitoring behaviour of the device

Cyber-defence service runs between hosts and devices information databases. Some of the principal operations of the component are detection and mitigation of cyber-attacks on networks and device level

Trust management system gathers the actions and the vulnerabilities of the IoT devices and calculates their trust score based on their behaviour and other characteristics. This component is basically connected with the defence operations of the device

Cyber-Trust registration module is part of the administration portal which is responsible for the registration of various actors (users, devices, organizations)

Network architecture and assets repository component is providing a set of tools allowing to get information on a network's architecture (including the topology and the security defences deployed therein), assets and their values, etc.

Smart device agent is a component which is running on the device and will inform the users for the device's health status (such as vulnerabilities detection, firmware updates, etc.). The users will be informed via alerting channels (like mobile app messages). Also, the device agent will have rule-based configuration options for detecting anomalies

Smart gateway agent is a component which is running on network gateway and is using machine learning packet analysis (MLPA) in order to identify anomalies. The inspected anomalies are feeding with other data components, like the monitoring and cyber-defence service

Intelligent intrusion response system (iIRS) is a Cyber-Trust component running on a network gateway at the user premises. This component will serve as a range of smart gateways and is responsible for continuous monitoring of the smart home's security status and the computation of possible mitigation actions to sophisticated cyber-attacks

Profiling service is the primary interface with the Cyber-Trust backend components and the responsible component for gathering and collecting information from the smart device agents (SDAs) and the smart gateway agents (SGAs) deployed on an ISP's network

Crawling service encompasses all the methods of harvesting the data available on the surface/deep/dark web and store them for further analysis

Table 5.3 Cyber-crime investigation steps depending on the type of the attack based on an amplified analysis (order or type of steps may vary across jurisdictions; thus, the overview is indicative)

DDoS attacks	Malware attacks
Report the incident to the national LEA	Victim should report the incident to LEA. A police preliminary inquiry is being conducted
Victim submits any piece of evidentiary item/data as its disposal for the assistance of the criminal investigation (log files, IP address, digital evidence) at its own initiative or as per LEA's request	Victim should submit any piece of evidentiary item/data at its disposal for the assistance of the criminal investigation (log files, IP addresses, digital evidence), at its own initiative or as per LEA's request.
LEA creates the case file by securing the digital evidence and conducting any other investigating actions that are considered necessary	LEA creates the case file by securing the digital evidence and conducting any other investigating actions that are considered necessary
Secured digital evidence is being sent to the FSD for forensic examination	Secured digital evidence is being sent to the FSD for forensic examination
The case file is being submitted to the pertinent prosecutor's office for further criminal evaluation	The case file is being submitted to the pertinent prosecutor's office for further criminal evaluation
The case file is being sent back to LEA with the prosecutor's order for conducting preliminary examination	The case file is being sent back to LEA with the prosecutor's order for conducting preliminary examination.
FSD, after the conclusion of the forensic examination, redacts a laboratory expertise report and exports any digital artefacts pertinent to the case for submission to LEA in order for any further investigative actions to take place	FSD, after the conclusion of the forensic examination, redacts a laboratory expertise report and exports any digital artefacts pertinent to the case for submission to LEA in order for any further investigative actions to take place
LEA further investigates any digital footprints/electronic traces that have occurred through the preliminary procedure (e.g. e-mail addresses, IP addresses)	LEA further investigates any digital footprints/electronic traces that have occurred through the preliminary procedure (e.g. e-mail addresses, IP addresses)
Digital traces that correspond to the same state where the procedure was initiated are being communicated to the respective provider company, and further information may be requested (including personal data of subscribers)	Digital traces that correspond to the same state where the procedure was initiated are being sent to the respective provider company, and further information may be requested (including personal data of subscribers)

(*continued*)

Table 5.3 (continued)

DDoS attacks	Malware attacks
If the digital traces correspond to EU countries, a formal request will be issued based on the EU evidence exchange and transfer scheme (e.g. European Investigation Order) or bilateral agreements between Member States	If the digital traces correspond to EU countries, a formal request will be issued based on the EU evidence exchange and transfer scheme (e.g. European Investigation Order) or bilateral agreements between Member States
If the digital traces correspond to non-EU countries, a formal request for legal assistance, based on multilateral or bilateral agreements, will be launched	If the digital traces correspond to non-EU countries, a formal request for legal assistance, based on multilateral or bilateral agreements, will be launched

5.4   CYBER-TRUST FOR LEAS

5.4.1   LEAs in Cyber-Trust Platform

Based on the needs of the LEAs, Cyber-Trust on the basis of the necessary legal requirements will enhance the communication and exchange process between the service providers and the LEAs during the criminal proceedings following a cyber-attack. Furthermore, the platform wants (i) to create a secure and fast communication channel between LEAs and ISPs based on the Blockchain; (ii) given the volatile character of cyber-crime, to enhance the process of information transfer from an ISP to a requesting LEA; and (c) to prevent, detect, and respond to cyber-threats against smart homes and mobile/cellular devices rapidly and efficiently. The main module for the LEAs is the development and integration of the Cyber-Trust Blockchain. As a prominent solution, and following an extensive research, the Hyperledger Fabric was chosen. It is the most advanced and maintained framework while having a very active community of developers allowing to provide support for new features and functionalities quite frequently [23]. The information of interest for LEAs that will be stored on the Blockchain will be the hash values and a very limited selection of metadata relating to actual information about abnormal/malicious activities (e.g. cyber-attack against devices secured by Cyber-Trust). The actual information will be stored off-chain in order to meet the legal requirements imposed by the EU legal framework (e.g. the right to erasure) and only for the time allowed by the national data retention framework.

Moreover, storing all this information on-chain would not be technically beneficial, provided that the ever-growing size of the Blockchain would have a great impact on the cost of maintenance.

5.4.2   Blockchain for LEAs

Cyber-Trust is building a Blockchain (CTB) which will be integrated in the system. It will support numerous functionalities, such as authority management, devices' critical information (e.g. firmware, configuration files), URL of the latest patches of each device under the supervision of Cyber-Trust, exchange of information, and metadata of evidence. The key point is that blockchain offers trust, integrity, transparency, accountability, and secure data which are key elements of the chain of custody procedure. The LEAs will exploit the attributes offered by the Blockchain based on the current legal framework and best practises. It is comprised of four core components: (i) the front-end user interface (UI, see Fig. 5.3) which gives the capability to end-users to view, invoke, or query blocks, transactions, chain codes, and so forth in the CTB; (ii) the Blockchain node ensuring the communication of the authorized participants with the CTB network; (iii) the trusted exchanged logs existing in the Blockchain which keeps the

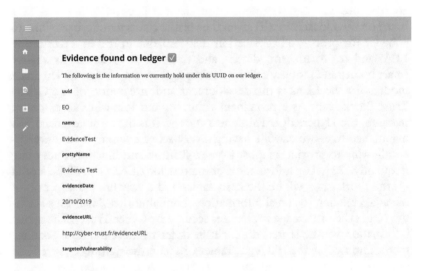

Fig. 5.3   Metadata of evidence stored on the DLT

historical record of facts about the exact time that an evidence piece was created and the way its ownership was transferred from one entity to another; and (iv) the forensic evidence DBs (off-chain database).

In order to guarantee security and validity of information in Blockchain processes, three confidentiality levels were created within Cyber-Trust; they are, namely, (i) public information, (ii) private information that can't be shared among partners, and (iii) private information that can be shared among partners. The confidentiality level is defined based on DLT data model [24].

It is important to briefly describe how the information that might hold forensic evidence is handled within Cyber-Trust, before detailed description of the CTB is provided. Assuming that a deviant behaviour is detected by the SGA/SDA, the minimum necessary amount of the respective data and metadata is collected and stored off-chain in the premises of the ISP for the limited time period allowed per jurisdiction, if such provision exists. Due to the legal limitations with regard to the approach proposed and implemented by Cyber-Trust, only the metadata are published on the Blockchain. The metadata which are going to be communicated through the Blockchain (on-chain) reach the third level of confidentiality, which means that this specific information is considered private information and can be shared among Cyber-Trust partners. Moreover, the metadata encompasses possible forensic evidence and other information useful to LEAs during criminal proceedings. Data is private and only visible to the creator of the data [25].

The CTB will also offer a communication channel between two or more interested parties (e.g. between LEA and ISP). Based on the decision made by LEAs, the communication between interested parties will be stored in the DLT. This information will be stored as private information that cannot be shared on both of the peers owned by the communicants. LEAs' main interest is DLT to be utilized in order to provide integrity and ownership of requests and, at the same time, to preserve the transparency and history of communications, for establishing a chain of custody.

As it is not possible to store information and data related to forensic evidence indefinitely and the time period is subject to national legislation, the Time to Live (TTL) feature was added. This will allow the automatic deletion of the data in specified (by the LEA) timeframe. As the timeframe of the TTL may differ in different legal frameworks, the TTL is passed in the DLT through the API with other parameters at the creation of the data. The API is based at the REST format built with Loopback in

JavaScript. This API is responsible to do the link between the components built by the Cyber Trust consortium and the smart contract of the DLT. In the mock-up, we used TTL of 5 years for the data shared between the peers, as an indicative timeframe, but the time can be adjusted to the particular needs of each jurisdiction. This means that after 5 years of storage inside the CTB, the data will be automatically deleted.

In order to successfully handle the chain of custody, CTB records and preserves the chronological order of (i) attacks' evidence and (ii) handling of this evidence (e.g. access history, read/write actions) [26, 27]. To ensure that only the appropriate members have access to the on-chain evidence and efficient permissions, a consensus algorithm has been implemented (Hyperledger Fabric) [23].

Towards that direction and according to the aforementioned (including the end-user requirements), CTB solution is based on the following principles:

- A private and permissive ledger is used.
- Only the metadata of the data which allegedly related to a cyber-attack will go on-chain.
- Access to the on-chain metadata will be provided only after LEA's request, which would satisfy all the legal requirements per jurisdiction.
- The actual forensic evidence will remain off-chain on the ISP side and will be accessed only after all the legal requirements have been satisfied.
- Time to Live feature is used for providing a way for the data to be deleted when no longer necessary for the criminal proceedings or based on the national legislation requirements for data retention.

5.4.3   LEAs User Interface (UI)

As mentioned before, during the gathering of end-user's requirements and platform's specifications, a large number of internal mini workshops with LEA experts occurred. The LEA experts, involved in the mini workshops, mainly were cyber-crime investigators and digital evidence examiners, with a small proportion of non-LEA digital forensic experts and Blockchain experts. The UI is characterized by simplicity and immediacy. More specifically, figures of user interface and attributes' usability are described below (Figs. 5.4, 5.5, 5.6, 5.7):

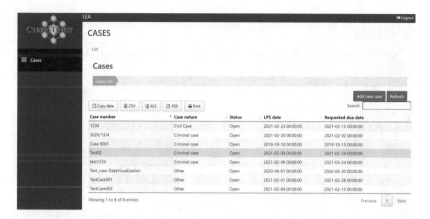

Fig. 5.4 List of cases of interest to law enforcement agencies

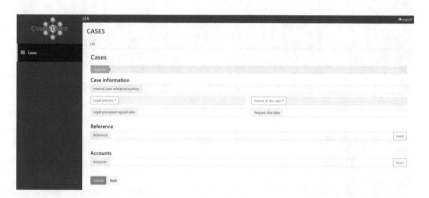

Fig. 5.5 Add a new case in the List of Cases in Law Enforcement Agencies (LEAs) User Interface (UI)

STEP A.   LEA user enters the URL link of the interface.

STEP B.   LEA user registers to the platform with specific username and password.

STEP C.   Users could see their cases list, with information regarding ID number of the cases, with the status of their cases (open/close), with reference of the case, and the date on which the case opened.

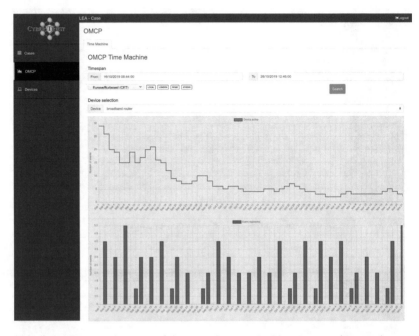

Fig. 5.6   The visualization of device activity in OMCP time machine and activity events

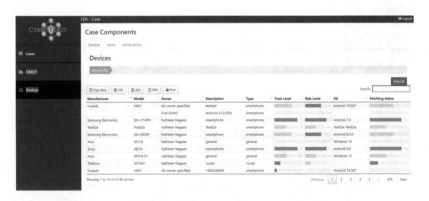

Fig. 5.7   List of devices which existed in the specific case

STEP D. Users have the capability to copy the list of cases with the aforementioned details, and the capability to print it in .pdf, .csv, and .xls forms.

STEP E. The user can also create a new case and refresh the case list.

STEP F. Users have already chosen the specific case from the list of cases, and then they choose whether they want to visualize the Operational Manager Control Panel (OMCP) Time Machine or the Devices.

STEP G. When users select the OMCP tab, they can select (a) the days duration, (b) the timespan in which they want to receive monitored information about the device, and (c) the specific device that they want to monitor.

STEP H. When users select the Device tab, they can see or add a device from the list of devices that existed in the account. Moreover, they have the capability to copy data from the list of devices, and the capability to print the data in. PDF, CSV, and XLS formats.

5.5 Conclusions

In a nutshell, we can say without any doubt that the development of LEAs methods and tools are encouraging advanced management and proactiveness with respect to cyber-attacks, by enabling communication between LEAs and ISPs. Also, the involvement of ISPs and smart homes in the Cyber-Trust platform enhances and emphasizes the public and individual cyber-security, respectively. Moreover, the end-user requirements are contributing to the improvement of system accuracy and applications towards cyber-threat mitigation. Emphasis should be given to the early identification of the potential hazards through the filtering of end-user's requirements. Finally, in implementation and validation phase, all the requirements should be cross-checked in order to validate users' needs alongside with platform's operations.

Acknowledgements

 Cyber-Trust project has received funding from the European Union's Horizon 2020 research and innovation programme under grant agreement No 786698. The content of this chapter reflects only the author's view, and the European Union is not responsible for any use that might be made of such content.

REFERENCES

1. Cyber-Trust: D2.3 Cyber-Trust Use Case Scenarios, [Online] available: https://cyber-trust.eu/wp-content/uploads/2020/02/D2.3.pdf. Accessed 18 Feb 2020, [p. 02].
2. IBM: Define and implement an IoT Security Strategy, IBM, [Online] available: https://www.ibm.com/internet-of-things/trending/iot-security. Accessed 01 Jul 2019, [p. 02].
3. IBM Security connect, manage and analyze IoT data with Watson IoT Platform, Watson Internet of Things, IBM, [Online] available: https://www.ibm.com/internet-of-things/solutions/iot-platform, 2019. Accessed 01 Jul 2019, [p. 02].
4. Murphy J. (2016). Enhanced Security Controls for IBM Watson IoT Platform, IBM Watson IoT Platform IBM [Online] available: https://developer.ibm.com/iotplatform/. Accessed 01 Jul 2019, [p. 02].
5. J. Clark, IBM and Whirpool: an innovative partnership, Internet of Things blog, IBM, [Online] available: https://www.ibm.com/blogs/internet-of-things/whirlpool/, 2016. Accessed 01 Jul 2019, [p. 02].
6. IBM X-Force Red Vulnerability Management Services, IBM Security, IBM, [Online] available: https://www.ibm.com/security/services/vulnerability-scanning, 2019. Accessed 01 Jul 2019, [p. 03].
7. Motorola solutions, CYBERSECURITY: Protect critical network infrastructure from cyber threats, Motorola, [Online] available: https://www.motorolasolutions.com/en_xp/managed-support-services/cybersecurity.html. Accessed 01 Jul 2019, [p. 03].
8. Prasad R. IoT Infrastructre, Empowered by F5's IoT solution", F5, [Online] available: https://www.f5.com/company/blog/iot-infrastructure-empowered-by-f5s-iot-solution. Accessed 01 Jul 2019, [p. 03].
9. Intel IoT Security and Scalability on Intel IoT Platform, [Online] available: https://www.intel.com/content/www/us/en/internet-of-things/iot-platform.html. Accessed 01 Jul 2019, [p. 03].
10. AWS Cloud Services, [Online] available: https://aws.amazon.com/iot/. Accessed 01 Jul 2019, [p. 03].
11. Ericsson IoT platform, [Online] available: https://www.ericsson.com/en/internet-of-things/iot-platform, Accessed 30 Jul 2019, [p. 03].
12. Seriot, Secure and Safe Internet of Things [Online] available: https://seriot-project.eu/, 2018, Accessed 01 Jul 2019, [p. 03].
13. SOFIE, Secure Open Federation for Internet Everywhere, [Online] available: https://www.sofie-iot.eu/, 2018. Accessed 01 Jul 2019, [p. 03].
14. SOFIE, State of the Art in Blockchain Technology and IoTSystems, [Online] available: https://www.sofie-iot.eu/news/state-of-the-art-in-blockchain-technology-and-iot-systems, 2018. Accessed 01 Jul 2019, [p. 03].

15. REACT, REactively Defending against Advanced Cybersecurity Threats, [Online] available: http://react-h2020.eu/, 2019. Accessed 04 07 2019, [p. 04].
16. Astrid, AddreSsing ThReats for virtuallseD services, [Online] available: https://www.astrid-project.eu/, 2019. Accessed 04 Jul 2019, [p. 04].
17. SPEAR, Secure and PrivatE smArt gRid, [Online] available: https://www.spear2020.eu/, 2017. Accessed 04 Jul 2019, [p. 04].
18. SecureIoT, [Online] available: https://secureiot.eu/, 2019. Accessed 04 Jul 2019, [p. 04].
19. CHARIOT, [Online] available: https://www.chariotproject.eu/About#LivingLabs, 2017. Accessed 04 Jul 2019, [p. 04].
20. SEMIoTICS framework, [Online] available: https://www.semiotics-project.eu. Accessed 04 Jul 2019, [p. 04].
21. GHOST project, [Online] available: https://www.ghost-iot.eu/ghost-project. Accessed 04 Jul 2019, [p. 04].
22. Cyber-Trust, D2.4 Cyber-Trust end-user requirements, [Online] available: https://cyber-trust.eu/wp-content/uploads/2020/02/D2.4.pdf. Accessed 18 Feb 2020, [p. 05].
23. Hyperledger Fabric, [Online] available: https://www.hyperledger.org/projects/fabric. Accessed 04 Jul 2019, [p. 09], [p. 11].
24. Cyber-Trust, D7.2 CYBER-TRUST distributed ledger architecture, [Online] available: https://cyber-trust.eu/wp-content/uploads/2020/02/D7.2.pdf. Accessed 18 Feb 2020, [p. 09].
25. Cyber-Trust, D7.3 CYBER-TRUST authority and publishing management, [Online] available: https://cyber-trust.eu/wp-content/uploads/2020/02/D7.3.pdf. Accessed 18 Feb 2020, [p. 10].
26. Brotsis S. (2019). Blockchain solutions for forensic evidence preservation in IoT environments. In: *2019 IEEE Conference on Network Softwarization (NetSoft)*, Paris, France, 2019, pp. 110–114, [p. 11].
27. Cyber-Trust, D7.5 CYBER-TRUST information and evidence storage, [p. 11]

CHAPTER 6

Cyber Ranges: The New Training Era in the Cybersecurity and Digital Forensics World

Athanasios Grigoriadis, Eleni Darra,
Dimitrios Kavallieros, Evangelos Chaskos,
Nicholas Kolokotronis, and Xavier Bellekens

6.1 Introduction

Nowadays, cybersecurity is considered as the key factor that an organization can be affected by a security incident. As there is a huge growth of cyber-attacks, successfully thwarting cybersecurity threats has become crucial up to the level of protecting people's everyday lives and the effect they

A. Grigoriadis (✉) · E. Darra
Center for Security Studies, Athens, Greece
e-mail: a.grigoriadis@kemea-research.gr; e.darra@kemea-research.gr

D. Kavallieros
Center for Security Studies, Athens, Greece

University of Peloponnese, Tripoli, Greece
e-mail: d.kavallieros@kemea-research.gr; d.kavallieros@uop.gr

B. Akhgar et al. (eds.), *Technology Development for Security
Practitioners*, Security Informatics and Law Enforcement,
https://doi.org/10.1007/978-3-030-69460-9_6

may have on a large number of businesses every year. For this reason, cybersecurity training can play a vital role in avoiding and defending against cyber-attacks and for securing systems, networks, and data, since they will teach all businesses' stuff involved how to protect against cybersecurity threats. Additionally, a cyber range (CR) is a micro-environment that offers tools and services to support the establishment of cybersecurity training courses and cybersecurity exercises to enhance the resilience and increase of cybersecurity capabilities of organizations. Furthermore, a cyber range allows the reproduction and execution of information technology (IT), operational technology (OT), and/or hybrid systems in a real or simulated environment. The diversity of cyber ranges across different sectors leads to the imperative need to review their current status focusing on the infrastructure they utilize and other relevant approaches having been taken into account for their development.

In this chapter, 27 CR environments are presented in Sect. 6.2, giving a short description of their capabilities, mission, and configuration features they embed and, in some cases, the highlighted advantages. The IT, OT, and Hybrid approaches of the CRs are presented in Sect. 6.3. Section 6.4 provides a full overview of the components of CRs, whereas Sect. 6.5 presents the operational impact of cyber ranges elements. The example of FORESIGHT approach, which brings together the state-of-the-art features of cyber ranges, is given in Sect. 6.6. Finally, a summary of this chapter is presented.

6.2   State-of-the-Art of Cyber Ranges

A CR can be categorized by the supporting sector, such as government (incl. military and LEAs oriented), academic, and commercial; its development depends on the design of several features such as flexibility, scalability, isolation, interoperability, effectiveness, access, service-based access, scoring and evaluation, and risk evaluation.

E. Chaskos · N. Kolokotronis
University of Peloponnese, Tripoli, Greece
e-mail: e.chaskos@uop.gr; nkolok@uop.gr

X. Bellekens
University of Strathclyde, Glasgow, UK
e-mail: xavier.bellekens@strath.ac.uk

6.2.1 Government, Military, and LEAs Oriented

It is well defined in the literature that there are plenty of CRs that have been indicated and identified in the government and military organizations determining the importance of cybersecurity training. Below some of the most known and often used cyber ranges for government and military organizations are presented.

6.2.1.1 Department of Defence (DoD) Cybersecurity Range

DoD CR environment is capable to support exercises, training, testing, evaluation, and education especially for the military parts [1]. Some of the modules embedded in this CR are the traffic generator, configurable user emulation, malware, spyware, and botnets emulation.

6.2.1.2 Arizona Cyber Warfare Range

The Arizona Cyber Warfare Range is a live-fire cyber warfare range created to rapidly train/upskill cybersecurity talent with hands-on learning [2]. Cyber Warfare Range drives innovation in cybersecurity techniques, technologies, and training across the United States and allied countries. This range is a privately funded non-profit entity and is 100% volunteer-driven.

6.2.1.3 Hybrid Network Simulation (HNS) Platform

HNS platform is an all-in-one cyber range that makes use of a turnkey platform for technical and operational preparedness of civilian and military cyber defenders. It includes multiple operations like up-to-date cyber range features and scenarios and a hyper-realistic and dedicated environment outside production systems for training in a red-blue team environment [3].

6.2.1.4 ManTech

ManTech is built on as an Infrastructure-as-a-Service model, coupled with more than a dozen tools. It instantly provides a precise emulation of any network environment, regardless of size, at any level of fidelity. Users have the ability to create replicas of existing systems and network structures and then to simulate those environments with realistic traffic, automated users, and even malware [4].

6.2.1.5 *École Navale CR*

École Navale's CR environment is used to support training and education for students, military, and researchers. It mainly focuses on naval systems, supervisory control and data acquisition (SCADA), and navigation equipment and makes use of the same programmable logic controllers (PLC) which are used on real ships. Furthermore, it is very flexible with virtual capacities. It is worth mentioning that services are still under construction [5].

6.2.1.6 *Airbus CR*

Airbus CR environment may support training and education for companies and the military. Specifically, it includes complete training stack, trainer console, training scenarios advanced customization tools, malware forensics, network security, penetration testing, certification, capture-the-flag benefits, user-friendly interfaces, and simplified management of the virtual environment. Some of the most worth-mentioned services are live traffic generation, individual and team training, multistep scenarios of threat, quick design and deployment of network infrastructures through a user-friendly interface, and customizable catalogue of assets and cyber-attacks [6].

6.2.2   Academic

There have been several efforts from universities to simulate the effects of computer network attacks. These approaches are primarily used for training and research from the students. An extensive list and description of the CRs are described below as follows.

6.2.2.1 *KYPO Cyber Range*

This CR is a realistic environment for cyber-training and support for cyber-testing, research, and training for students and researchers [7]. As hosted in the cloud, it includes several capabilities of web access, role-based access, user-specific content, dynamically creation and destruction of the virtual environments, and large target network replication for multiple and simultaneous usages. Some of the most important features include the complete training stack, training scenarios, advanced customization tools, malware forensics, network security, penetration testing, certification, and the benefit of capture-the-flag (CTF) environment.

6.2.2.2 Augusta University CR

The Augusta University CR environment is able to support exercises and training for education and research [8]. It is a training methodology that can lead to certified courses keeping in.

6.2.2.3 US Cyber Range

US CR is a scalable, cloud-hosted infrastructure that provides users with a virtual environment for realistic, hands-on cybersecurity labs and exercises [9]. One of the key features of this CR is that it is defined as an immersive environment because students can practice what they've learned in hands-on laboratory exercises. Furthermore, it is a cloud-hosted infrastructure that can be accessed from any device either from school or home, and instructors can deploy cloud-based virtual environments to students using a simple point-and-click interface. Additionally, having administrative access to the instance of cyber range software, customers create accounts for faculty and other cyber range users.

6.2.2.4 Austrian Institute of Technology Cyber Range

The Austrian Institute of Technology's CR is an environment for sharing knowledge in the cybersecurity domain for critical infrastructure providers, industry, research, and the public sector [10]. Advanced training exercises and competition on different levels, visualization, industrial control systems, digital networks, and critical infrastructures focus on cybersecurity research and development.

6.2.2.5 Saros Technology

Saros CR uses manufacturers like Cisco and Ixia's virtual machines to create real infrastructure for developing real cyberthreat scenarios [11]. Fully virtualized, the Virtual Test Lab is available on-demand, reusable, easily replicated, and self-service. Each cloud is executed as its own single file and is treated just like a document so it can be edited, copied, shared, and backed up.

6.2.2.6 European Space Agency (ESA) CR (by RHEA Group)

This CR is focused on cybersecurity and computer network defence for ESA [12]. The embedded services are instantiation of a full mission environment, mission control systems, pre-launch, launch, IoT, ground and satellite simulators, and operations and development network. It claims to be the perfect environment to support training and education.

6.2.2.7 Virginia CR

This CR provides an environment to increase the number, and the preparedness, of students entering the cybersecurity workforce in operations hosted in the cloud, web access, role-based access, user-specific content, dynamic creation and destruction of the virtual environments, and large target networks which can be replicated for multiple and simultaneous usages [13].

6.2.2.8 THE Michigan CR

The Michigan CR aims to strengthen Michigan's cyber defences by mitigating the growing number of cyberthreats and providing a more secure environment that promotes economic development. This can be accomplished by nurturing a cybersecurity industry that leverages Michigan's unique advantages, which include educational institutions, a large IT workforce, the manufacturing base, and federal cooperation with the security industry [14].

6.2.3   Commercial

Several commercial cybersecurity simulation products exist in the market. In the list below, there are the general-purpose CRs, without a specific academic or military orientation but with the possibility to be used by multi-domain users.

6.2.3.1 IXIA Cyber Range

This CR is used to provide an environment to train the participants of an organization to combat modern cyberthreats using a variety of IXIA's products. It can offer a service, flexible, scalable, application and threat intelligence, visualization modules, security information and event management (SIEM), and traffic generator. It can also provide complete training stack, trainer console, training scenarios, advanced customization tools, various training scenarios, capture-the-flag environment, and cybersecurity competitions [15].

6.2.3.2 Palo Alto Networks Cyber Range

It is used to train the participants of an organization to combat modern cyberthreats and enhance their prevention, detection, and response skills through hyper-realistic network simulation exercises [16]. It can also provide an isolated and realistic environment with network traffic-generator capabilities, application traffic-generator, and the support of multiple courses.

6.2.3.3 IBM Cyber Range
IBM CR delivers an environment in order to offer a training experience in a cyber-incident [17]. The objective is to exercise a rapid-response thinking in a pressured environment, understand how security solutions work together and experience on how the teams work together.

6.2.3.4 CybExer Cyber Range
CybExer CR is an environment that supports training and education for companies, military, and LEAs [18]. It offers a complete training stack, trainer console, training scenarios, advanced customization tools, malware forensics, network security, penetration testing training, certification, and CTF environment.

6.2.3.5 Raytheon Cyber Range
Raytheon CR provides an environment to support training and education for all different companies [19]. It offers a network environment emulation for air traffic control, power grids, water supplies, security operations centre (SOC) capabilities, scalable and agile architecture, automation, and interconnection with external hardware.

6.2.3.6 CYBERBIT Cyber Range
This CR is able to provide a hyper-realistic simulated training environment to enterprises, governments, and academic institutions [20]. This CR prepares the security team for the attack, by providing a complete training stack, trainer console, training scenarios, advanced customization tools, malware forensics, network security, penetration testing, certification courses, and capture-the-flag environment.

6.2.3.7 Breaking Point
Commercial appliances from breaking point are advertised as providing CR capabilities [21]. Their products provide traffic generation and a strike pack of network security and malware attacks in a single rack-mountable appliance. Traffic generation is highly configurable up to and including layer 7 of the Open Systems Interconnection (OSI) model. Breaking point uses simulation to achieve its high scalability. Large network topologies

involving hundreds of thousands of hosts can be simulated in a single appliance.

6.2.3.8 RGCE

The Realistic Global Cyber Environment (RGCE) utilizes modern ways to combine virtualization techniques, physical devices, and business-specific systems [22]. It is also possible to create tailored environments for an organization's specific training, exercise, or research and development needs. RGCE provides individual training for cybersecurity specialists, security analysts, and pen-testing operators, cybersecurity training and capability development for organizations and teams, ready-made business sector organization environments, digital forensics, and incident response training and exercises.

6.2.3.9 Berkatweb

Berkatweb is a cyber range with features with various security challenges and the ability of modelling and simulation of dynamic exercises and a fully customizable API framework [23]. This CR provides an extensible virtualized platform for cybersecurity training, modelling, simulation, and advanced analytics. It offers a secure environment in which to assess network and system attack and defend strategies as well as supplies a proven training path, helping your organization improve its cyber resilience and maturing. The CR's architecture gives users the ability to test, evaluate, and train for next-generation threats, similar to training on a traditional weapons test range.

6.2.3.10 CYBERGYM

CYBERGYM emulates complex cyber-attack scenarios in OT and IT environments [24]. Apart from that, it is a common cyber range model which provides training for all departments, specializing in active cyber defence, event mitigation, and crisis management. The three teams participate in the training sessions including the red team, the blue team, and the white team. Utilizing the red team throughout the training provides unique insights into a hacker's mindset and point of view. The blue team is faced with real attacks they have to identify, defend, and harden their environment against using the necessary methods and tools, while the white team manages the training and debriefing process, reviews the blue team's performance, and provides recommendations.

6.2.3.11 CyberCENTS

The CENTS® platform solutions provide a relevant, integrated, Live-Virtual-Constructive (LVC) cyber range environment for demonstration, training, exercising, tool development, and testing full-spectrum cyber-space capabilities [27]. The CENTS solution permits closed-network engagements or use of a virtual private network (VPN), engagements with multiple interconnected environments. Each CENTS unit has an IEEE RFC-compliant traffic generation that features dynamic traffic flows and protocols that can be manipulated to follow a customer profile. The emulated elements of the environment (e.g. users, traffic, attacks, the Internet) interact with the virtualized and physical elements of the system providing true-life system response. This CR supports social media services and multilayer and dynamic websites. All IP addresses and website URLs resolve in the cyber range's domain name system (DNS). All virtualized Internet IP space uses real-world geo-IP addresses. The cyber scenarios can be executed in automatic or manual mode.

6.2.3.12 Silensec Cyber Range

The key benefit of Silensec CR includes cloud technology improvement on how to scale up to thousands of concurrent users and virtual environments. It is available as a service or hosted as a highly secure on-premise cyber range, capabilities for integration with IoT and supervisory control and data acquisition/industrial control system (SCADA/ICS) environments. Additionally, it supports individual and team-based cyber exercises, not to mention the competence-based scoring and assessment system embedded [28].

6.2.3.13 Cisco Cyber Range

The Cisco Cyber Range is offered as a service [38]. It is a training course that aims to train the participants to combat modern cyberthreats. Cisco CR is based on real-world scenarios and provides a war-gaming environment that allows participants to play the role of both the attacker and the defender, in order to learn the latest methods of vulnerability exploitation and the use of advanced tools and techniques to mitigate the threats. The Cisco CR provides real-life experience of reacting to and defending against complex cyber-attacks, including advanced persistent threats (APTs). It provides training in security methodologies, operations, and procedures using a variety of security tools and techniques.

6.3   IT, OT, AND HYBRID APPROACHES
OF CYBER RANGES

Cyber ranges offer an environment for teams to train collectively, improve their cyber defence skills, and gain critical insight into a variety of stakeholder actions within every organization. This tends to improve teamwork across the enterprise and the communication skills between stakeholders as it gives teams a better understanding of what other departments or people are responsible for. This is critical to building a successful incident response team, and it's difficult to obtain that experience through conventional training simulations. Teams are trying to support an IT environment, an OT environment, or most often a combination of them, a hybrid one. In the table below, a categorization of CRs is depicted depending on the approach of the network environment and their involved assets:

CR approach	Description	Implementation
IT	IT cyber ranges provide a complete and tested framework to help IT security organizations improve their overall security posture. Used by companies and governments around the world, cyber ranges offer hyper-realistic simulated training and testing scenarios, which dramatically improve cybersecurity performance while providing tools for simulating various network setups, attack scenarios, and traffic patterns	As network attacks have been very common now, understanding their mechanism and knowing how to detect and respond to it is essential to the cybersecurity of every enterprise. Distributed denial of service (DDoS) SYN flood, DNS amplification attack, and man-in-the-middle attack will be taught in the hacker's perspective and the defender's perspective to equip the students with an all-rounded understanding of the attacks and prevention approaches
OT	Hands-on training simulations that are ultra-realistic and safe regarding that OT services and machines rely on	It is about the creation of a replica environment of an operational ecosystem of an organization in order to exercise on it. That means the hardware and software detect or cause a change through the direct monitoring and/or control of physical devices, processes, and events in the organization. Further to that, it is related to testing, fine-tuning, and perfecting the response to cyber incidents. The result is to gain interactive hands-on learning by leveraging the OT incident response and building an integrated cyber response strategy

CR approach	Description	Implementation
Hybrid	Traditional IT security training is largely ineffective because it relies on sterile, mostly theoretical training. To get the security teams prepared to face today's multidimensional IT and OT security challenges, the focus should be in a technology-driven environment that mirrors organizations own, facing real-life threats. In other words, it could be defined as hyper-realistic hybrid simulation	The potential of simulation-based training, as compared to traditional training, is substantial. Organizations can not only train people but also test processes and technologies in a safe environment. Furthermore, security teams can be trained as individuals or as a group, to improve their teamwork. With the help of simulation, a team can experience high-fidelity threat scenarios while training and improve their capabilities, rather than encountering these threats for the first time during the actual attack. This results in a dramatic improvement in their performance

6.4   Components of Modern Cyber Ranges

Modern cyber ranges are referring to the ones that need realistic, industry-relevant content as well as trainees' tools to practice governance activities in emulated networks. This will help all trainees to better understand how to address a threat in real-life scenarios. It is well-known that cybersecurity attacks require teams to combat them, and for that reason cyber ranges are able to allow the team training and engagement for professionals to gain a better understanding of what it really needs to be taken into consideration in order to stop evolving threats.

6.4.1   Artificial Intelligence (AI) and Machine Learning

With advances in AI and machine learning, cyber ranges are considered vital to leverage such technologies, especially in the last few years. The collected intelligence from the information gathering component and the collected network-flow samples from various attacks will feed the threat data visualization tools, threat simulation tools, and threat forecasting tools regarding AI and machine learning new capabilities and "zero-day" features. Big data and machine learning can enhance this process by learning how to automatically detect unusual patterns in web traffic. The huge raise of encrypted traffic makes machine learning valuable because of its capability of monitoring previously unseen encrypted network traffic [25]. Additionally, a threat forecasting module can help to improve the security posture prior to an attack regarding such kind of attacks. The holistic view with the gathered intelligence accompanied by the data analysis of the pattern and trends of the attacks and the network flow analysis of AI and machine learning will improve the cybersecurity awareness of trainees (e.g. LEAs). In the end, they will be able to build a threat modelling for their needs of proactive data-driven by big data analysis [26].

6.4.2   *Information Gathering and Sharing*

An important element of the planning stage of a cyber range training scenario is to ensure that scenarios are realistic and up to date. In order to achieve this, it is necessary to receive information from multiple sources. such as the process of sharing data on attacks, malware, malware indicators, indicators to compromise, research statistics, incident reports, or even suspicious actions and methodologies. Results and conclusions from threat hunting and research procedures related to the training scenario under development should also be included [29]. Cooperation and information sharing are considered to be a key important training objective during a cyber defence exercise or training in a cyber range as dependencies between each other systems, similar networks, and similar attacks should foster cooperation and information sharing between the blue teams [31]. It should be mentioned that information sharing is also included in the major aspects of special scoring during the evaluation phase. There are various modules and extensions used for threat information sharing framework which are used for collecting, processing, and exporting high-quality indicators of compromise (IOCs). This allows a security analyst or a player – blue team leader to collect and standardize structured and unstructured threat intelligence. Applying threat intelligence to security operations enriches alert data with additional confidence, context, and co-occurrence. This means that users and trainees can apply research from third parties to security event data to identify similar, or identical, indicators of malicious behaviour and enrich a scenario with it.

6.4.3 Gamification and Serious Gaming

In the context of training and military strategy, games and simulations are well-known to Roman military commanders as tools to visualize and manipulate small physical representations of battlefields as Smith R. said in [33]. Nowadays, various methodologies have endeavoured to introduce gaming components in cybersecurity education using cyber ranges. These methodologies vary from using simple games for beginners and non-experts to cybersecurity training ecosystems for cybersecurity professionals from the frame of cybersecurity sectors. Through gamification, trainees can obtain the appropriate practical skills and the corresponding to an incident or a special cyber-attack occasion. A problem-solving mechanism through lab environments enables the participants to grasp the problem's full details but also the key decision-makers to find a solution, in a cyber-security incident, which increases cyber-resilience and their creative-thinking methods. Most of the game genres are applicable to cybersecurity training as well [26].

Serious gaming is simulations that are often adopted by those organiz-ing cyber war games, involving the drilling or training of military and security personnel. These types of games are a cyber version of different activities well-known in the military and/or LEAs. One type of cyber exer-cise that can utilize a full-scale simulation is an activity known as "capture the flag". CTF activities are often selected for large, international exer-cises, such as Locked Shields [36]. CTF is a form of war gaming where participants are divided into red and blue teams, with red teams playing the part of the aggressor or hacker and blue teams defending. Depending on the nature of the scenario, blue teams may be required to work together or independently to achieve game goals. In gameplay, teams are awarded points depending on how deep they penetrate a defended network or how swiftly they respond to and remedy an incident or attack.

6.4.4 Evaluation Module

The exercise life cycle ends with an evaluation. As cyber range is the test-bed for cybersecurity and cyber defence exercises, an evaluation should be conducted after the exercise training in a specific cyber range that consists of a collection of:

- Evaluation (after action) workshop
- Feedback survey (after-action report)
- Scoring subsystem

The most visible part of this phase is the evaluation workshop attended by the blue team (CR defenders' team). The red team (CR attackers' team) prepares an overview of its success in attacks against particular teams and best practices related to the attacks used in the exercise. Both teams benefit from data collected and entered into the scoring subsystem. Furthermore, the green team (CR technical team) stores all collected logs during the exercise of other teams if needed. Feedback provided by the blue teams in the survey before the evaluation workshop is also incorporated. The evaluation workshop shows the exercise scenario and timeline from the perspective of the red team and the white team. It is the only opportunity when the learners can authoritatively learn about attacks used by the red team. They can discuss their approach in particular situations and phases. Until this point, they were only able to see the results of their experimentation during the exercise without an explanation of why something happened. It is therefore recommended not to underestimate this part of the exercise and deliver analysis and lessons that will have value to the learners. For instance, a handout with best practices for system hardening might be useful in the daily routine of the participants [30]. During the evaluation phase, a member of the design team or another selected staff member should develop an after-action report that determines the functionality of the tested systems or components [32]. The introduction to the after-action report should document background information about the test such as the scope, objectives, and tests. The after-action report should also document observations made by the test team during the test and recommendations for enhancing the IT plan that had its components or systems tested, along with associated procedures and components. The after-action report should also include a list of test participants and may provide information from any participant surveys that were distributed during the hotwash (workshop) to solicit feedback.

Not only during the exercise life cycle but especially during the evaluation workshop and the after-action report, evaluation tools should be used. The main tool is a scoring subsystem which is used for the penalization of blue teams' life cycle. During this phase, scenario-specific data is used to define scoring rules. Attack plans, objectives, and penalty values are set according to the expected goals of the exercise and learners' skills

[31]. The scoring system is considered an essential part of CR exercises and provides feedback and the option for comparison to the technical blue teams (BTs). The scoring system is a mixture of graded procedures by human evaluators and an automated process by CR systems. The scoring aspects that can be measured are based on the red team (RT) reporting, analysis of yellow team observations, and decisions of white team members. Detailed scoring rules should not be released to the BTs or players in order to avoid them focusing only on how to get higher scores. The following categories are proposed to be measured [31]:

- Availability of services
- CR usability
- Successful red team attacks
- Situation reporting
- Responding to injects
- Requests for support to GT
- Special scoring

where the last category includes penalties, bonus points for outstanding performance, and information sharing, amongst others.

6.5   OPERATIONAL IMPACT OF CYBER RANGE ELEMENTS

The advantages of using exercises such as simulations as training tools have been well-known especially to military and security personnel for centuries. Such activities provided practice for soldiers in preparation for real situations and actual combat.

6.5.1   Impact of Training in Cybersecurity/Defence

From the perspective of organizational learning, there is no fundamental difference between a simulated event and a real incident [35]. Conducting cyber range exercises can help with validating policies, plans, and procedures and with training, improving current tools, or rolling out new equipment; testing information and communications technology (ICT); and identifying gaps in resources. As a result, this kind of exercises is of benefit to a broader range of actors and organizers than simply military and security law enforcement organizations [33]. Cyber range training can be carried out by small, individual entities such as single ministries or private firms or, in the case of large multinational simulations, exercises which can involve a multitude of actors from different areas of the security nexus, such as private corporations, government ministries, utility providers, and military units. Additionally, cyber ranges tend to be highly technical in nature with a focus on testing technological capabilities and resources. This is an important aspect of national cyber defence and cybersecurity in combination with full-scale simulations in procedural scenarios. Conflict in cyberspace, while new and technically challenging, still conforms to traditional models of conflict. As do defenders of other domains, defenders of cyberspace strive to minimize the fog of war (is the uncertainty in situational awareness experienced by participants in military operations), either deliberately or intuitively. However, the volume, velocity, and variety of operations in the cyber defence domain, coupled with enormous attack surfaces and the low cost to adversaries of mounting a cyberattack, make the goal of minimizing both information ambiguity very difficult with the tools available. The findings, training applications, and user interface improvements made through serious games, gamification research, and cyber ranges have the potential to greatly decrease fog of war while increasing operational readiness and efficacy in cyber defence space [34].

6.5.2   Impact of Training in Digital Forensics

Digital forensics is necessary for law enforcement and investigation but also has applications in commercial, private, or institutional organizations. All activity conducted on an individual's computer systems and on a company network leaves a digital trace, which can range from web browser history caches and cookies all the way to document metadata, deleted file fragments, email headers, process logs, and backup files [37]. With cyber ranges, trainees learn to dissect and analyse real forensic cases. During such specific training, not only the technical aspects of digital forensics are considered but also the legal and organizational aspects.

The main purpose of the digital forensics' simulation environment is the ability to compile detailed forensic reports by trainees for use in both organization and in a court of law. A hands-on lab with basic forensics process and methodology scenarios demonstrates the ability to use forensics tools and analyse artefacts such as Windows file systems, Windows registry, memory images, Windows logs, and (optionally) network packet captures. This lab has frequently complicated configurations of multiple computers networked to each other, to a common server, to network devices, or a combination of these. Securing a scene and collecting digital evidence in these environments may pose challenges to the first responder. Improperly shutting down a system may result in lost data, lost evidence, and potential civil liability. The first responder may find a similar environment to train and learn in safe similar locations, in order to be ready to deal with real case scenarios. These kinds of environments need to be built according to true needs and expectations along with the proper standards. The main objective of digital forensics training in a cyber range is to present the trainees with the principles of digital forensics and evidence gathering. Furthermore, there is an intention to establish a common knowledge of the requirements regarding evidence admissibility in the court of law along with the server-centric approach to evidence gathering as a valuable source for further legal proceedings as well as for establishing patterns of malicious activity. The patterns are then used to quickly identify similar events from the past in the future as they take place. A CR digital forensics exercise also gives an overview of popular malware characteristics, methods of identification, and tools that may be used at the real scene especially in a situation where the case is supported by LEAs.

Law enforcement agencies perform an essential role in achieving our nation's cybersecurity objectives by investigating a wide range of cybercrimes, from theft and fraud to child exploitation, and apprehending and prosecuting those responsible. It is important to develop and execute a cyber range training plan to maximize the readiness to conduct high-impact criminal investigations and disrupt/defeat cybercriminals. The added value is to prioritize the training of technical experts, develop standardized methods, and broadly share cyber response best practices and tools through cyber range training. Criminal investigators and network security experts should be trained for a deeper understanding of the technologies malicious actors are using and the specific vulnerabilities they are targeting with updated and well-designed scenarios. Complementary cybersecurity and law enforcement capabilities are critical to safeguarding and securing cyberspace.

6.6   FORESIGHT Paradigm

The EU H2020 FORESIGHT project aims to develop a federated cyber range solution in order to enhance the preparedness (prevention, detection, reaction, and mitigation) of cybersecurity professionals at all levels (from junior to senior) by delivering a realistic training and simulation platform that brings together unique cybersecurity aspects from the aviation, power grid, and naval ecosystems. Hybrid scenarios will also be implemented by introducing IoT-simulated devices (e.g. sensors) to the ecosystems. FORESIGHT proposes the development and deployment of a beyond the state-of-the-art federated cyber-training environment, able to cater for multi-domain cyber training scenarios. To realistically and practically achieve this task, FORESIGHT proposes the design, development, and deployment of an internetworked federated controller that will provide gateway functionality between the multi-domain cyber ranges and will also provide virtual machines management and deployment capabilities. In this respect, FORESIGHT will enable the creation of a large-scale federated cyber range environment that can deliver multi-domain cyber training scenarios. Such kind of hybrid scenarios will include sub-scenarios across a range of domains (aviation, smart power grid, naval, or similar). In order to develop such scenarios to "feed" cyber range, current approaches and standards shall be considered in order to specify the best practices to be followed, improvements that are to be made, and constraints that are to be met, concurrently avoiding traps and obstacle that were faced in past implementations. The main driver of the system architecture will be the user requirements, ensuring that the system to be implemented meets the user needs and is able to provide the capabilities required. End-users from the above-mentioned domains replicate and configure critical systems of the airport/power grid/ports on the cyber range modelling and replicate IT and OT networks in order to build a near-real environment through a cyber range. The training will improve its cyber defence strategy and train its cybersecurity teams.

6.7   Conclusions

Cybersecurity training is growing in an appropriate way that is able to prevent and handle cyber breaches. New strategic methods such as anomaly detection, threat intelligence, simulation tools, big data analysis, threat analysis, forensic evidence collection, and gamification are needed in order to provide sufficient situational awareness of cybersecurity threats to companies, governments, and researchers. Cyber ranges can provide a continuous training environment using state-of-the-art methodologies, techniques, and a multi-domain training program directly guiding cybersecurity experts and professionals to create ways to implement and integrate security measures. In respect to the above, this chapter gives a well-defined overview of the existing cyber ranges, focusing specifically on LEAs and defence stakeholders. Finally, the FORESIGHT approach has been described giving a projection of the future capabilities and potentials about cyber ranges design, implementation, and execution of training which target the involvement of the end-users.

Acknowledgements The FORESIGHT project has received funding from the European Union's Horizon 2020 research and innovation programme under grant agreement No. 833673. The results reflect only the author's view, and the agency is not responsible for any use that may be made of the information it contains.

REFERENCES

1. DoD CR [Online]. Available: https://www.hqmc.marines.mil/docscr/. Accessed 27 Nov 2019.
2. Arizona Cyber Warfare Range [Online]. Available: https://www.azcwr.org/. Accessed 27 Nov 2019.
3. HNS platform [Online]. Available: https://www.hns-platform.com/. Accessed 27 Nov 2019.
4. ManTech [Online]. Available: https://www.mantech.com/capabilities/cyber. Accessed 27 Nov 2019.
5. École Navale CR [Online]. Available: https://www.chaire-cyber-navale.fr/. Accessed 27 Nov 2019.
6. AirBus CR [Online]. Available: https://airbus-cyber-security.com/products-and-services/prevent/cyberrange/. Accessed 27 Nov 2019.
7. Kypo CR [Online]. Available: https://www.kypo.cz/en. Accessed 27 Nov 2019.
8. Augusta Cyber Range [Online]. Available: https://cyber.augusta.edu/georgia/. Accessed 27 Nov 2019.
9. US Cyber Range [Online]. Available: https://www.uscyberrange.org/. Accessed 27 Nov 2019.
10. AIT [Online]. Available: https://www.ait.ac.at/en/research-topics/cyber-security/cyber-range/. Accessed 27 Nov 2019.
11. Saros [Online]. Available: http://www.saros.co.uk/products/it/cyber-range/. Accessed 27 Nov 2019.
12. Rhea CR [Online]. Available: https://www.rheagroup.com/services/cyber-and-physical-security. Accessed 27 Nov 2019.
13. Virginia CR [Online]. Available: https://www.virginiacyberrange.org/. Accessed 27 Nov 2019.

14. The Michigan CR [Online]. Available: https://www.merit.edu/cyberrange/. Accessed 27 Nov 2019.
15. Ixia CR [Online]. Available: https://www.ixiacom.com/solutions/cyber-range. Accessed 27 Nov 2019.
16. Palo Alto [Online]. Available: https://www.paloaltonetworks.com/solutions/initiatives/cyberrange-overview. Accessed 27 Nov 2019.
17. IBM [Online]. Available: https://www.ibm.com/security/services/managed-security-services/security-operations-centers. Accessed 27 Nov 2019.
18. Cybexer [Online]. Available: https://cybexer.com/. Accessed 27 Nov 2019.
19. Raytheon [Online]. Available: https://www.raytheon.com. Accessed 27 Nov 2019.
20. Cyberbit [Online]. Available: https://www.cyberbit.com/solutions/cyber-range/. Accessed 27 Nov 2019.
21. Breaking Point [Online]. Available: https://www.ixiacom.com/products/network-security-testing-breakingpoint. Accessed 27 Nov 2019.
22. RGCE [Online]. Available: https://jyvsectec.fi/cyber-range/. Accessed 27 Nov 2019.
23. Berkatweb [Online]. Available: https://www.berkatweb.com/cyber-range/. Accessed 27 Nov 2019.
24. Cyber-Gym [Online]. Available: https://www.cybergym.com/. Accessed 27 Nov 2019.
25. Cisco Threats Report [Online]. Available: https://www.cisco.com/c/en/us/products/security/security-reports.html. Accessed 03 Dec 2019.
26. Cyber-security training: A comparative analysis of cyber ranges and emerging trends [Online]. Available: https://pergamos.lib.uoa.gr/uoa/dl/frontend/file/lib/default/data/2864976/theFile. Accessed 02 Dec 2019.
27. CyberCents [Online]. Available: https://cybercents.com/cyber-ranges/cents/. Accessed 27 Nov 2019.
28. Silensec Cyber Range [Online]. Available: https://cyberranges.com/#cyber-range-securing-cyber-space. Accessed 27 Nov 2019.
29. [MISP – Open Source Threat Intelligence Platform & Open Standards for Threat Information Sharing [Online]. Available: http://www.misp-project.org. Accessed 02 Dec 2019.
30. Lessons learned from complex hands-on defense exercises in a cyber range [Online]. Available: https://ieeexplore.ieee.org/document/8190713. Accessed 02 Dec 2019.
31. Locked Shields 2018 After Action Report. Accessed 02 Dec 2019.
32. NIST: Guide to Test, Training, and Exercise Programs for IT Plans and Capabilities [Online]. Available: https://www.nist.gov/publications/guide-test-training-and-exercise-programs-it-plans-and-capabilities. Accessed 04 Dec 2019.

33. Smith, R. (2010). The long history of gaming in military training. *Simulation and Gaming, 41*, 6–19. https://doi.org/10.1177/1046878109334330.
34. CSS Cyber Defense Report, Cybersecurity and Cyber defense Exercises [Online]. Available: https://css.ethz.ch/content/dam/ethz/special-interest/gess/cis/center-for-securities-studies/pdfs/Cyber-Reports-2018-10-Cyber_Exercises.pdf. Accessed 04 Dec 2019.
35. Prior, T., & Roth, F. (2016). *CSS study: Learning from disaster events and exercises in civil protection organizations (CSS Study), risk and resilience reports.* Zurich: Center for Security Studies.
36. CCDCOE Cyber Defense Exercise [Online]. Available: https://ccdcoe.org/exercises/locked-shields/. Accessed 04 Dec 2019.
37. Role and impact of digital Forensics in cyber crime investigations [Online]. Available: https://www.researchgate.net/publiction/331991596_ROLE_AND_IMPACT_OF_DIGITAL_FORENSICS_IN_CYBER_CRIME_INVESTIGATIONS/link/5c9a39c445851506d72d8fdc/download. Accessed 04 Dec 2019.
38. Qiu, P., Cisco Cyber Range [Online]. Available: https://www.cisco.com/c/dam/global/en_hk/assets/event/cisco_connect_2015/pdf/4-3.pdf. Accessed 17 Dec 2019.

PART II

Serious and Organized Crime (SOC)

CHAPTER 7

COPKIT: Technology and Knowledge for Early Warning/Early Action-Led Policing in Fighting Organised Crime and Terrorism

Raquel Pastor, Franck Mignet, Tobias Mattes, Agata Gurzawska, Holger Nitsch, and David Wright

7.1 Introduction

Incorporation of ongoing development of the intelligence policing concept is essential for dealing with current threats [1]. There is a need to focus on developing priorities built on multiple factors, including intelligence analysis. The main goal should be to get ahead of threats and

R. Pastor (✉)
Ingeniería de Sistemas para la Defensa de España, Madrid, Spain
e-mail: rpastor@isdefe.es

F. Mignet
Thales Netherlands, Delft, The Netherlands
e-mail: franck.mignet@nl.thalesgroup.com

T. Mattes · H. Nitsch
Bavarian University for Policing, Fürstenfeldbruck, Germany
e-mail: tobias.mattes@polizei.bayern.de; holger.nitsch@pol.hfoed.bayern.de

B. Akhgar et al. (eds.), *Technology Development for Security
Practitioners*, Security Informatics and Law Enforcement,
https://doi.org/10.1007/978-3-030-69460-9_7

121

criminal behaviour by proactively identifying indicators and taking action based upon that knowledge ([2], p. 43). In this context, information becomes intelligence via the intelligence cycle, not only directly by a step by step process, but a multidirectional one.

The term "intelligence-led policing" (ILP) originates in parallel in Great Britain and the USA ([3], p. 5). ILP originally focuses on key criminal activities. Once crime problems are identified and quantified through intelligence assessments, key criminals can be targeted for investigations and prosecution ([3], p. 8). The core parts of the well-known variation of ILP called "problem-oriented policing" are analysing and studying upcoming crime phenomena associated with an evaluation of the carried-out police measures. It requires assessing each new problem and developing a tailored response ([4], p. 14).

This model is sometimes considered to be synonymous with the SARA (scanning, analysing, responding and assessing) process. This is a broader analytical model used in various fields, not only policing. Nonetheless, the SARA model can be applied to collecting and applying intelligence. Scanning may be viewed as part of the collection process. Analysis and assessment are part of the intelligence process, and response is the outcome of the intelligence process [5].

Clark adapted the classic approach of the intelligence cycle to new needs. The goal was to redefine the intelligence process in such a way that all of the parts of the intelligence cycle come together as a network. It is a collaborative process where all involved stakeholders are integral and information does not always flow linearly [6, 7].

The Australian Crime Commission extended the classical intelligence approach by the factor of so-called indications and warnings system (I&W system). The I&W system orients collection and warning on specified issues to service potential policy decisions, by using a process-based analytical method to consider the impact of observed threats, events and trends on the criminal justice environment. The approach provides a synthesis of different information supplied by agencies, combined with open

A. Gurzawska · D. Wright
Trilateral Research Ltd, London, UK
e-mail: agata.gurzawska@trilateralresearch.com;
david.wright@trilateralresearch.com

source data. When firm indications of strategic change emerge, an alert will be produced [8]. Especially the part of the strategic early warning methodology was further evolved by the Criminal Intelligence Service Canada during the development of their "Strategic Early Warning System for organised and serious crime (SEWS)" [9]. This approach is more focused on strategic policing, but neglects the operational side.

Despite various developments, intelligence-led policing lacks the interaction between long-term high-level strategic intelligence and operational intelligence usable as a decision support in case investigation. Due to the increased complexity of intelligence work, there is a need for new approaches that on the one hand combine existing and new methods and on the other hand support the combination of intelligence resulting from analysis carried out at strategic and operational levels. Such cross-fertilisation must be supported by the fusion and exchange of information from different data and information sources and the incorporation of domain knowledge available at the two levels of analysis.

To improve their capacity to anticipate, law enforcement agencies (LEA) need concepts (supported by tools) that extend intelligence-led policing for all types and time scales of intelligence. We suggest to extend ILP approaches by creating a continuum of "from early warning to early action" (EW/EA). In this new approach, the knowledge of possible new usage of technology coming from strategic long-term analysis is made available to the case analyst and can help him or her gain a better situation awareness on a given case, improving the quality of tactical decisions. Conversely, the case analyst will recognise earlier new characteristics in a given case, thus enriching the information available for the strategic analysts able to provide improved and earlier warnings. Such an approach narrows the classical temporal and operational gap between analysis and decision-making.

EW explains how crimes are evolving, identifying "weak signals", warnings and new trends, and provides a basis for assisting decision-makers, at both strategic and operational levels, in order to develop EA (preparedness, mitigation, prevention and other security policies).

Knowledge extraction and sharing is at the base of this EW/EA approach. In the next sections, some of the most relevant capabilities and characteristics of the technologies and techniques necessary for that development are presented. Moreover, the ethical and data protection aspects arising from the proposed approach are discussed.

7.2   RELEVANT CHARACTERISTICS
OF THE TECHNIQUES USED

The technologies used to realise the technical capabilities required to support the intelligence cycle will depend on the exact functional requirements, the nature of data or information (structured or not, type of medium, the semantic, etc.) and the goal of the processing step. The variety of data types and processing goals involved in the analysis of cybercrime requires a large number of processing and artificial intelligence (AI) techniques. When choosing a technique, its performance for the problem at hand will be important, but the characteristics of the EW/EA methodology and of the cybercriminality suggest that other properties are important as well. Recalling that the EW/EA methodology can realise its full potential by leveraging knowledge, Sect. 7.2.1 will discuss how certain techniques can facilitate the incorporation of domain knowledge. The tools able to fulfil the required capabilities mostly rely on AI and, in particular, machine learning (ML) techniques to provide decision support functions to LEAs. The rapid adaptation of criminal actors to new LEA (detection) capabilities and policies has as a result that models are likely to become inaccurate or obsolete rapidly. This could be even truer in cybercriminality, where the fast evolution in digital technologies in society provides a constant flow of new opportunities and the immateriality of the target space enables faster adaptation. Techniques relying on easily interpretable models can facilitate the understanding of causes of the loss of accuracy or obsolescence of an AI component (e.g. the new criminal behaviour) and, in addition, contribute to the satisfaction of the rising demand for accountability in application of AI for policing. Section 7.2.2 will introduce how AI techniques can be "explainable" and support accountability. Finally, Sect. 7.2.3 will expand on the requirements for technologies that can be used to realise an information system supporting the flexibility and the information sharing that are key to the EW/EA methodology.

7.2.1   Approaches to Incorporation of Knowledge

The incorporation of prior knowledge in AI systems has been present since the 1970s. For instance, expert system consisted of formalised expert knowledge and a reasoning engine, without automatic learning capability. The incorporation of prior knowledge in ML techniques came in the early

1980s to design more efficient intelligent agents and, separately, to compensate for insufficient training data from which to learn [10]. There are different moments at which knowledge can be incorporated in automated systems, at usage time and at design time.

During usage, the expert can use his knowledge to overcome automation difficulties. For instance, appropriate interactivity can allow a specialised LEA operator to respond to detection counter-measures during the collection of data from the dark net market. The expert can also act as a "teacher", providing feedback or correction to a learning system. For instance, semantic analysis of texts written by cybercriminals (such as adverts) is rendered difficult by the informal type of language and its fast-changing pace. Techniques such as on-line learning and re-enforcement learning are able to use feedbacks to reach higher performance.

At design time, expert knowledge can contribute to the construction of models that are used for reasoning or classification tasks, for instance, to assess a situation. As a basic example, one can imagine an advert selling a set of personal data. Assessing whether the data sold is actual (not yet having been used) is an important factor to decide on further law enforcement actions. A technical component estimating the likelihood (a form of classification task) that the data sold is actual can be a useful tool, using as inputs information extracted from the advert (for instance, the claims of the seller, the quantity of data, the price, the reputation of the seller, etc.). Such a component requires the construction of a model allowing the computation of the said likelihood from the input information. The importance of prior knowledge in constructing the model can range from light influence (with a large amount of knowledge being automatically learned using ML techniques applied to a suitable relevant and sizeable dataset) to very strong influence or even completely determine the model (such as in traditional expert systems). Note that knowledge (or intuition) regarding the particular patterns in the problem at hand always guides the choice of an AI technique (following the "no free lunch theorem"). A minimal incorporation of knowledge is to propose a large set of characteristics of the advert that could be useful (proposing a feature set), a frequent approach when deep learning (although recent research [11] is proposing means to incorporate more domain knowledge in deep learning approaches). The ML process will compute which characteristics are most useful and their quantitative contribution to the desired estimation. Providing a smaller set of characteristics regarded as the most useful could constitute a stronger intervention in the ML process. The expert could further constrain the ML process by making explicit the relationships that exist between input characteristics, for instance, indicating that the amount of data sold influences directly the price but not (directly) the seller's reputation. The ML process is then only responsible for the computation of the strength of the relationships. Baumgartner [12], among others, applies this strategy to construct Bayesian Network models for the profiling of criminals with limited data.

The injection of domain knowledge in the process of creating models can offer opportunities for using ML approaches when data is limited, a useful characteristic in the context of cybercriminality. However, constraining with incorrect knowledge can prevent the learning process from finding the models with high performance. The complexity and cost of formalising the knowledge can be another drawback.

7.2.2   *Interpretability of Techniques*

Research aiming at making the results computed by AI systems easier to understand by human beings has been developed under the label of explainable AI (XAI). There is no formal definition of XAI, but DARPA provides a description of the goals of their XAI project, i.e., to "create a suite of machine learning techniques that (i) produce more explainable models, while maintaining a high level of learning performance (prediction accuracy); (ii) enable human users to understand, appropriately trust, and effectively manage the emerging generation of artificially intelligent partners" [13]. A full review of the technical challenges associated with making AI explainable is beyond the scope of this paper, but this section provides some insights. The concepts of XAI are still in development. In fact, the definition of what constitutes an "explanation" of a computational model and which category of "users" is addressed are subjects of ongoing research. In the ML community, the trend seems to be to consider a partial causal explanation, of a specific result as an "explanation" [14]. In the field of ML, the term "interpretable" is often preferred to "explainable". The characteristics of the techniques used often determine the approach to interpretability. When each technical step followed in the technique is interpretable in the represented domain ("the real world phenomenon"), one can speak of transparent AI. Other techniques can use a post hoc interpretability approach, aiming at giving qualitative insights on the relationships between the (specific) inputs and the (specific) conclusions. The LIME approach [15] for deep neural networks (formerly seen as a textbook example of black-box AI) is an example of post hoc interpretability approach. In between, hybrid AI approaches, in which a function is realised by combining a black-box approach and a knowledge-based approach, are (partially) interpretable.

A specific case of incorporation of domain knowledge was mentioned in Sect. 7.2.1, in which the expert makes explicit which features are related with each other. In the resulting model, the learned strengths relate already interpretable features. The learned model is therefore generally interpretable, often even transparent (depending on the algorithms used to reach the conclusion). Interestingly, approaches in which the subject matter knowledge strongly (e.g. more than choosing input features) constrains the ML process may result in higher interpretability.

Recalling the discussion in Sect. 7.2.1, if the used techniques combine transparent models with an easy way to incorporate (new) domain knowledge, the creation of an improved AI component could be also facilitated by partially reusing existing components. Therefore, if performance is sufficient for the problem at hand, techniques supporting incorporation of subject matter knowledge and techniques that are interpretable offer specific advantages.

7.2.3 Requirements for the Information System

The EW/EA approach relies heavily on the capability to reuse knowledge, which implies that the knowledge can be efficiently stored, retrieved and shared. In the domain of policing, sharing is always paired with the constraints of keeping the data secure and often with ethical, legal and privacy-related challenges. Complaints from LEA personnel regarding the challenges encountered in sharing data are commonplace. Appropriate information systems should allow fine-grained publishing and fine-grained, role-based and decentralised access control of intelligence. Furthermore, from the point of view of the analyst, the realisation of relevant functions will require the dynamic combination of diverse components, for instance, a component performing extraction of information from adverts, followed by a component assessing characteristics of the adverts. Data lakes are often presented as an enabler of workflows and sharing for big data applications. However, when applied across organisations (or even "units"), migrating to data lake architectures implies moving data to a (somewhat centralised) infrastructure resulting in (some) loss of control and, for legacy systems, significant costs. Alternatively, techniques, handling computational components and data sources as dynamically composable services and encapsulating existing systems [16] could provide the level of interoperability enabling analytics application on disparate data from LEAs.

7.3   ETHICAL, DATA PROTECTION AND RELATED ASPECTS

Data analytics tools have much promise and have received much attention. Nevertheless, there are legal (particularly regarding data protection), ethical and societal pitfalls to avoid when adopting these approaches. It has been argued that technological solutions could result in LEAs moving into wider and deeper aspects of social and public policy [17]. Therefore, legal, ethical and societal challenges posed by the deployment of such tools are examined and approaches that have the potential to mitigate risks and negative consequences are proposed.

7.3.1   Legal, Ethical and Societal Challenges

While there is a clear desire to become more data-driven or intelligence-led, privacy is one of the main concerns regarding data-driven policing tools. One of the most important questions related to the use of such tools is what types of data and how the data available to the police ought to be used by LEAs [18] and who ought to have an access to them. More specifically, to what extent is the digital footprint of citizens (e.g. their movements with public transport or social media) private and can it be used unconditionally [19]? What are the legal limits of profiling individuals in society [20]? There are significant ethical, legal and societal constraints governing police use of data-driven policing tools and data [21]. There is a range of critical positions all of which point to the absence of and need for an adequate legal framework that can guide the development and use of AI-based policing tools. There are differences emerging from the national and/or institutional context. The procedures vary depending on the country and institution. Moreover, legal accountability, regulation and transparency are lacking [22].

In terms of ethical and societal challenges, the academic community and civil society organisations (CSOs) criticise AI-based policing tools, particularly predictive policing tools, due to (1) organisation and societal reasons, such as the lack of transparency (e.g. in terms of potential bias) and information among the public and LEAs with regard to how predictive algorithms reach their conclusions [23], which data sources are used, how the analyses work, how they are used and who has access to this data within and outside the police [24], and (2) scientific or technical reasons, for instance, related to the lack of robust evaluation of the forecast (how effective they are in contributing to the uncertainty and bias) [32] concerning not only predictive policing but also knowledge discovery and assessment.

Moreover, such tools have been criticised for their inaccuracy, automation bias and discriminatory results [25]. Critics argue that AI-based policing tools may lead to stigmatisation of people (e.g. vulnerable or minority groups and individuals) and places (hot spots) [18, 26]. Factors such as postal codes, age, sex, employment status and social and family situation can be used as proxies for race, as some are more highly correlated with vulnerable and minority groups [27]. As a result, they may perpetuate ethnic stereotypes or confirm historic biases [28].

These arguments are not ill-founded. Currently, existing predictive policing tools such as PredPol and COMPAS regularly raise controversies in terms of inaccuracy, discrimination and stigmatisation, automation bias and transparency. Therefore, the potential for enormous benefits is coupled with considerable risks [24]. Hence a number of approaches to data-driven policing tools addressing legal, ethical and societal challenges is proposed.

Firstly, the development and use of the data-driven policing tools in the EU context should be guided by three major European legal acts: (1) the Charter of Fundamental Rights of the EU, (2) the General Data Protection Regulation (GDPR) [31] and (3) the Law Enforcement Directive, which set forth numerous provisions that need to be respected, such as fundamental human rights including dignity, non-discrimination, privacy and data protection. Policing tools should follow six principles established by the GDPR and the Law Enforcement Directive, namely, lawfulness, fairness and transparency (personal data processing); purpose limitation; data minimisation; accuracy; storage limitation; and integrity and confidentiality. Since there are significant differences emerging from the national and institutional contexts, this study proposes that tools are designed in a flexible way to respect these differences and offer a modular approach to ensure adaptability to various regimes.

Secondly, policing tools using AI should also consider ethical guidelines regarding the use of AI. Even though there are no specific principles for the use of AI in policing, such tools could follow the principles drafted by the High-Level Expert Group on AI (AI HLEG), established by the

European Commission. The European strategy puts forward trust as a prerequisite to ensure a human-centric approach to AI (European Commission) [29]. Since AI-based policing tools aim to support LEAs in fighting crime, the tools should reflect and support ethical standards and principles for the performance of their duties. These values may differ from country to country; nevertheless, the core principles should include accountability, integrity, openness, fairness, impartiality, respect, honesty, objectivity, lawfulness, proportionality, due discretion and public interest (e.g. College of Policing [30]).

Lastly, policing tools using AI should respect societal values and serve the common good and benefit humanity, through supporting LEAs in fighting crime. At the same time, citizens have the right to know about the tools that are used by LEAs. In order to ensure a higher level of trust and understanding of the AI-based policing tools, their functions, objectives and potential impacts (both positive and negative), LEAs should ensure a high level of transparency and develop communication with the public.

7.4   Conclusions

LEAs, with their available resources, that are often scarce, need to fight organised crime and terrorism. The use of new, innovative approaches to tackle this problem becomes necessary and of the utmost importance. For LEAs, it is crucial to improve their management of the technology that can help them on this task.

With this chapter, important capabilities have been identified to support the EW/EA methodology developed in COPKIT, with many of these capabilities relying on AI techniques. Among the many techniques required, those supporting incorporation of subject matter knowledge and techniques that are interpretable offer specific advantages for this methodology. This is proposed as solution for analysts, at different levels, from investigative to strategic, to support them in all steps of the intelligence cycle, from data collection to knowledge and intelligence building, including forecasting to anticipate to future threats, underpinning the different decision-making processes that support from investigations to strategic planning while considering security, ethical, legal and societal aspects at all stages.

Acknowledgement

 This chapter is based on research, inter alia, undertaken in the context of the EU-funded COPKIT project, which has received funding from the European Union's Horizon 2020 research and innovation programme under grant agreement no. 786687. The views expressed in this chapter are those of the authors alone and are in no way intended to reflect those of the European Commission.

REFERENCES

1. Huber, N. (2019). Intelligence-Led Policing for Law Enforcement Managers. Available on https://leb.fbi.gov/articles/featured-articles/intelligence-led-policing-for-law-enforcementmanagers. last accessed 15 July 2020.
2. Baker, T. (2009). *Intelligence-led policing: Leadership, strategies, and tactics.* Flushing, NY: Looseleaf Law Publications.
3. Smith, A. et al. (1997). Intelligence-Led Policing: International Perspectives on Policing in the 21st Century. Available online: https://www.ialeia.org/docs/ILP_intl_perspectives.pdf. Last access 15 July 2020.
4. Goldstein, H. (2003). On further developing problem-oriented policing: The Most critical need, the major impediments, and a proposal. In J. Knutsson (Ed.), *Crime prevention studies:Vol. 15. Problem-oriented policing. From innovation to mainstream* (pp. 13–57). Monsey, Devon: Criminal Justice Press; Willan Pub.
5. Peterson, M. (2005). Intelligence-Led Policing: The New Intelligence Architecture. Available online: https://polis.osce.org/file/4811/download?token=rUkdXg7n. Last accessed 15 July 2020.
6. Clark, R. (2004). *Intelligence analysis: A target-centric approach.* Washington, D.C.: CQ Press.
7. Jardines, E. (2005). Using open source effectively: Hearings before the Subcommitee on Intelligence, Information and Terrorsim Risk Assement; Committee on Homeland Security. Available on https://www.govinfo.gov/content/pkg/CHRG-109hhrg24962/html/CHRG-109hhrg24962.htm. Last accessed 15 July 2020.
8. Quarmby, N. (2003). Futures Work in Strategic Criminal Intelligence: Research Paper presented at the Evaluation in Crime and Justice: Trends and Methods Conference. Available on http://docplayer.net/2599648-Futures-work-in-strategic-criminal-intelligence-neil-quarmby-australian-crime-commission.html. Last access 15 July 2020.

9. Criminal Intelligence Service Canada (CISC) Strategic Early Warning for Criminal Intelligence: Theoretical Framework and Sentinel Methodology. (2007). Available online https://pdfs.semanticscholar.org/8107/ce7733c5422 3f058c1c594de3ff583b70dfc.pdf?_ga=2.2826352.1761634195.1594833830-2147085485.1594833830, last accessed 15 July 2020.

10. Duda, R. O., Hart, P. E., & Stork, D. G. (2001). *Pattern classification*. New York: Wiley.

11. Salakhutdinov, R. (2019). Integrating Domain-Knowledge into Deep Learning. In Proceedings of the 25th ACM SIGKDD International Conference on Knowledge Discovery & Data Mining (KDD '19). ACM, New York, NY, USA, 3176–3176. https://doi.org/10.1145/3292500.3340416.

12. Baumgartner, K., Ferrari, S., & Palermo, G. (2008). Constructing Bayesian networks for criminal profiling from limited data. *Knowledge-Based Systems, 21*, 563–572. https://doi.org/10.1016/j.knosys.2008.03.019.

13. Gunning, D. (2017). Explainable artificial intelligence (XAI), Darpa/I2O, Program Update November 2017. Available at: https://www.darpa.mil/attachments/XAIProgramUpdate.pdf, last Accessed 26 July 2019.

14. Mittelstadt, B., Russell, C., & Wachter, S. (2019). Explaining explanations in AI. In *Proceedings of the conference on fairness, accountability, and transparency (FAT* '19)* (pp. 279–288). New York, NY, USA: ACM.

15. Tulio Ribeiro, M., Singh, S., & Guestrin, C. (2016). *""why should I trust you?": Explaining the predictions of any classifier" in proceedings of the 22nd ACM SIGKDD international conference on knowledge discovery and data mining (KDD '16)* (pp. 1135–1144). New York, NY, USA: ACM.

16. Pavlin, G., Quillinan, T., Mignet, F. & de Oude, Patrick. (2013). "Exploiting Intelligence for National Security" in Strategic Intelligence Management (Eds Babak Akhgar, Simeon Yates), (pp. 181–198).

17. Policy Connect: Building ethical data policies for the public good. (2019). Available online: https://www.policyconnect.org.uk/sites/site_pc/files/report/1214/fieldreportdownload/raa35577ipcldatatechethicsreportlsin-glepagesl0519.pdflast accessed: 15 July 2019.

18. Norwegian Board of Technology. (2015). Predictive policing can data analysis help the police to be in the right place at the right time?.

19. Edwards, L., & Urquhart, L. (2016). Privacy in public spaces: What expectations of privacy do we have in social media intelligence? *International Journal of Law and Information Technology, 24*(3), 279–310.

20. Lammerant, H., & De Hert, P. (2016). Predictive profiling and its legal limits: Effectiveness gone forever. In *Exploring the boundaries of big data* (pp. 145–173). Amsterdam University Press/WRR.

21. Babuta, A. (2017). Big data and policing: An assessment of law enforcement requirements, expectations and priorities. In *Royal United Services Institute for Defence and security studies*.

22. Sanders, C. B., & Sheptycki, J. (2017). Policing, crime and 'big data'; towards a critique of the moral economy of stochastic governance. *Crime, Law and Social Change, 68*(1–2), 1–15.
23. Couchman, H. (2019). Policing by Machine: Predictive policing and a threat to our rights. Liberty. Available on https://www.libertyhumanrights.org.uk/sites/default/files/LIB%2011%20Predictive%20Policing%20Report%20WEB.pdf, last accessed 11 July 2019.
24. Barrett, Lindsey, Reasonably Suspicious Algorithms. (2017). Predictive policing at the United States Border. 41 N.Y.U. *Review of Law & Social Change,* 327–363.
25. Ferguson, A. G. (2012). Predictive policing and reasonable suspicion. *Emory LJ, 62,* 259.
26. Rinik, C., Oswald, M., & Babuta, A. (2019). Machine learning algorithms and police decision-making: Legal. *Ethical and Regulatory Challenges.*
27. Kirkpatrick, K. (2017). It's not the algorithm, it's the data. *Communications of the ACM, 60*(2), 21–23.
28. Ethics Committee, West Midlands Police and Crime Commissionaire: Notes of meeting held Wednesday 03 April 2019 (2019). Available online: https://www.westmidlandspcc.gov.uk/ethics-committee/, last accessed 11 July 2019.
29. European Commission. Policy: Artificial Intelligence, available on https://ec.europa.eu/digital-single-market/en/artificial-intelligence, last accessed 26 November 2019.
30. College of Policing. Code of Ethics: A Code of Practice for the Principles and Standards of Professional Behaviour for the Policing Profession of England and Wales. (2014). Available on https://www.college.police.uk/What-we-do/Ethics/Documents/Code_of_Ethics.pdf, last accessed 11 July 2019.
31. GDPR Regulation (EU) 2016/679 of the European Parliament and of the Council of 27 April 2016 on the protection of natural persons with regard to the processing of personal data and on the free movement of such data, and repealing Directive 95/46/EC (General Data Protection Regulation) (Text with EEA relevance).
32. Robinson, D. and Koepke, L., 2016. Stuck in a Pattern. *Early evidence on predictive policing.*

CHAPTER 8

Detection of Irregularities and Abnormal Behaviour in Extreme-Scale Data Streams

Konstantinos Demestichas, Theodoros Alexakis,
Nikolaos Peppes, Konstantina Remoundou,
Ioannis Loumiotis, Wilmuth Muller,
and Konstantinos Avgerinakis

8.1 Introduction

Crime always has been an issue of outmost importance for every society. In 2017, a total of 205 foiled, failed and completed terrorist attacks were reported in the EU [1]. The aforementioned number represents a sharp

K. Demestichas (✉) · T. Alexakis · N. Peppes · K. Remoundou · I. Loumiotis
Institute of Communication and Computer Systems, Athens, Greece
e-mail: cdemest@cn.ntua.gr; talexakis@cn.ntua.gr; npeppes@cn.ntua.gr;
kremoundou@cn.ntua.gr; i_loumiotis@cn.ntua.gr

W. Muller
Fraunhofer Institut of Optronics System Technologies and Image
Exploitation – IOSB, Karlsruhe, Germany
e-mail: wilmuth.mueller@iosb.fraunhofer.de

K. Avgerinakis
Catalink Ltd, Nikosia, Cyprus
e-mail: koafgeri@catalink.eu

B. Akhgar et al. (eds.), *Technology Development for Security
Practitioners*, Security Informatics and Law Enforcement,
https://doi.org/10.1007/978-3-030-69460-9_8

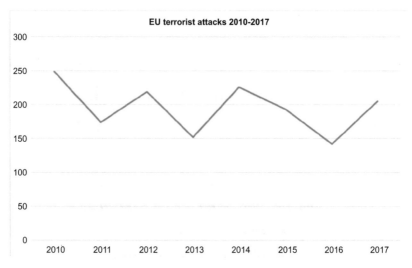

Fig. 8.1 Number of failed, foiled and completed terrorist attacks in the EU 2010–2017

increase around 45% compared to 2016 terrorist attacks and stopped a downward trend which had begun on 2014 (Fig. 8.1).

In the direction of providing increased security, the European Commission developed the European Agenda on Security which sets terrorism, organized crime and cybercrime as interlinked cross-border challenges in which EU countries must have a common strategical approach and coordinated action [2].

Post-study of criminal attacks reveal common patterns among them [3], such as radicalized individuals with history in organized crimes as perpetrators. This underlines the increasing necessity of LEAs to combine, prioritize and analyse heterogeneous massive data streams. Thus, there is the need to integrate sociological, psychological and linguistic models alongside with ICT tools in order to aid predictive policing methods and extract precursors or predictors of abnormal behaviour. The upward trend in cybercrime activities such as attacks on information systems, forms of online fraud and forgery, dissemination of illegal online content and more renders the need even more pertinent. According to cyberterrorism experts, approximately 90 percent of terrorist activity on the Web takes place through the usage of social networking tools [4]. Despite the LEAs' efforts to counter fight terrorism propaganda through social networks, the lack of linguistic capabilities and expertise as well as the differences in assessment of content are being exploited by terrorists in order to infest social media with their outlaw messages [5]. Another relative challenge is that data generation considering crimes is massive and mostly in semi-structured or unstructured data format. Taking into account that it is very difficult to analyse semi-structured or worse unstructured data formats using traditional data mining techniques such as SQL databases and common statistical analysis, Big Data analytics appear as a promising and feasible solution. Additionally, Big Data provide powerful tools by means of social networks analysis, semantic technologies as well as the utilization of advanced linguistic models. Therefore, LEAs must develop and integrate future-proof solutions and tools which will empower them with supreme analytical and predictive capabilities against terrorists, organized crime groups (OCGs) and individuals.

The remainder of the chapter is organized as follows: Sect. 8.2 presents an overview of state-of-the-art research projects in EU; Sect. 8.3 provides a brief review about the available technology; Sect. 8.4 presents the proposed architecture; while Sect. 8.5 concludes the chapter.

8.2 STATE-OF-THE-ART RESEARCH PROJECTS

The European Union and European Commission realize the necessity of research and development in order to promote state-of-the-art solutions in the field of public safety and security. In this light there are many research programs across Europe most of them totally or partially funded by European Commission. According to the H2020 strategic program, research and state-of-the-art approach is not just about creating new technologies and applying new tools but also requires understanding phenomena such as violent and abnormal behaviour [6]. So, it is clear that alongside with the emerging technologies such as Big Data analytics and machine learning, social and human sciences must be involved. This increases the complexity of all current tools and solutions and makes research mandatory in order to equip LEAs with powerful and future proof tools. Starting from crisis management, the beAWARE project demonstrates an integrated solution to support forecasting, early warnings, transmission and routing of the emergency data as well as the coordination between the first responders and the authorities.

beAWARE aims to provide decision support services to crisis management centres and make first responders and authorities more situational aware. In study [7] of beAWARE project are represented some of the main challenges and solutions such as the collection and integration of heterogeneous data from various sources in a common framework. In addition to crisis management, there are several innovative research projects running with the scope of fighting terrorism and organized crime. The TENSOR project aims to provide LEAs with a terrorism intelligence platform in order to act and plan fast for the prevention and the early detection of organized terrorist activities.

The TENSOR platform integrates a set of tools which allows the detection and gathering of various online data both from the Surface and the Dark Web. In the same direction, the RED-ALERT project brings data mining and predictive analytics tools to the next level, developing novel natural language processing, semantic media analysis, social network analysis, complex event processing and artificial intelligence technologies for online terrorist content. The study [8] is directly connected to the RED-ALERT project and focuses on social media terrorist content and how to remove it. Supplementary to RED-ALERT, the VICTORIA project aims to utilize the data streams from video sources in order to address the need for video analysis for investigation of criminal and terrorist activities. The latest terrorist attacks in London, Brussels, Barcelona, Paris, Berlin and Nice highlight the importance of video recordings. VICTORIA aims to deliver a TRL- 6 Video Analysis Platform (VAP) that accelerates video analysis with reliable results using Big Data tools as described in the paper of Alexander Schindler et al. [9]. One step further is the combination of heterogeneous data streams and the creation of a perpetually self-improving knowledge base according to a sophisticated and representational model which is also the basic idea and concept of MAGNETO project. MAGNETO intends to create a powerful set of tools that will be based on Big Data analytics, semantic reasoning and augmented intelligence well integrated in a common TRL-6 platform. Thus, MAGNETO targets to help LEAs to deal with large volumes and diversity of data in their fight against terrorism and organized crime in general.

Moreover, European Union tries to reduce not only the terrorism rates but also the crime rates generally. In this perspective, the CONNEXIONs project develops and demonstrates next-generation detection, prediction, prevention and investigation based on integration and correlation of multimodal data. CONNEXIONs uses augmented and virtual reality tools in order to construct crime scenes for post or pre-occurrence of a crime. Another major issue the last few years in the EU is the protection of public spaces where large crowds congregate; thus the LETS-CROWD project, which is in collaboration with CONNEXIONs project, aims to overcome the challenges associated with the effective implementation of European Security Models. Study [10] of the LETS- CROWD project focus on human-centred tools and solutions for real-time behaviour forecasting as well as risk assessment methodologies for soft targets and a policy-making toolkit for the long-term and strategic approach of these issues.

Aside from inner security and safety issues, border protection is another one major issue for the EU. Projects like ROBORDER and TRESSPASS aim to provide technologically advanced frameworks which support passenger risk assessment and decision services considering border patrolling and protection. For example, in ROBORDER a fully functional autonomous border surveillance system with unmanned mobile robots and vehicles will be developed which will incorporate multimodal sensors as well as route algorithms for optimal path detection as presented by Athanasios Kapoutsis et al. in [11]. Collected heterogeneous data are analysed and semantically integrated from authorities in order to provide accurate decision support services for border patrolling. TRESSPASS focuses on the utilization of LEA databases in order to contribute to risk-based passenger screening. In the context of border protection, stopping illegal trafficking inside and outside Europe is of high importance. To this end, certain advanced research projects focus on different aspects of illicit markets and illegal trafficking, e.g. the ANITA project. ANITA project and authors of [12] propose a method about textual similarity in video content as a solution which improves the investigation capabilities of LEAs by offering a toolset as well as techniques to efficiently address online illegal trafficking of falsified medicines, NPS, drugs and weapons.

8.3 Available Technologies in Crime Investigations and Future Trends

LEAs, practitioner analysts and investigators have always been equipped with tools and capabilities so they can prevent and fight crime as early as possible. These tools and capabilities are constantly evolving so they can follow and counteract to criminals' advanced methods. In the following paragraphs, we try to make a presentation of the available technologies and their future trends for crime prevention and investigation.

8.3.1   *Visual Intelligence*

Visual intelligence is one of the most celebrated technologies for crime investigation. It contains several algorithms, leverage deep learning and shallow representation technologies so as to tackle object and person detection, tracking human activity recognition as well as abnormal event detection, face detection and recognition and finally crisis event detection in surveillance and crawled visual data. Object detection and association can be performed by several algorithms such as Fast R-CNN [13], Faster R-CNN [14], YOLO [15], SSD [16], KCF [17], MDNET [18] and VITAL [19]. The same techniques can be used for face detection as well [16, 20], but for face recognition more information is required; therefore, frameworks like DeepFace [21] and FaceNet [22] with more sophisticate deep CNN algorithms are used. Going a step further, vehicle identification uses plate recognition [23] or DCNN [24] features, while activity localization and recognition use activity behaviour patterns of tracked people, objects or vehicles by computing their global spatiotemporal trajectories [25]. Having identification of people objects and vehicles as well as their activity localization and recognition can lead to abnormal behaviour detection which involves trajectory-based analysis on specific tracked objects [26] or statistical-based methods, such as PLSM [27] in order to discover motifs of abnormal activities that may occur in the scene.

A future approach on object detection and tracking can be a hybrid representation of shallow and deep representation features that will encode HOG and SIFT descriptors with a Fisher vector and represent them with a Deep CNN. As far as face detection and recognition are concerned, research should aim to leverage facial point detection and a combination of shallow features with a deep convolutional framework. The object detection representations would also be useful for vehicle detection and representation if combined with a plate DCNN scheme which could lead to a robust vehicle identification technique. Moreover, for action recognition, the goal-based descriptors [28] should be extended with spatiotemporal texture; Fisher vectors and a neural network would then transform these features to apply deep learning capability, while action localization will be deployed with a spatiotemporal saliency detector.

8.3.2 Semantic Integration and Technologies

LEAs and practitioner analysts gather data from a vast variety of sources which in most cases are semi-structured or unstructured. Thus, this information must be transformed to knowledge mainly by cognitive processes conducted by domain experts. In the direction of data and information fusion, the JDL model framework defines how to perform fusion as well as which processes and resources are involved [29]. Today's available tools for semantic reasoning and integration can be divided in three main categories: (i) tools based on mathematical class of random fields, (ii) logical models and frameworks (i.e. if-then rules) and (iii) Markov logic networks. Mathematical class of random fields is a solid and well-understood basis to represent stochastic processes with their random variables and the problem-specific dependencies among them. Logical models represent concepts, instances (objects in the world) and declarative knowledge (rules) in a consistent manner using if-then rules to achieve the fusion. Additionally, the use of OWL (Web Ontology Language) to represent semantic models in an object-oriented manner is a perfect fit for such a logic framework. Markov logic networks basically are the combination of knowledge graphs with stochastic modelling [30] which allows the computation of queries under uncertainties.

The application of information fusion comes with some challenges such as data association, adaptation of the logical model during its usage as well as the need for a high-quality statistical model. Future research should aim to adopt special techniques which incorporate relational knowledge during a probability mapping of the attributes of the related objects in the domain using Big Data analytics tools and technologies. Therefore, future proposals in the field of semantic reasoning and data fusion have to be easily adaptable according to changes in the environment; advanced algorithms, which compute the benefit for incorporating new concepts and rules into the actual model, can realize this.

8.3.3 Data Mining and Detection of Cybercriminal Activities

The World Wide Web is a huge data repository, and LEAs have a huge interest on how to crawl useful data from it. Common approaches focus on the efficiency of the crawling using different methodologies, but they do not take into account the structure of the sources of data. Such common approaches include Focused Crawling [31] or Path Ascending

Crawling [32]. As a counteract to data crawling tools from LEAs, cyber-criminals show a rising trend of utilization of various techniques which allow to thwart or at least to delay the detection of their nefarious activities. In order to tackle this and considering that a more flexible and scalable detection solution is needed, data mining-based solutions should be developed that will explore the suitability of pattern discovery, for which initial results seem to be rather promising [33].

Parallel to data mining, security agencies nowadays adopt more and more Big Data technologies in order to apply analytics over gathered information from various online sources such as social media, the Dark Web, etc. [34]. These technologies usually are implemented in the context of i) artificial neural network (ANN) models to predict national security problems in near real time [35], ii) data mining to reveal fraud [36] and iii) classification methods to predict deception in computer-mediated communications [37]. Big Data analytics drawing on online and other activities can be used to determine relevant behavioural indicators. This will help LEAs to detect outliers or anomalies, discover previously unknown associations and rules and continuously monitor data streams as a preventive measure. In this light, the target for future studies, as far as Web intelligence frameworks are concerned, is to leverage existing state-of-the-art data analytics tools, specifically assessing their impact on the information and data stream management.

8.4   PROPOSED ARCHITECTURE

This chapter demonstrates the proposed high-level system architecture which provides the required methods for LEAs to accelerate their investigations and remain careful in consideration of terrorist and cybercriminal threats by effectively integrating massive data streams.

The overall strategy of the current proposal comprises an iterative development methodology, where software updates are made available to LEAs and practitioners end-users for testing as well as thorough evaluation and validation purposes. Specifically, they are presented with details relevant to the main functional modules as well as the components that the system generally is consisted of, in a coherent manner. The details supply in a comprehensive way the structure of the system architecture based on the provided requirements together with the combination of the different components into a common proposed platform which will be deployed into LEAs' and practitioners' facilities. Additionally, the initial system-wide functional testing, evaluation and validation will be executed.

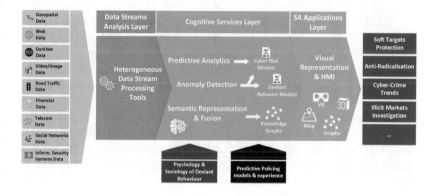

Fig. 8.2 Platform architecture

The main effort is being put on standardization of the platform's open architecture and its constituent components, interfaces as well data exchange formats. In order to succeed, extensive monitoring studying and contribution into activities related to standards of ISO/TC 292 (Security and Resilience) in the area of security is foreseen.

The platform could be considered as a multiple level system with a deeper view of system layers. The system architecture is designed in such a way so as to meet performance and resiliency requirements at scale. The main target of the platform development is to reach TRL-7. The main components of this architecture are shown in Fig. 8.2.

8.4.1 Visual Intelligence Modules

The visual intelligence modules is implemented by using face recognition and face detection algorithms in order to achieve specific identifications of persons contained in images and videos crawled from the Web, social media and/or footage from static or moving cameras. Furthermore, object recognition techniques are also applied in static or wearable surveillance cameras aiming to detect suspicious objects. The choice of the aforementioned techniques could also be extended in order to track people and objects throughout the video. Finally, suspicious or abnormal activities are being tracked by the adoption of crowd analysis and human action recognition with spatiotemporal localization.

The starting point for the prementioned implementations comprises deep convolutional network, DeepFace, SSD coupled with a YOLO architecture, KCF, goal-based descriptors and swarm intelligence for crowd analysis and abnormal activity detection. Therefore, libraries for computer vision and machine learning purposes targeted at dealing with real-time scenarios will be used.

The challenges that arise are data transmission failures and activities or objects that may be occluded. The contingency plan that had to be designed before the experimental implementation contains multiple transmitter and cameras in outdoors pilots, so that possible failures can be tackled.

8.4.2   Data Mining Modules for Crime Prevention and Investigation

Data mining is used to extract valuable information from the existing data. Crawling and mining take place by using crawl points, social media accounts and keywords, from which posts and sites are extracted and stored for analysis purposes. Eventually data from multiple sites are processed and exported into a common format that is available to the other components.

To develop and study crime patterns, we used the existing open-source Web and social media mining components integrated in the TENSOR and HOMER projects in conjunction with APIs provided by social media as well as extensive and scalable open-source Web crawlers software projects. However, Dark Web sites entrance difficulty and specific rate limits on social media access constitute a possible risk. The contingency plan pays attention to key Dark Web Sites and narrow social media access using specific keywords.

8.4.3   Semantic Information Representation and Fusion Modules

A module dedicated to data and information fusion is applied to the heterogeneous data that are collected from different sources, where the use of semantic technologies results to information transformation into valuable knowledge. The baseline of this module includes knowledge graphs, JDL model, OWL language and Markov logic networks along with the use of appropriate semantic information fusion models.

Main problems that arise include the uncertain, various and conflicting data and information, the requirement for solid training data as well the growing distinction between the model and the real world during the model's lifecycle. The use of a manually engineered model for the beginning of the training process, the selection of a confirmed model and the modification of the training frequency are defined as a contingency plan.

8.4.4 Trend Detection and Probability Prediction Modules for Organized Terrorism and Criminal Activities

To predict organized terrorism and criminal activities, Big Data analytics techniques are applied over the collected data in order to identify hidden trends inside the datasets. The deployment of analytics results to predictive models is the link between analytics and decision-making process.

Many widely used types of Artificial Intelligence Algorithms such as artificial neural networks, decision trees, pattern recognition and lifelong learning algorithms are developed by using the appropriate libraries as well as suitable predictive policing software. Some problems that occurred are the inaccurate results that were extracted from the prediction model and the false-positive alert that a model generated in some cases. A proposed contingency plan foresees the use of data sources of higher degree of diversity as well the creation of more sophisticated models for explaining deviant behaviours.

8.4.5 Detection Modules of Cybercriminal Activities

In addition to identifying anomalies, behavioural indicators and also revealing previously unknown associations and rules that are connected to cybercriminal activities, advanced Big Data analytics techniques, based on artificial neural networks and classification methods, are applied to the collected data.

Three common machine learning algorithms are used, i.e. K-means clustering, support vector machines and deep learning algorithms, and are developed with open-source libraries for numerical computation in order to achieve faster results. Moreover, the suitable ML algorithms for data mining tasks are used.

Risks which appeared in module presented in Sect. 8.4.4 could not only be the inaccurate results of the model, but also model results lose quality over time. The contingency plan gives attention to the selection of more complex model, fit the training frequency as well test and modify the model.

8.4.6   *Situation Awareness and HMI Modules*

All of the aforementioned functionality has to be demonstrated to the end-users in order to increase the situation awareness of the decision-makers. For this purpose, innovative visualization tools such as virtual and augmented reality technologies are used. The baseline comprises open-source libraries for visual analytics in addition to powerful, secure and flexible end-to-end analytics platforms for data visualization and representation purposes.

Possible problems that may arise and need to be overcome could be the inadequate offered visualization tools for some LEAs, the requirement for additional data views in certain use cases and the difficulty in using and handling the visualization environment. A proposed contingency plan includes the adaptation of the visualization environment, the creation of applicable solutions on top of the existing platform as well more detailed training courseware.

8.5   Conclusions

In this study, we made a brief presentation of state-of-the-art research projects in the domain of crime fight as well as involved technologies and tools. Moreover, we highlighted some of the currently used technologies and tools by LEAs and crime analysts as well as their future prospects. Considering the needs of LEAs for future-proof tools and expertise adoption so to fight and counteract in all types of crimes as well as the increasing resolve of criminals to adopt new technology, we presented a state-of-the-art platform accompanied with a detailed description of the required modules. The modules are integrated and deployed into a common architecture and next will be available to real end-users for detailed validation and evaluation. The architecture comprises all the necessary modules regarding extreme-scale data stream analytics, data mining processes, visual intelligence and machine learning algorithms, semantic and reasoning integration, probability prediction and eventually situation awareness modules and applications, all of them interconnected through open interfaces and a standardized platform. In order to reach TRL-7, an iterative approach will take place with the interaction of real end-users in conjunction with numerous software updates.

Acknowledgements This work has been performed in the context of the PREVISION project, which has received funding from the European Union's Horizon 2020 research and innovation programme under grant agreement no. 833115. The chapter reflects only the authors' view, and the Commission is not responsible for any use that may be made of the information it contains.

REFERENCES

1. Europol. (2018). *TE-SAT 2018: EU terrorism situation and trend report.* Publications Office of the European Union.
2. European Agenda on Security - Migration and Home Affairs - European Commission. (2016). Retrieved from https://ec.europa.eu/home-affairs/what-we-do/policies/european-agenda-security_en.
3. Joint statement of EU Ministers for Justice and Home Affairs and representatives of EU institutions on the terrorist attacks in Brussels on 22 March 2016 - Consilium. Retrieved from http://www.consilium.europa.eu/en/press/pressreleases/2016/03/24/statement-on-terrorist-attacks-in-brussels-on-22-march.
4. Terrorist groups recruiting through social media I CBC News. (2012). Retrieved from https://www.cbc.ca/news/technology/terrorist-groups-recruiting-through-social-media-1.1131053.
5. Europol. Internet Organised Crime Threat Assessment (IOCTA). (2017). Retrieved from https://www.europol.europa.eu/iocta/2017/index.html
6. Horizon 2020 – Work Programme 2018-2020, 14. Secure societies – Protecting freedom and security of Europe and its citizens – European Commission. (2019, July 02). Retrieved from https://ec.europa.eu/programmes/horizon2020/en/h2020-section/secure-societies---protect-ing-freedom-and-security-europe-and-its-citizens
7. Hilbring, D., Moßgraber, J., Hertweck, P., Hellmund, T., van der Schaaf, H., Karakostas, A., Kontopoulos, E., Vrochidis, S., Kompatsiaris, I., Gialampoukidis, I., & Andreadis, S. (2018). Harmonizing data collection in an ontology for a risk management platform. In *International conference on informatics for environmental protection (EnviroInfo)* (pp. 126–132). Munich.
8. van der Vegt, I., Gill, P., Macdonald, S., & Kleinber, B. (2019). Shedding light on terrorist and extremist content removal. In *GRNTT.* London: RUSI.
9. Schindler, A., Boyer, M., Lindley, A., Schreiber, D., & Philipp, T. (2019). Large scale audio- visual video analytics platform for forensic investigations of terroristic attacks. In I. Kom-patsiaris, B. Huet, V. Mezaris, C. Gurrin, W. H. Cheng, & S. Vrochidis (Eds.), *MultiMedia modeling. MMM 2019. Lecture notes in computer science* (Vol. 11296). Cham: Springer.
10. Dambra, C., Gralewski, A., & Arias, J. (2019). LETSCROWD: Dynamic risk assessment for mass gatherings. In *Proceedings of the 16th ISCRAM conference.* València, Spain.

11. Kapoutsis, A., Malliou, C., Chatzichristofis, S., & Kosmatopoulos, E. (2017). Continuously informed heuristic A∗-optimal path retrieval inside an unknown environment. In *2017 IEEE Interna- tional symposium on safety, security and rescue robotics (SSRR)* (pp. 216–222). Shanghai.

12. Gkountakos, K., Dimou, A., Papadopoulos, G. T., & Daras, P. (2019). Incorporating textual similarity in video captioning schemes. In *2019 IEEE international conference on engineering, technology and innovation (ICE/ITMC)* (pp. 1–6). Valbonne Sophia-Antipolis, France.

13. Girshick, R. (2015). Fast R-CNN. In *2015 IEEE international conference on computer vision (ICCV)* (pp. 1440–1448). Santiago.

14. Ren, S., He, K., Girshick, R., & Sun, J. (2015). Faster R-CNN: Towards real-time object detection with region proposal networks. *IEEE Transactions on Pattern Analysis and Machine Intelligence, 29*, 91–99.

15. Hu, P., & Ramanan, D. (2017). Finding tiny faces. In *The IEEE conference on computer vision and pattern recognition (CVPR)* (pp. 1522–1530).

16. Derpanis, K. G., Lecce, M., Daniilidis, K., & Wildes, R. (2012). Dynamic scene understanding: The role of orientation features in space and time in scene classification. In *Proceedings CVPR, IEEE computer society conference on computer vision and pattern recognition* (pp. 1306–1313). IEEE Computer Society Conference on Computer Vision and Pattern Recognition.

17. Henriques, J. F., Caseiro, R., Martins, P., & Batista, J. (2015). High-speed tracking with kernelised correlation filters. *IEEE Transactions on Pattern Analysis and Machine Intelligence, 37*(3), 583–596.

18. Nam, H., & Han, B. (2016). Learning multi-domain convolutional neural networks for visual tracking. In *IEEE conference on computer vision and pattern recognition* (pp. 4293–4302).

19. Song, Y., Ma, C., Wu, X., Gong, L., Bao, L., Zuo, W., Shen, C., Lau, R., & Yang, M. H. (2018). VITAL: VIsual tracking via adversarial learning. In *Proceedings 2018 IEEE/CVF con- ference on computer vision and pattern recognition (CVPR)* (pp. 8990–8999).

20. Jiang, H., & Learned-Miller, E. (2017). Face detection with the faster R-CNN. In *Proceedings 2017 12th IEEE international conference on automatic face & gesture recognition (FG 2017)* (pp. 650–657). Washington, DC.

21. Parkhi, O., Vedaldi, A., & Zisserman, A. (2015). *Deep face recognition*. British Machine Vision Conference (BMVC) 1, pp. 41.1–41.12.

22. Schroff, F., Kalenichenko, D., & Philbin, J. (2015). Facenet: A unified embedding for face recognition and clustering. In *Proceedings of the IEEE conference on computer vision and pattern recognition* (pp. 815–823). Boston.

23. Liu, X., Liu, W., Mei, T., & Ma, H. (2016). A deep learning-based approach to progressive vehicle re-identification for urban surveillance. In B. Leibe, J. Matas, N. Sebe, & M. Welling (Eds.), *Computer vision – ECCV 2016. ECCV 2016. Lecture notes in computer science* (Vol. 9906). Cham: Springer.

24. Liu, H., Tian, Y., Wang, Y., Pang, L., & Huang, T. (2016). *Deep relative distance learning: Tell the difference between similar vehicles.* 2016 IEEE conference on computer vision and pattern recognition (CVPR), pp. 2167–2175, Las Vegas.

25. Wang, H., & Schmid, C. (2013). *Action recognition with improved trajectories.* 2013 IEEE international conference on computer vision, pp. 3551–3558, Sydney, NSW.

26. Yang, W., Gao, Y., & Cao, L. (2013). TRASMIL: A local anomaly detection framework based on trajectory segmentation and multi-instance learning. *CVIU, 117*(10), 1273–1286.

27. Varadarajan, J., Emonet, R., & Odobez, J. M. (2012). A sequential topic model for mining recurrent activities from long term video logs. *IJCV, 103*(1), 100–126.

28. Tachos, S., Avgerinakis, K., Briassouli, A., & Kompatsiaris, I. (2017). Mining discriminative descriptors for goal-based activity detection. *Computer Vision and Image Understanding, 160,* 73–86.

29. Liggins, M., II, Hall, D., & Llinas, J. (2008). *Handbook of multisensor data fusion: Theory and practice* (2nd ed.). Boca Raton: CRC Press.

30. Richardson, M., & Domingos, P. (2006). Markov logic networks. *Machine Learning, 62*(1-2), 107–136.

31. Chakrabarti, S., van den Berg, M., & Dom, B. (2000). Focused crawling: A new approach to topic- specific web resource discovery. *Computer Networks, 31*(11–16), 1623–1640.

32. Cothey, V. (2004). Web-crawling reliability. *Journal of the American Society for Information Science and Technology, 55*(14), 1228–1238.

33. Cabaj, K., Mazurczyk, W., Nowakowski, P., & Zorawski, P. (2018). Towards distributed network covert channels detection using data mining-based approach. In *Proceedings of criminal use of information hiding (CUING) workshop co-located with ARES 2018.*

34. Davis, P. K., Walter, L. P., Brown, R. A., Douglas, Y., Parisa, R., & Voorhies, P. (2013). *Using be- havioral indicators to help detect potential violent acts: A review of the science base.* Santa Monica: RAND Corporation.

35. Bueno de Mesquita, B. (2011). Applications of game theory in support of intelligence analysis. In B. Fischoff & C. Chauvin (Eds.), *Intelligence analysis: Behavioral and social scientific foundations* (pp. 57–82). Washington, D.C: National Academies Press.

36. Li, S. H., Yen, D., Lu, W., & Wang, C. (2012). Identifying the signs of fraudulent accounts using data mining techniques. *Computers in Human Behavior, 28*(3), 1002–1013.

37. Zhou, L., Burgoon, J., Twitchell, D., & Qin, T. J., Jr. (2004). A comparison of classification methods for predicting deception in computer-mediated communication. *Journal of Management Information Systems, 20*(4), 139–165.

CHAPTER 9

Visual Recognition of Abnormal Activities in Video Streams

Konstantinos Gkountakos, Konstantinos Ioannidis,
Theodora Tsikrika, Stefanos Vrochidis,
and Ioannis Kompatsiaris

9.1 Introduction

The massive streams of visual information captured by CCTV surveillance and body-worn cameras cannot be easily monitored by human operators, particularly in the field of law enforcement. To assist law enforcement officers in their daily tasks and to improve their operational and investigation capabilities, several tools have been developed in order to automatically process and analyse such video streams and subsequently alert the human operators when events of interest, such as any abnormal activities, take place. Abnormalities can be considered as non-normal states, unknown states, everything abnormal, deviant, or outliers. This work

K. Gkountakos (✉) · K. Ioannidis · T. Tsikrika · S. Vrochidis · I. Kompatsiaris
Centre for Research and Technology Hellas, Information Technologies Institute,
Thessaloniki, Greece
e-mail: gountakos@iti.gr; kioannid@iti.gr; theodora.tsikrika@iti.gr;
stefanos@iti.gr; ikom@iti.gr

151
B. Akhgar et al. (eds.), *Technology Development for Security
Practitioners*, Security Informatics and Law Enforcement,
https://doi.org/10.1007/978-3-030-69460-9_9

focuses on such systems that aim to recognise actions of interest performed by humans or vehicles and categorise each action to one of existing pre-defined categories. Leveraging the significant advancements in deep learning neural networks, state-of-the-art action recognition methods are based on convolutional neural networks (CNNs) and recurrent neural networks (RNNs) [10, 12]. Moreover, the architectures of such activity recognition systems typically consist of two parts: the feature extractor and the classifier. To this end, this work proposes an end-to-end activity recognition framework that extracts visual features from video streams and classifies them to predefined activities. The proposed framework is evaluated using the VIRAT [8] dataset and the activities considered in the TRECVID Activities in Extended Video (ActEV) evaluation series [3].

The main contributions of this work are the proposal of a complete end-to-end activity recognition framework based on deep learning neural networks, the investigation of early and late fusion techniques in the context of this framework, and the extensive evaluation experiments using the VIRAT dataset. Moreover, since some of the ActEV activities are fine-grained, we group similar activities together so as to consider coarser-grained activities that are likely to be of more interest to general activity-based recognition systems; we have thus performed evaluation experiments using both the finer- and the coarser-grained activities.

The remainder of the chapter is structured as follows. Section 9.2 discusses related work and relevant datasets, Sect. 9.3 presents the proposed framework, Sect. 9.4 describes the experimental setup and presents the evaluation results, and Sect. 9.5 concludes this work.

9.2   Related Work

State-of-the-art activity recognition methods are based on deep learning techniques. Simonyan et al. [9] proposed a 2D convolution-based architecture that takes into account the visual and stacked optical-flow features and generates a two-stream neural network that can learn simultaneously the motion and the appearance of the input video. Ji et al. [5] proposed a 3D convolution-based approach in order to extract spatio-temporal features, while Tran et al. [12] also trained a 3D convolutional neural network. Hara et al. [4] extended previous works that make use of 3D convolutional kernels with filter size equal to 3×3×3 by using varied kernel sizes and very deep convolutional neural networks. They also concluded that the Kinetics [6] dataset, consisting of more than 300,000 videos that depict 400 human-related activities, can be widely employed for training and testing activity recognition systems, similarly to the wide use of the ImageNet [2] dataset for training object detection systems.

Apart from Kinetics, several other datasets have been built for the activity recognition problem. HMDB-51 [7] is one of such dataset that consists of more than 6766 videos, with a mean duration of approximately 3 seconds, categorised into 51 human activities extracted from movies. The ActivityNet [1] is another such dataset consisting of around 20, 000 videos categorised into 200 human activities. Finally, both the videos of the VIRAT [8] dataset and their annotations are provided by the National Institute of Standards and Technology (NIST – https://www.nist.gov/) in the context of the TRECVid Activities in Extended Video (ActEV – https://actev.nist.gov/) evaluation series.

9.3   ACTIVITY RECOGNITION FRAMEWORK

This work follows the supervised learning paradigm for human-related activity recognition that employs a deep neural network architecture, namely, the 3D ResNet neural network [4]. This 3D convolutional-based architecture achieves faster processing and can thus perform human activity recognition in (near) real time while using simultaneously (batch) frame processing. In particular, the architectures with 18, 50, and 101 layers as described in [4] have been deployed.

The 3D-ResNet-18 architecture consists of basic blocks, with each block consisting of two 3D convolutional layers followed by batch normalisation and ReLU (rectified linear unit) activation layers, as depicted on the left part of Fig. 9.1. The other two architectures (3D-ResNet-50 and 3D-ResNet-101) follow the bottleneck blocks approach (see right part of Fig. 9.1), where each bottleneck block consists of three 3D convolution layers followed by batch normalisation and ReLU activation layers, with the convolution kernels being 1×1×1 for the first and third convolution layers and 3×3×3 for the middle one.

Finally, it should be noted that the weights of the Kinetics dataset [6] were pre-loaded for all architectures. The Kinetics dataset was selected since it covers a large number of human activity classes (400 classes) and also contains videos that were not collected from sources in specific domains (e.g. movies, soccer games, etc.), but videos from diverse data sources uploaded on YouTube.

Fig. 9.1 3D-ResNet
basic and bottleneck
blocks [4]. "to"
3D-ResNet basic and
Bottleneck blocks (as
illustrated by [4])

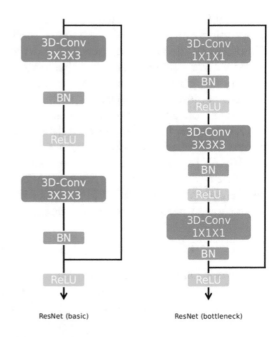

ResNet (basic)   ResNet (bottleneck)

9.4   Experiments

This section reports on the experimental evaluation of the proposed activity recognition framework by presenting first the datasets used in our experiments (Sect. 9.4.1), then the overall experimental setup (Sect. 9.4.2), and finally the evaluation results of our experiments (Sect. 9.4.3).

9.4.1   Dataset

In order to evaluate the proposed method, we selected the dataset provided by NIST under the ActEV evaluation series. This dataset was selected since it contains several human activities and vehicle actions that can be considered as abnormal in particular contexts. In particular, ActEV considers activities where one or more people generate movements or interact with objects (or groups of objects), such as other people (P) and vehicles (V). Specifically, ActEV defines and clearly annotates 18 human activities and vehicle actions listed in Table 9.1. The ActEV dataset consists of a total of 2446 annotated activities in its training and validation sets extracted

Table 9.1 ActEV activities official declaration

#	Activity name	Objects acts	Description
1	Closing	(P, V) or (P)	A person closing the door to a vehicle or facility
2	Closing trunk	(P, V)	A person closing a trunk
3	Entering	(P, V) or (P)	A person entering (going into or getting into) a vehicle or facility
4	Exiting	(P, V) or (P)	A person exiting a vehicle or facility
5	Loading	(P, V)	An object moving from person to vehicle
6	Open trunk	(P, V)	A person opening a trunk
7	Opening	(P, V) or (P)	A person opening the door to a vehicle or facility
8	Transport heavy carry	(P, V)	A person or multiple people carrying an oversized or heavy object
9	Unloading	(P, V)	An object moving from vehicle to person
10	Vehicle turning left	(V)	A vehicle turning left or right is determined from the POV of the driver of the vehicle
11	Vehicle turning right	(V)	A vehicle turning left or right is determined from the POV of the driver of the vehicle
12	Vehicle U-turn	(V)	A vehicle making a U-turn is defined as a turn of 180 and should give the appearance of a "U"
13	Pull	(P)	A person exerting a force to cause motion toward
14	Riding	(P)	A person riding a "bike"
15	Talking	(P)	A person talking to another person in a face-to-face arrangement between $n + 1$ people
16	Activity carrying	(P)	A person carrying an object up to half the size of the person
17	Specialised talking phone	(P)	A person talking on a cell phone where the phone is being held on the side of the head
18	Specialised texting phone	(P)	A person texting on a cell phone

from 118 videos of the VIRAT (release 1.0 and 2.0) dataset (http://virat-data.org/). The training set consists of 64 videos that contain 1338 recognised activities, while the validation set consists of 54 videos that contain 1128 recognised activities. The test set will not be considered as its annotations are not publicly available. The distribution of the activities both for the training and validation sets is depicted in Fig. 9.2. As it can be observed, ActEV is a challenging dataset, as it is highly unbalanced.

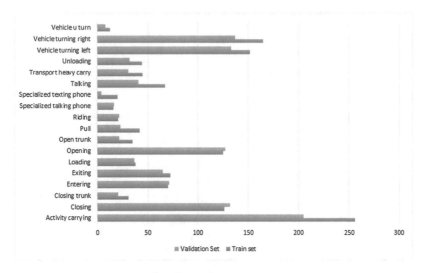

Fig. 9.2 ActEV dataset activities distribution

As some of the ActEV activities are rather fine-grained, we have also grouped similar activities together so as to consider coarser-grained activities that are likely to be of interest to more general activity-based recognition systems (e.g. recognition of vehicle-relevant activities). Table 9.2 lists these so-called super-activities, while Fig. 9.3 depicts the distribution of these super-activities for the training and validation sets, which is also highly unbalanced, similarly to before.

9.4.2   Experimental Setup

The aim of the evaluation experiments was to assess the effectiveness of the activity recognition system, and therefore they focused on processing and analysing only the parts of the video streams where some form of activity had been observed. To this end, first, the frames from all videos were extracted; to be more specific, one every four frames was extracted. Then, only the frames that depict an activity were considered and were stored in a valid format (.png).

The same training strategy was followed for each experiment. Specifically, the batch size was set to 32, the number of total epochs was set to 200, and stochastic gradient descent [11] was used as an optimiser

Table 9.2 ActEV activities grouped to "super-activities"

#	Activity name	ActEV dataset activities
1	Vehicle	1. Vehicle turning left
		2. Vehicle turning right
		3. Vehicle U-turn
		4. Riding
2	Talking	Talking
3	Person exits	Exiting
4	Person enters	Entering
5	Person carrying activities	1. Loading
		2. Transport heavy carry
		3. Unloading
		4. Activity carrying
6	Person interacts with phone	1. Specialised talking phone
		2. Specialised texting phone
The following activities are not taken into account		
1		Closing
2		Closing trunk
3		Open trunk
4		Opening
5		Pull

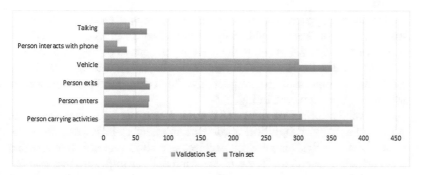

Fig. 9.3 ActEV dataset super-activities distribution

with an initial learning rate equal to 0.1. A "reduce on plateau" strategy was applied in order to create a learning rate schedule with max patience equal to 10 epochs. This strategy allows to reduce the learning rate by a factor once learning stagnates; if no improvement is seen for a "patience" number of epochs, the learning rate is reduced. Furthermore, five

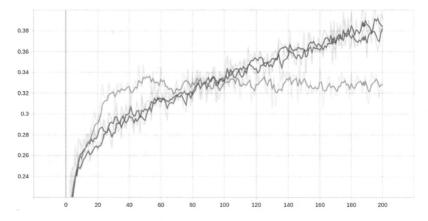

Fig. 9.4 Accuracy during training of ResNet-18(blue), ResNet-50(orange), and ResNet-101(red) with respect to the number of epochs

different scale factors were used for data augmentation [1.0, 0.84, 0.70, 0.59, 0.49], while a corner cropping strategy was also applied; this refers to the random selection of a cropped box from the four corners and the centre.

The training process was monitored for a complete evaluation by utilising the TensorBoard application downloaded from the TensorFlow[1] repository. Figure 9.4 presents the accuracy per epoch during training and denotes the 3D-ResNet architecture consisting of 18, 50, and 101 layers with blue, orange, and red, respectively. The correspondingly losses during training are depicted in Fig. 9.5.

The validation set of the ActEV dataset was used for evaluating the proposed activity recognition framework in order to investigate how the depth of a 3D-ResNet network architecture affects its effectiveness. To this end, we applied two different experimental settings, one that considers the 18 activities of the ActEV dataset and one that considers the 6 super-activities. Regarding the super-activities, we apply both late and early fusion. For the late fusion, the accuracy of each super-class comprises the summation of the subclasses' predictions during testing, whereas for early fusion, the super-activities are merged during training (i.e. a single training set is created for each super-activity by merging the training sets of its sub-activities).

---

[1] https://github.com/tensorflow/tensorboard

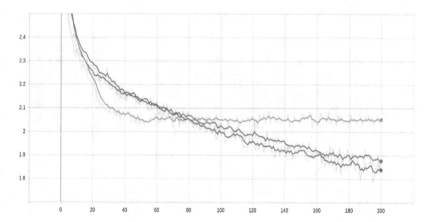

Fig. 9.5 Cross-entropy loss during training of ResNet-18(blue), ResNet-50(orange), and ResNet-101(red) with respect to the number of epochs

Precision@N is used as the basic evaluation criterion which allows us to show the accuracy of the framework for different numbers of retrieved activities where $N \in \{1, ..., 18\}$ in the case of ActEV activities and $N \in \{1, ..., 6\}$ in the case of super-activities. Precision@1 indicates the percentage of videos where the top prediction by our framework corresponds to the correct activity shown in the video. Hence, Precision@18 for the ActEV activities and Precision@6 for the super-activities should always be equal to 1, as the framework is bound to predict correctly if it simply provides all available activities. In addition, confusion matrices are also presented.

9.4.3   Results

This section presents the results for the different ResNet architectures both for the 18 activities and also for the 6 super-activities; in the latter case, the results listed below correspond to the late fusion, whereas the results for the early fusion are presented at the end of this section.

ResNet-50 results. Figure 9.6 presents the Precision@N using the ResNet-50 architecture. Precision@1 equals to 28% when all 18 activities are considered and 51% in the case of super-activities. As expected, coarser-grained activities can be more easily identified. Figures 9.7 and 9.8 present the confusion matrices of the prediction activities both for the 18 activities

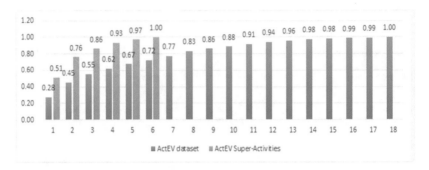

Fig. 9.6 Precision@N, ActEV, and super-activities trained using ResNet-503

Ground truth \ Predicted	Activity carrying	Vehicle turning right	Entering	Talking	Vehicle turning left	Exiting	Opening	Loading	Specialized texting phone	Transport heavy carry	Riding	Vehicle u turn	Specialized talking phone	Closing	Closing trunk	Open trunk	Pull	Unloading
Activity carrying	127	11	3	0	27	0	1	0	0	1	0	0	0	29	0	0	0	0
Vehicle turning right	31	26	2	8	44	0	5	0	0	2	0	0	0	19	0	0	0	0
Entering	25	6	2	2	7	0	8	0	0	4	0	0	0	16	0	0	1	0
Talking	10	3	0	17	1	0	1	0	0	1	0	0	0	8	0	0	0	0
Vehicle turning left	27	20	2	12	48	0	1	0	0	0	0	0	0	22	0	0	1	0
Exiting	21	2	0	1	7	0	5	0	0	0	0	0	0	29	0	0	0	0
Opening	27	0	1	6	15	0	15	0	0	4	0	0	0	56	0	0	3	0
Loading	14	1	0	0	5	0	2	1	0	4	0	0	0	9	0	0	1	0
Specialized texting phone	1	0	0	0	1	0	0	0	0	0	0	0	0	2	0	0	0	0
Transport heavy carry	9	2	0	6	3	0	1	0	0	7	0	0	0	3	0	0	0	0
Riding	17	0	1	1	1	0	1	0	0	0	0	0	0	1	0	0	0	0
Vehicle u turn	0	2	0	0	6	0	0	0	0	0	0	0	0	0	0	0	0	0
Specialized talking phone	7	1	2	0	4	0	0	0	0	0	0	0	0	3	0	0	0	0
Closing	23	2	0	2	18	0	20	1	0	1	0	0	0	64	1	0	0	0
Closing trunk	7	0	0	1	2	0	1	0	0	0	0	0	0	10	0	0	0	0
Open trunk	6	0	0	0	2	0	1	0	0	3	0	0	0	10	0	0	0	0
Pull	16	0	0	0	1	0	0	1	0	2	0	0	0	0	0	0	3	0
Unloading	9	1	0	1	4	0	3	0	0	0	0	0	0	13	0	0	0	0

Fig. 9.7 Confusion matrix using ActEV dataset trained on ResNet-50

Fig. 9.8 Confusion matrix using super-activities dataset trained on ResNet-50

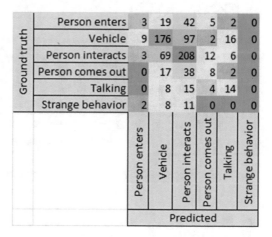

	Person enters	Vehicle	Person interacts	Person comes out	Talking	Strange behavior
Person enters	3	19	42	5	2	0
Vehicle	9	176	97	2	16	0
Person interacts	3	69	208	12	6	0
Person comes out	0	17	38	8	2	0
Talking	0	8	15	4	14	0
Strange behavior	2	8	11	0	0	0

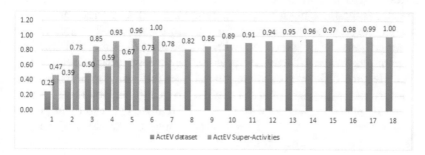

Fig. 9.9 Precision@N, ActEV, and super-activities trained using ResNet-18

and the 6 super-activities. A detailed examination indicates that the unbalanced characteristics of the ActEV dataset lead the model to a dominated learning state adapted to the activity with the highest occurrence ("activity carrying"). On the other hand, in the super-activities dataset, the number of false negatives and false positives has been reduced and disengaged from a dominating activity.

ResNet-18 Results. Figure 9.9 presents the Precision@N using ResNet-18 architecture. Precision@1 has decreased to 25%, compared to the 28% achieved by the ResNet-50 architecture for the 18 activities. Regarding the super-activities, Precision@1 has also decreased from 51% to 47%.

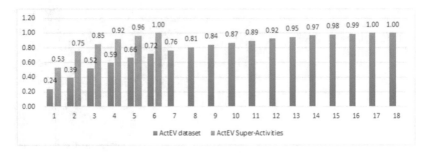

Fig. 9.10   Precision@N, ActEV, and super-activities trained using ResNet-101

ResNet-101 Results. Finally, the results of the experiments for the ResNet-101 neural network architecture are depicted in Fig. 9.10. As the results indicate, a higher capacity neural network can learn more accurately the classification problem. Specifically, the ResNet-101 architecture outperforms the previous ones when considering the super-activities, but the results for the 18 activities dataset are even lower than the ResNet-50 architecture. A detailed examination indicates that many of these 18 activities are closer (in terms of visual content) to each other, and thus, a higher capacity neural network which tries to differentiate between them aggressively results in lower Precision@1, even though the Precision@5 remains similar to the ResNet-50 results.

Early Versus. Late Fusion. In addition to the late fusion experiments presented above, we also carried out early fusion experiments for the case of super-activities.

To compare the effectiveness of the two approaches, we select the ResNet-101 architecture as it achieves the best performance in the case of super-activities. Figure 9.11 depicts the Precision@N both for early and late fusion. Specifically, early fusion increases the system performance for all N except for Precision@1. Furthermore, Fig. 9.12 compares the confusion matrices for early and late fusion and indicates that although the Precision@1 is lower when applying early fusion, the value of the error of misclassified activities is smaller and Precision@N for $N > 1$ is higher.

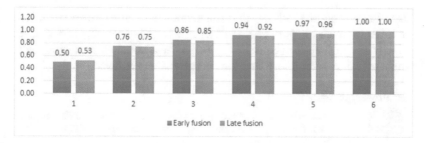

Fig. 9.11 Precision@N both for early and late fusion using ResNet-101

		Late fusion						Early fusion					
Ground truth	Person enters	3	23	39	6	0	0	2	18	40	9	2	0
	Vehicle	1	203	76	9	10	1	4	177	91	11	17	0
	Person interacts	2	76	204	15	1	0	4	87	197	5	5	0
	Person comes out	2	19	34	9	1	0	8	23	23	8	3	0
	Talking	1	15	16	2	7	0	1	9	16	2	13	0
	Strange behavior	0	8	11	2	0	0	0	10	9	2	0	0
		Person enters	Vehicle	Person interacts	Person comes out	Talking	Strange behavior	Person enters	Vehicle	Person interacts	Person comes out	Talking	Strange behavior
		Predicted						Predicted					

Fig. 9.12 Confusion matrices both for early and late fusion using ResNet-101

9.5 CONCLUSIONS

This work presented a framework for recognising activities in video streams. Specifically, the framework makes use of 3D convolutional filters in order to learn the spatio-temporal representation of activities. The framework was evaluated using the challenging ActEV dataset and also a second dataset that was created using the same data and which merges the ActEV activities into super-activities in order to evaluate the proposed framework in a more general activity-based recognition domain. The experimental results indicate that our framework can capture coarse level representations as it performs satisfactorily in the super-activities dataset. Finally, the early fusion approach proved to be advantageous in contrast to the late fusion when more than one activity were retrieved.

Acknowledgements

 This research has received funding from the European Union's H2020 research and innovation programme as part of the CONNEXIONs (H2020-786731) project.

BIBLIOGRAPHY

1. Caba Heilbron, F., Escorcia, V., Ghanem, B., & Carlos Niebles, J. (2015). Activitynet: A large-scale video benchmark for human activity understanding. In *Proceedings of the IEEE conference on Computer Vision and Pattern Recognition (CVPR)* (pp. 961–970). IEEE.
2. Deng, J., Dong, W., Socher, R., Li, L. J., Li, K., & Fei-Fei, L. (2009). Imagenet: A large-scale hierarchical image database. In *Proceedings of the IEEE conference on Computer Vision and Pattern Recognition (CVPR)* (pp. 248–255). IEEE.
3. Awad, G., Butt, A. A., Curtis, K., Lee, Y., Fiscus, J., Godil, A., ... & Quenot, G. (2020). Trecvid 2019: An evaluation campaign to benchmark video activity detection, video captioning and matching, and video search & retrieval. arXiv preprint arXiv:2009.09984.
4. Hara, K., Kataoka, H., & Satoh, Y. (2018). Can spatiotemporal 3D CNNs retrace the history of 2D CNNs and ImageNet? In *Proceedings of the IEEE conference on Computer Vision and Pattern Recognition (CVPR)* (pp. 6546–6555). IEEE.
5. Ji, S., Xu, W., Yang, M., & Yu, K. (2012). 3D convolutional neural networks for human action recognition. *IEEE Transactions on Pattern Analysis and Machine Intelligence (PAMI), 35*(1), 221–231.
6. Kay, W., Carreira, J., Simonyan, K., Zhang, B., Hillier, C., Vijayanarasimhan, S., Viola, F., Green, T., Back, T., Natsev, P., et al. (2017). The kinetics human action video dataset. *arXiv preprint arXiv:1705.06950.*
7. Kuehne, H., Jhuang, H., Garrote, E., Poggio, T., & Serre, T. (2011). HMDB: a large video database for human motion recognition. In *Proceedings of the IEEE International Conference on Computer Vision (ICCV)* (pp. 2556–2563). IEEE.

8. Oh, S., Hoogs, A., Perera, A., Cuntoor, N., Chen, C. C., Lee, J. T., Mukherjee, S., Aggarwal, J., Lee, H., Davis, L., et al. (2011). A large-scale benchmark dataset for event recognition in surveillance video. In *Proceedings of the IEEE conference on Computer Vision and Pattern Recognition (CVPR)* (pp. 3153–3160). IEEE.
9. Simonyan, K., & Zisserman, A. (2014). Two-stream convolutional networks for action recognition in videos. In *Proceedings of the international conference of advances in Neural Information Processing Systems (NIPS)*, (pp. 568–576), NeurIPS foundation.
10. Singh, D., Merdivan, E., Psychoula, I., Kropf, J., Hanke, S., Geist, M., & Holzinger, A. (2017). Human activity recognition using recurrent neural networks. In *Proceedings of the international Cross-Domain conference for Machine Learning and Knowledge Extraction (CD-MAKE)* (pp. 267–274). Springer.
11. Sutskever, I., Martens, J., Dahl, G., & Hinton, G. (2013). On the importance of initialization and momentum in deep learning. In *Proceedings of the International Conference on Machine Learning (ICML)* (pp. 1139–1147).
12. Tran, D., Bourdev, L., Fergus, R., Torresani, L., & Paluri, M. (2015). Learning spatiotemporal features with 3d convolutional networks. In *Proceedings of the IEEE International Conference on Computer Vision (ICCV)* (pp. 4489–4497). IEEE.

CHAPTER 10

Threats and Attack Strategies Used in Past Events: A Review

Ioannis Daniilidis and Symeon Andriotis

10.1   Introduction

10.1.1   Background

Almost two decades after the 9/11 terrorist attacks in the USA, the landscape of terrorist threats and attack strategies has been continuously evolving through time and space, thus constituting a major global security challenge for governments, but also a regional concern for decision-makers and the general public.

During the last years, attacks in the European territory conducted by al-Qaeda and Daesh-inspired or affiliated groups (i.e. Paris in 2015 and Brussels in 2016) increased concerns regarding the threat posed by violent

I. Daniilidis (✉)
Center for Security Studies, Athens, Greece
e-mail: i.daniilidis@kemea-research.gr

S. Andriotis
Hellenic Police, Athens, Greece
e-mail: s.andriotis@hellenicpolice.gr

167

B. Akhgar et al. (eds.), *Technology Development for Security Practitioners*, Security Informatics and Law Enforcement, https://doi.org/10.1007/978-3-030-69460-9_10

jihadist individuals and groups. At the same time, Europe seems to be facing a significant threat by far-right groups, with right-wing motivated attacks comprising now a priority in most of the security portfolios of the EU Member States.

With these in mind, this chapter presents an overview of the key global trends and patterns in terrorism, delineating current and emerging terrorist threats and relevant attack strategies, with emphasis on explosives.

10.1.2   Purpose and Contents of the Chapter

The overall purpose of this chapter is to shed light upon the dynamic nature of terrorism over the past decades, concentrating on the threat raised by the use of explosives.

To this end, the chapter will provide an overview of the terrorist threat, by adopting a historical and legal perspective, as well as of the past and currently implemented terrorist attack strategies.

Hence, the last chapter will draw upon the main findings of the previous sections and report the deduced conclusions on emerging trends and patterns, which could further be the basis of plausible scenarios and predictions of future incidents.

10.2   Review of Terrorist Threats

Initially, a conceptual definition of terrorism is being approached, in tandem with the delineation of its evolutionary path until the current historical juncture. Subsequently, a series of contemporary developments in modern terrorism are being illustrated, encompassing the so-called Foreign Terrorist Fighters (FTFs) threat as well as the terrorist landscape in the European territory.

10.2.1   Defining Terrorism

At EU level, the definition of the term terrorist offences is specified in the Directive 2017/541 of the European Parliament and of the Council of 15 March 2017 on combating terrorism that all EU Member States have transposed in their respective national legislations.

In particular, based on this Directive, terrorist offences are defined as:

- Intentional acts which, given their nature or context, may seriously damage a country or an international organisation when committed

- Having the aim of seriously intimidating a population, or unduly compelling a government or international organisation to perform or abstain from performing an act, or seriously destabilising or destroying the fundamental political, constitutional, economic or social structures of a country or an international organisation

Nonetheless, despite the existence of working definitions, few terms and concepts in modern political discourse present such a plethora of conceptual approaches as in the case of terrorism. The lack of a commonly accepted definition can be attributed to the fact that terrorism is a highly subjective term, with a strong political tone, depending on the subject's experiences and personal views since "the same kind of action [...] will be described differently by different observers, depending when and where it took place and whose side the observer is on" [1].

10.2.2   Origins and Typologies of Terrorism

Terrorism seems to be as old as human history. However, modern terrorism is considered to have originated with the French Revolution, when the term "terror" was first coined (1795) to refer to a policy systemically used to protect the fledgling French republic government against counterrevolutionaries. Thenceforth, modern terrorism has become a very dynamic concept, dependent to some degree on the political and historical context within which it is employed [2].

David Rapoport's theoretical scheme of the "terrorist waves" constitutes an endeavour to shed light on the evolution of modern terrorism. According to the American academic, since the end of the nineteenth century, there have been four "terrorist waves", which he describes as "Anarchist", "Anti-colonial", "New Left", and "Religious" [3]. More specifically:

- Anarchism comprises the first of Rapoport's waves. Between 1880 and 1905, anarchist terrorists assassinated the Austrian empress, the king of Italy, French and American Presidents, as well as dozens of citizens, accused as being part of the bourgeoisie. Although the pursued international revolution did not materialise, anarchists exerted a significant influence, most notably through the so-called propaganda-by-the-deed in which acts of individual heroism sought to elicit similar chains of reaction [4].

- The Anti-colonial wave emerged in 1930 and reached its peak in 1950. The violent groups, that composed it, were integrated in the population and aimed at combating foreign domination, leading to the eventual withdrawal of colonial forces. This wave laid the foundations for the conversion of terrorism in the late 1960s from a mainly local phenomenon to a global security issue [5].
- The New Left was largely composed of members of the upper middle class. Its central aim was the emergence of a new, socially just, and anti-authoritarian society situated on socialist principles, but in a distance from the ongoing version of socialism in the Eastern coalition countries [4]. The basic strategy was to incite the sociopolitical overthrow from the urban areas by waging spectacular attacks against governmental targets and "systemic agents".
- The onset of the Religious wave dates back to 1979, a year marked by the Iranian revolution, the USSR invasion of Afghanistan, and the capture of the great mosque in Mecca by Sunni Muslims, while, according to the Muslim calendar, 1979 was the beginning of a new century [4]. Murders and hostages comprised common practices of the third wave, but "suicide attacks" were the most impressive and innovative tactics, with Islamist terrorists being internationally networked. During this fourth wave, a terrorist organisation appeared, with apparently "pioneering" methods of recruiting and operating in the history of terrorism – al-Qaeda.

On the other hand, focusing on the basis of their source of motivation and ideological background [6] categorises terrorist organisations as (a) jihadist, (b) right-wing, (c) left-wing and anarchist terrorism, (d) ethnonationalism and separatism, and (e) single-issue.

10.2.3   Key Developments in Modern Terrorism

The contemporary elements that compose the nature of terrorist activity are briefly addressed through the exploration of the notion of "new terrorism", the depiction of the concerns associated with the FTFs threat, along with the overall impact of terrorism at European level during the past five decades.

Table 10.1   Fundamental elements of new and old terrorism [7]

	New terrorism	*Old terrorism*
Aims	Religiously inspired, absence of ideological rigour	Predefined set of political, social, and/or economic objectives
Methods	Mass civilian attacks; excessive violence	"Legitimate" targets; rules of engagement
Targets	Civilians, infrastructure, officials; soft and – less frequently – hard targets	Symbolic targets (e.g. embassies, banks) or persons representing authoritarianism; hard targets
Structure	Global network and agenda	Hierarchical structure

10.2.3.1 The Profile of "New Terrorism"

Notwithstanding the sometimes-broad use of the term, "new terrorism" consists an additional point of disagreement among researchers, with its origin dating back to the 1990s and 2000s [7]. Being adopted during the period of the terrorist attacks of 11 September 2001 and at the heyday of al-Qaeda terrorist organisation, new terrorism bears a number of partially distinct characteristics (Table 10.1):

- Religiously motivated operational action is a constituent component.
- The attacks from the new terrorist organisations are more lethal, given the consolidation of methods of action, such as suicide attacks.
- The theatre of operations is characterised by international scope and impact, as a result of a globalised network of terrorist actors.
- New terrorism can be approached through the lenses of conducting asymmetric/non-conventional war operations between terrorist organisations and nation states.

Following the terrorist attacks of 11 September 2001 and the ensuing "War on Terror", the structural identity of terrorism had acquired a significantly decentralised form. The decentralisation of terrorism can be interpreted as the result of the influence of factors that contributed to the formation and consolidation of terrorist organisations with an extremely high degree of "diffusion".

10.2.3.2 Foreign Terrorist Fighters (FTFs)

According to the UN Security Council and its Resolution 2178, Foreign Terrorist Fighters (FTFs) are defined as "… nationals who travel or attempt to travel to a State other than their States of residence or nationality, and other individuals who travel or attempt to travel from their territories to a State other than their States of residence or nationality, for the purpose of the perpetration, planning, or preparation of, or participation in, terrorist acts, or the providing or receiving of terrorist training, including in connection with armed conflict".

Since the Syrian conflict began in 2011, thousands of EU nationals have travelled or attempted to travel in conflict zones in Iraq and Syria to join insurgent terrorist groups, such as ISIS/Daesh. This influx of the so-called Foreign Terrorist Fighters (FTFs) to Syria and Iraq seems to had reached, in 2018, a number of more than 40,000 individuals originating from around 110 countries, of which it has been estimated that around 30% have already returned to their place of origin [8].

Currently, the issue of the FTFs remains high on the political agenda at both Member State and EU level inasmuch as it touches upon a broad spectrum of policies, related to the prevention of radicalisation, information exchange at EU level, criminal justice responses to returnees, as well as disengagement/deradicalisation inside and outside prisons [8].

10.2.3.3 The Terrorist Landscape in Europe

The 9/11 attacks have been a key point in redefining the role of terrorism and helping to raise awareness in terms of international security issues. In fact, the global terror attacks have led to an intensive effort to exercise internal control and vigilance in the fight against terrorism. At the same time, new forms of "cross-border coalitions" were established between countries, with an emphasis on the use of military and civilian power and the overriding aim of ensuring world peace and security [9].

Focusing on the last decades of terrorist activities [10], the attacks of 11 September 2001 signalled the shift towards religiously inspired terrorism and jihadism. Since then, Europe has witnessed by large-scale attacks, such as in Madrid on 11 March 2004, when 191 people were killed and 1755 injured, as well as in London on 7 July 2005, when 50 were killed and 700 injured.

From 2014 onwards, Daesh joined al-Qaeda as a new salafi-jihadist terrorist group. Between 2014 and 2016, Europe was the place of several major terrorist attacks, like (Fig. 10.1):

- The January 2015 attack against the satirical newspaper Charlie Hebdo and other targets in Paris.

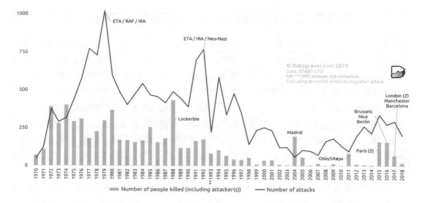

Fig. 10.1 Terrorism in Europe (1970–2018). (Source: START GTD [10])

- The November 2015 massacre in the Bataclan theatre and other targets in Paris.
- The 2016 attacks at the Brussels Airport and Maalbeek metro station in Belgium.
- The 2016 Nice vehicle attack, as well as the similar attack on a Christmas market in Berlin the same year.

Nevertheless, alongside the salafist-jihadist terrorist threat, Europe faces a significant rise of the extreme right-wing ideology and extremist action. In particular, extreme right-wing attacks fluctuated from 9 in 2013 to 21 in 2016 and 30 in 2017, namely, the highest number of right-wing attacks in Europe since 1994 [11].

10.3   REVIEW OF TERRORIST ATTACK STRATEGIES

The inherently fluid nature of terrorism contributes substantially to the upgrade of several terrorist organisations into "lifelong learning entities" [12]. In order to ensure the survival of the organisation, this "learning process" requires the gradual alteration of key elements and tactics of action. The formulation and effectiveness of the latter are crucially underpinned by technological innovations, along with the ability to manipulate democratic institutions and values.

10.3.1   Lone Actors and Organisational Structure

A key element in the structural transformation of modern terrorism is the emergence of a variety of operational actors. Among them, the salient role of the "lone wolves" can be identified, along with their implemented strategy – the so-called leaderless resistance.

A lone wolf – terrorist – can be considered as a lone actor, namely, an autonomous perpetrator, who aims to have an impact on the wider community, acting without direct support during the planning, preparation, and execution phase of the attack, and whose decision to act is based on inspiration rather than direct guidance from peers [13].

In other words, a lone wolf comprises an autonomous operational actor, often integrated into the community and capable of "self-activation" at any time. Usually, there is no direct link to a terrorist organisation (training, funding, etc.), while a lone wolf is usually driven by political and/or religious motives.

On the other hand, another element concerning individualised terrorist activities is the notion of "leaderless resistance". Leaderless resistance could be seen as a "confrontational strategy" that encourages involvement in acts of political violence, which are independent of any hierarchical structure or support network [14]. In this way, individuals or small cells can fight against an established power through independent acts of violence, without being centrally coordinated and with limited or non-existent communication between them.

10.3.2   Rationalism and Decision-Making Model

In many instances, an overall decision-making model for terrorist action is centred on the theory of "rational choice", which seems to partially interpret the behavioural patterns of terrorist organisations, especially when lone actors are deployed.

In general, rationalism in terrorist activities could be seen as a process whereby the decision-maker performs a cost-benefit analysis, in order to select the most beneficial course of action by effectively offsetting the risk and the costs inherently involved while achieving its goal and objectives.

As such, although in many cases jihadist terrorists are portrayed by the media as maniacs and mentally disturbed killers, they are in fact "disturbingly normal" persons, as they make careful calculations before committing heinous crimes, as well as evaluation of their effectiveness and impact [5].

10.3.3   Modern Technology and Online-Digital Environments

The structural changes observed in modern terrorism would not have been so evident without the contribution of a technological revolution, in particular through the usage of the so-called "new media". New media and their multiple applications facilitated the transformation of the pyramidal organisational structures into horizontal networks, in which numerous members are linked by advanced means of communication [15].

By using state-of-the-art communication tools, terrorist organisations succeeded in being developed beyond a narrowly structured network of terrorist organisations. On the contrary, the current diversity of terrorist actors with a common ideological background tends to be described as a "network of networks", paving the way for a global interconnection of heterogeneous entities [16].

Hence, an important parameter for modern terrorist organisations is their involvement in the digital world, especially through the use of the Internet. Indeed, since the early period of al-Qaeda, its online presence was seen as a significant mechanism for the transition to an era of terrorism, characterised by the active role of digital extremist communities with a high degree of resilience.

10.3.4   Explosives as Weapons of Choice

In the vast majority of recent terrorist attacks in the EU, such as in Paris in 2015, Brussels in 2016, as well as Manchester and Parsons Green in 2017, home-made explosives have been extensively used. The chemicals to produce them, known as explosives precursors, can be found in a number of products, like detergents, fertilisers, lubricants, and water treatment chemicals with the EU strengthening rules regulating the purchase of these substances [8].

The explosives used in the Paris and Brussels attacks were TATP (triacetone triperoxide), a home-made explosive which is for years the explosive of choice for terrorists [17]. TATP can be made from commonly available ingredients – it is difficult to detect – while instructions on how to produce it can be found on the Internet.

Moreover, the explosive-related jihadist plots are usually aimed at soft targets and mass gathering locations. In contrast, targets chosen by anarchist extremist groups are mostly state, financial, military, or law enforcement targets, with the anarchist groups utilising simple improvised incendiary devices (IIDs) filled with flammable liquids or IEDs filled with easily accessible explosive materials, such as pyrotechnic mixtures [6].

On the other hand, commercial explosives are more difficult for terrorists to acquire, while military explosives are even harder to access, even if explosive remnants of war and illicit trafficking in explosives are still presenting a significant threat to the EU [18].

10.4   Emerging Threats and Attack Strategies in Terrorism

This last part of the chapter provides a synopsis of threats and attack strategies that have already started to manifest themselves in the EU internal security environment, eventually concentrating on the (mis)use of explosives and technology.

10.4.1   Trends and Patterns in Modern Security Environment

- Returning FTFs are always highlighted as a potential threat, taking into consideration the possibilities of recruiting vulnerable individuals and other members of diasporas, thus contributing to the formation of a radicalised European network.
- A large number of recent terrorist attacks was carried out through the use of improvised weapons, like knives and vehicles, namely, weaponised "daily objects" that do not require special prior training or extensive logistical support. Possibly, this lack of sophistication and the sometimes-spontaneous nature of actions make the majority of the incidents involving knives and/or vehicles, as weapons of choice, hard to detect let alone to predict.
- The spread of extremist ideologies that further lead to an increased polarisation in society remains of deep concern among EU Member States, including violent Islamist, right-wing and left-wing ideology, as well as how they fuel each other.
- The Daesh military defeat in Syria/Iraq had a significant negative impact on the group's digital capabilities, which, nonetheless, maintained its online presence thanks to unofficial supporter networks and media outlets [6].

- Despite the lack of capacity of Daesh to coordinate and conduct external attacks, it maintains the intent to perform such operations, potentially using "sleeper cells" [19].
- Radicalisation in prison remains a considerable challenge, given the potential security risks posed by those convicted of terrorist offences and/or those radicalised in prison that are released.
- The so-called "hawala" informal banking system, along with the misuse of credit systems, non-profit organisations, and small-scale business ventures constitute means of fundraising and financing of terrorism.

10.4.2   The Explosives Threat

In Europe, the unlawful use of explosives is highly related not only to groups or lone actors linked to jihadist terrorism but also to many organisations and individuals with radical right-wing and left-wing ideologies. In particular, a number of current and emerging trends, with regard to the use of explosives for terrorist purposes, are as follows [6]:

- A shift is noted from the previously predominant use of TATP to a broader range of home-made explosives (HME), such as black powder, chlorate mixtures, and fertiliser-based mixtures.
- An increased use of pyrotechnic mixtures (mainly fireworks) is observed, which are obtained legally or, more often, illegally.
- Attempts to use IEDs in combination with chemical or biological toxins were identified in 2018, something that was also promoted in jihadist propaganda and IED-making manuals.
- Knowledge transfer, in terms of HME and IED production, is enhanced through the use of online, and often encrypted, social networks and forums, while there is an increasing trend of receiving bomb-making knowledge from readily available online open sources (i.e. pyro/explosive enthusiast sites and forums).

10.4.3   Misuse of Technological Advances

New modi operandi and criminal activities may be enabled by advanced technologies like online trade in illicit goods, virtual currencies, alternative banking platforms, and encrypted communication technologies. In particular, the following technological advances seem to raise important concerns among experts, while their application could critically influence almost every area of crime, including terrorist activities and the use of explosives [20]:

- Artificial Intelligence can transform the security landscape by becoming a tool for conducting cyberattacks, target selection, production and spreading of false information (fake news, deep fakes, etc.), as well as handling AI drones and self-driving vehicles.
- Darknet and cryptocurrencies that comprise key facilitators for trade in illicit goods, with decentralised darknet markets enabling vendors and customers to carry out transactions with high degree of anonymity.

The criminal abuses of 3D printing technology can obtain an even more complex nature with the development of programmable matter (PM) technology and its use in 4D printing.

Acknowledgement

 The work reported has been performed during the course and funded by the European Union's Horizon 2020 research and innovation project EXERTER: Security of Explosives pan-European Specialists Network (Grant Agreement: 786805). The content of this chapter "Threats and attack strategies used in past events" reflects only the authors' views, and the European Union is not liable for any use that may be made of the information contained herein.

BIBLIOGRAPHY

1. Teichman, J. (1996). How to define terrorism. In C. A. Gearty (Ed.), *Terrorism*. Aldershot: Dartmouth.
2. Cronin, A. K. (2003). Behind the curve: Globalization and international terrorism. *International Security, 27*(3), 30–58.
3. Rapoport, D. C. (2002). The four waves of rebel terror and September 11. *Anthropoetics, 8*(1), 1–17.
4. Neumann, P. R. (2016). *Radicalized: New jihadists and the threat to the west*. London: Bloomsbury Publishing.
5. Hoffman, B. (2006). *Inside terrorism*. New York: Columbia University Press.
6. Europol. (2019). *TE-SAT 2018: EU terrorism situation and trend report*. Publications Office of the European Union, p 13, 64

7. Simon, S., & Benjamin, D. (2000). America and the new terrorism. *Survival,* *42*(1), 59.

8. European Parliament. (2018). The return of foreign fighters to EU soil: Ex-post evaluation. Brussels: European Parliamentary Research Service, pp 5–6. Retrieved from: https://www.europarl.europa.eu/thinktank/en/search.html?word=foreign+fighters

9. Das, R. (2016). The burden of insecurity: Using theories of international relations to make-sense of the state of post-9/11 politics. Online article. Retrieved from: http://relationsinternational.com/burden-insecurity-using-theories-international-relations-make-sense-state-post-911-politics

10. START (National Consortium for the Study of Terrorism and Responses to Terrorism). (2018). *Global terrorism database (GTD).* https://www.datagraver.com/case/people-killed-by-terrorism-per-year-in-western-europe-1970-2015 with data deriving from START.

11. Jones, S. G., Toucas, B., & Markusen, M. B. (2018). *From the IRA to the Islamic state: The evolving terrorism threat in Europe.* Washington, DC: Center for Strategic & International Studies, p iv.

12. Ganor, B. (2015). *Global alert: The rationality of modern Islamist terrorism and the challenge to the liberal democratic world.* New York: Columbia University Press, p 1.

13. Ellis, C., Pantucci, R., van Zuijdewijn, J. R., Bakker, E., Gomis, B., Palombi, S., & Smith, M. (2016). *Lone-actor terrorism analysis paper* (Countering lone-actor terrorism series no. 4). London: Royal United Services Institute for Defence and Security Studies, p iv.

14. Joosse, P. (2007). Leaderless resistance and ideological inclusion: The case of the Earth Liberation Front. *Terrorism and Political Violence, 19*(3), 351. https://doi.org/10.1080/09546550701424042

15. Tucker, D. (2001). What is new about the new terrorism and how dangerous is it? *Terrorism and Political Violence, 13*(3), 9. https://doi.org/10.1080/09546550109609688

16. Mockaitis, T. R. (2007). *The "new" terrorism: Myths and reality.* London: Greenwood Publishing Group, p 55.

17. Europol. (2020). TE-SAT 2019: EU Terrorism Situation and Trend Report. Publications Office of the European Union, p 20–21. Retrieved from https://www.europol.europa.eu/activities-services/main-reports/european-union-terrorism-situation-and-trend-report-te-sat-2020

18. Europol (2019). Do criminals dream of electric sheep? How technology shapes the future of crime and law enforcement. Publications Office of the European Union, p 10–15. Retrieved from https://www.europol.europa.eu/newsroom/news/do-criminals-dream-of-electric-sheep-how-technology-shapes-future-ofcrime-and-law-enforcement

19. Burton, F., & Stewart, S. (2008). The 'lone wolf' disconnect. Terrorism Intelligence Report-STRATFOR – Online article. Retrieved from https://worldview.stratfor.com/article/lone-wolf-disconnect
20. Europol. (2019). *Do criminals dream of electric sheep? How technology shapes the future of crime and law enforcement.* Publications Office of the European Union. Online article, https://edition.cnn.com/2017/06/21/europe/brussels-train-station-attack/index.html

CHAPTER 11

Syntheses of 'Hemtex' Simulants of Energetic Materials and Millimetre Wave Characterisation Using the Teraview CW400 Spectrometer: Fundamental Studies for Detection Applications

Hemant J. Desai, Richard Lacey, Daniel O. Acheampong,
Anthony Clark, Philip Dixon, Matthew Hogbin,
Robert Hudson, Sam Pollock, Usman Waheed,
and Hannah Whitmore

11.1 INTRODUCTION

Explosives or energetic materials are hazardous, expensive and difficult to handle safely. Appropriate simulants are desirable in order to overcome these barriers for various purposes. These include training, for both staff and canine or animal olfaction; acceleration of development of novel detection systems to

H. J. Desai (✉) · R. Lacey · R. Hudson
Defence Science and Technology Laboratory, Salisbury, UK
e-mail: hdesai@dstl.gov.uk; rjlacey@dstl.gov.uk; rjhudson@dstl.gov.uk

© Crown 2021   181
B. Akhgar et al. (eds.), *Technology Development for Security*
Practitioners, Security Informatics and Law Enforcement,
https://doi.org/10.1007/978-3-030-69460-9_11

prevent acts of terrorism; and provision of materials for daily calibration/verification of instruments and to aid achievement of regulatory accreditation.

With the advances in active and passive millimetre wave systems for threat detection, there is a need for accurate characterisation of threat materials and simulants with respect to these technologies. A modular process for developing simulants has been created, and formulations have been tested to characterise fundamental properties in the millimetre and submillimetre wave regions.

Characterisation within the frequency range of 25–250 GHz has been conducted using a Teraview CW400 spectrometer. This frequency range covers the majority of active and passive systems, providing data upon which material characterisation can be based.

Many millimetre wave characterisation instruments fail to provide such a large frequency range. Low cost and fast data acquisition make the CW400 an ideal candidate for characterisation work.

Millimetre wave characterisation suffers from the following limitations:

- Low power resolution without cryogenic cooling.
- Complex detection techniques.
- Sample thickness must be large so that uncertainties due to inhomogeneity in target materials are minimised.
- Very little reference data available.

The Teraview CW400 spectrometer was first used to determine the real and imaginary parts of the relative permittivity of reference liquids, to validate the system. Once validated, a characterisation study of a Semtex simulant was undertaken.

11.2   Theory

Active millimetre wave and submillimetre wave detection technologies are sensitive to the complex dielectric constant (or relative permittivity) of concealed explosives. Simulants designed to match the dielectric

D. O. Acheampong · A. Clark · P. Dixon · M. Hogbin · S. Pollock
U. Waheed · H. Whitmore
Home Office, London, UK
e-mail: Daniel.Acheampong@dft.gov.uk; Tony.Clark@defra.gov.uk;
philippe@physics.org; matthew.hogbin@homeoffice.gov.uk;
sam.pollock@iconal.com; u.waheed16@imperial.ac.uk;
Hannah_Whitmore@MEEI.HARVARD.EDU

properties (refractive index and absorption coefficient) at the operating frequencies of these systems should be adequate as a simulant for tests. The structure of the material may also be important when the probing wavelength matches the scale of the features in the material (e.g. pelletised explosives or cracks and voids in solid explosives). To match a specific threat material, this could be either a single simulant or a set of simulants, each designed for use in part of the frequency range.

For passive millimetre wave systems, the simulants will need to match some additional physical properties. In addition to dielectric properties, passive millimetre wave imaging is also sensitive to the thermal properties of the explosive, such as the thermal conductivity and specific heat.

The dominant interaction of electromagnetic waves at millimetre wavelengths in most materials is Debye relaxation. Due to frictional forces between dipoles, realignment with an electric field is not instantaneous. This response time is what causes the millimetre wave to be perturbed [1]. The specific interaction with the material by the millimetre waves is governed by the real and imaginary parts of the permittivity (ε_r), which is a complex function expressed as:

$$\varepsilon_r = \varepsilon_r' + i\varepsilon_r''$$

where ε_r' is the real part of permittivity indicating the polarisation response of a material to the electric field [2] and ε_r'' is the imaginary part of permittivity and measures how dissipative a material is to an electric field.

Broadly, the refractive index ($n \approx \sqrt{\epsilon'}$) determines the reflection coefficient from the material surface(s), and the loss tangent $\left(\tan\delta = \dfrac{\varepsilon_r''}{\varepsilon_r'} \right)$ determines the degree of attenuation experienced as the wave propagates through the material.

The CW400 spectrometer measures the change in optical path length, and therefore the phase difference through the material which can be used to derive the refractive index and real part of the dielectric constant since ($n \approx \sqrt{\epsilon'}$) as follows:

$$n = 1 + \frac{\varphi c}{2\pi fL}$$

where φ = phase difference measured, f = millimetre wave frequency, L = thickness of dielectric and c = speed of light.

For relatively low loss materials, the loss can be approximated by the Beer-Lambert law and used to derive the following [3]:

$$\varepsilon_r^{''} = \frac{\alpha \lambda \sqrt{\varepsilon_r^{'}}}{\pi}$$

where α is the exponential loss factor and λ is the wavelength of the incident wave. The exponential loss factor α is calculated from the measured amplitudes of the incident and transmitted waves through the material [4].

11.3   EXPERIMENTAL

11.3.1   *Materials*

Simulant materials were prepared in-house. The ingredients were purchased mostly from Aldrich unless stated otherwise. Basic Yellow 40 (BY40) was purchased from Keystone UK. Their synthesis procedures are reported below. Once the materials were made, they were kept in sealed anti-static bags and the contents handled with nitrile gloves.

11.3.2   *Synthesis Procedure*

The original Hemtex simulants were made to mimic Semtex-1H and consisted of:

- Hexamine (as a replacement for RDX)   200 g
- Pentaerythritol (as a replacement for PETN)   200 g
- Polystyrene-co-butadiene   30 g
- Tributyl citrate   30 g
- n-Dioctyl phthalate   30 g
- N-Phenyl-2-naphthyl amine   5 g
- Sudan IV   5 g

The main principle was to remove energetic moieties and to be as close as possible in terms of material properties and atomic ratios. Hence, the nearest synthesis precursors were considered first as ingredients. Adjustments were then made to alter properties or provide fluorescence for trace application as deemed appropriate. This created a modular method. A table of the ingredients of the 6 Hemtex simulant formulations used here is shown in Table 11.1. The Hemtex009 and Hemtex010 formulations were created to compare any batch-to-batch variation and were more pliable than the other formulations. The Hemtex013, Hemtex014, Hemtex016 and Hemtex025 were chosen to provide different physical properties, especially malleability.

The general procedure to make simulants tested here is as follows.

Hexamine crystals were ground to an approximate grain size of table salt or sugar (as a method to determine particle size distribution was not available) so that a more uniform mix would form. The polymer (polystyrene-co-butadiene) was dissolved at room temperature, using dichloromethane (DCM), approximately 300 cm^3 (Aldrich, used as received) by stirring in a large bunged conical flask. The dissolved polymer was mixed with the remaining ingredients, as shown above, using a commercial heating food mixer (Kenwood Chef). The mixture was stirred thoroughly for 8 hours, in total, and the solvent was gradually evaporated off in a fume cupboard by heating the mixture to 40 °C during and following mixing. A card-ice/acetone cold trap was used to retrieve as much of the solvent as possible before suitable disposal. After the composition was found to be of a similar texture to Semtex, it was then heated to 100 °C, to remove any trapped solvent for a further 1 hour. The final material was pliable like a real plastic explosive. The pliability could be adjusted by using the PEG polymer which made the composition softer and easier to manipulate.

The solutions of 0%, 0.1%, 1% and 10% chlorobenzene in cyclohexane by volume were prepared for a parallel liquids characterisation study. The chemicals were 99.9% pure and were purchased from Aldrich Chemical Co.

11.3.3 Characterisation

The Teraview CW400 spectrometer produces coherent continuous waves sweeping from 10 GHz to 1.8 THz. Such a large frequency sweep is uncommon and made possible due to the unique method of controlling frequency using a temperature differential between two lasers (Fig. 11.1).

Two class 3B DFB lasers with wavelength of ~850 nm are frequency-offset using Peltier heaters that incrementally change the central wavelength. The laser beams at ω_1 and ω_2 are coupled together to produce a beat frequency ω_{THz} which is split to both the emitting and receiving modules [6]

Table 11.1 Compositions of 6 Hemtex explosive simulant formulations characterised using the Teraview CW400 spectrometer

Unique identifier code	Nominal weights	RDX replacement	PETN replacement	Plasticiser	Plasticiser	Polymer	Polymer	Stabiliser (nitro ester)	Solvent	Other chemical
Hemtex009	~2 kg	Hexamine	Pentaerythritol	Di-n-octyl phthalate	Tributyl citrate	Polystyrene co butadiene	Poly (ethylene glycol) (PEG)	N-Phenyl-2-naphthylamine	DCM	
Hemtex010	~2 kg	Hexamine	Pentaerythritol	Di-n-octyl phthalate	Tributyl citrate	Polystyrene co butadiene	Poly (ethylene glycol) (PEG)	N-Phenyl-2-naphthylamine	DCM	
Hemtex013	~500 g	Hexamine	Erythritol	Di-n-octyl phthalate	Tributyl citrate	Polystyrene-co-butadiene	–	N-Phenyl-2-naphthylamine	DCM	
Hemtex014	~500 g	Cynamic acid	Pentaerythritol	Di-n-octyl phthalate	Tributyl citrate	Polystyrene co butadiene	–	N-Phenyl-2-naphthylamine	DCM	
Hemtex016	~500 g	Phosphonitrillic chloride	Pentaerythritol	Di-n-octyl phthalate	Tributyl citrate	Polystyrene-co-butadiene	–	N-Phenyl-2-naphthylamine	DCM	
Hemtex025	~500 g	Urea 200 g	Pentaerythritol 100 g	Di-n-octyl phthalate 30 g	Tributyl citrate 30 g	Polystyrene-co-butadiene 30 g	–	N-Phenyl-2-naphthylamine 5 g	DCM	Basic yellow 40 100 g

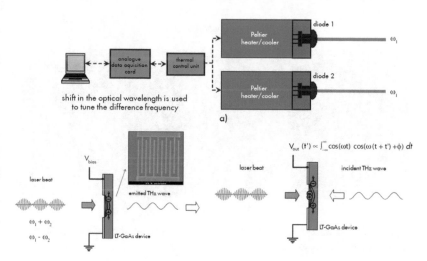

Fig. 11.1 Schematic diagram of how the lasers are offset and how the CW wave is produced and detected [5]

$$\omega_{THz} = \omega_1 - \omega_2$$

where ω = angular frequency.

The THz beatnote incident on the LT-GaAs substrate causes scintillation to occur. By use of a bias voltage connected to a bow-tie antenna, an electric field between the two plate-ends creates a dipole effect [7]. This causes continuous terahertz waves to be emitted.

Detection is made possible by using homodyne detection. The optical probe beam and the incoming wave interact with the semiconductor substrate connected to the antenna. The system uses a lock in detection scheme based on modulation of the emitter bias voltage which allows for the current to be converted into a measurable signal. A fibre stretcher delay line is used to vary the relative phase of the emitter pump and receiver probe beams, enabling vector measurements of the THz signal.

The current setup of the system includes PMMA lenses that collimate the output from the system and then focus it onto the detector after it passes through the sample. The simulation below shows the effect the lenses have on the wavefronts emitted from the receiver module. Wavefronts with minimal curvature ensures minimal change in amplitude when passing through the dielectric. Furthermore, lenses ensure a

PMMA lenses

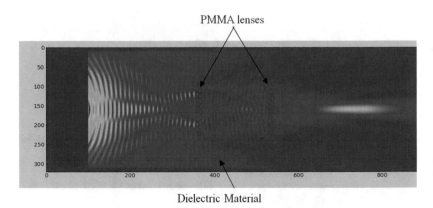

Dielectric Material

Fig. 11.2 Diffraction pattern of the continuous wave at 30 GHz, as it passes through the PMMA lenses, and the dielectric material

predictable diffraction pattern (Fraunhofer diffraction) observed in the receiver plane with and without a dielectric. In comparison, without lenses the variation of power received on the detector creates significantly greater variability within results (Fig. 11.2).

11.4   Results and Discussion

11.4.1   Liquid Characterisation Results

The objective of the experiment was to validate the CW400 with materials already characterised as reference materials. Reference data [8] was obtained from BCR Project 43 organised by National Physical Laboratories (NPL).

A PTFE sample holder was used to contain the liquid during characterisation. This material is strong and robust, resistant to chemicals and low loss, so the effect on the measured results should be negligible (Fig. 11.3).

Comparing reference data to measured data for low and medium loss materials (Figs. 11.4 and 11.5), we estimate a 1% error in ε_r' and 20% in ε_r'' in the low loss materials and an improved accuracy as the concentration of chlorobenzene, and hence the loss, is increased. For high loss materials, measurements for ε_r' and ε_r'' were accurate to 0.5% and 2%, respectively, when compared to NPL reference data [8]. We are therefore confident in our subsequent results.

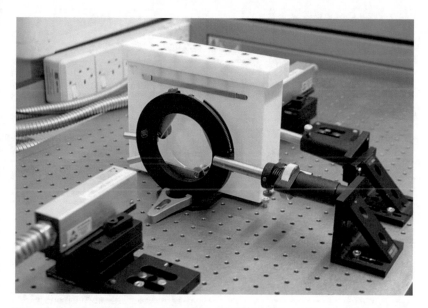

Fig. 11.3 Current arrangement of the CW400 system. Liquid is placed inside the solid PTFE block, which is hollow inside, and designed to work similarly to a cuvette. The emitter is to the left of the image and the receiver to the right. Both modules are placed at the focal length of the lenses ~14 cm

11.4.2 Simulant Characterisation Results

This experiment characterised six different formulations of Hemtex. These materials were characterised by placing them between two polyethylene slabs separated by spacers. The slabs were compressed to keep the sample under constant and equal pressure. This ensures faces are perpendicular to the optical bench and surface defects are minimised. Using spacers also ensures the thickness of the material is known accurately when under compression.

Figures 11.6 and 11.7 shows measurements made with the CW400 on the different batches of the simulant material detailed above.

Age, malleability and concentration of plasticiser are all variables that may affect millimetre wave measurements, and so it is necessary to repeat and vary these types of experiments.

The simulants were compared with Semtex as the threat material which varies from ~1.6 to 1.75 in refractive index according to the literature [9, 10] possibly due to batch-to-batch variation. The Hemtex simulants

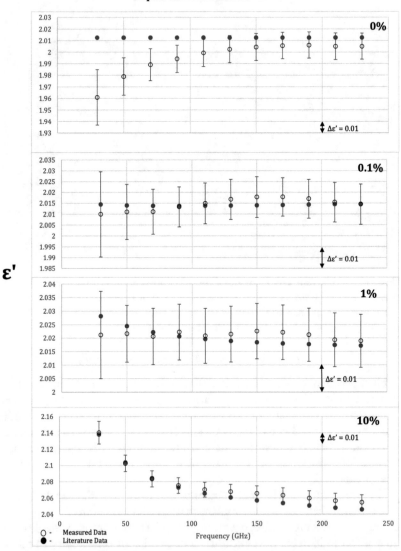

Fig. 11.4 Comparison of measured data against NPL reference data [8] for different concentrations of chlorobenzene in cyclohexane. Graph compares ε_r' against frequency for 0%, 0.1%, 1% and 10% chlorobenzene in cyclohexane, respectively. Literature data [8] for ε_r' accurate to ±0.2% [8]. Error bars [1] for measured data represent standard error with 95% confidence. Note the data points represent the raw data; the connecting lines are for indication purposes

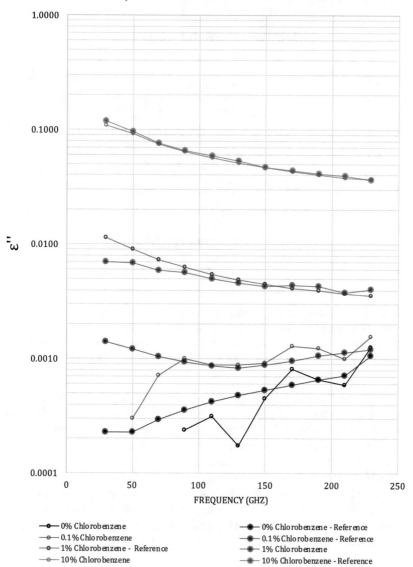

Fig. 11.5 Comparison of measured data against NPL reference data [8] for different concentrations of chlorobenzene in cyclohexane. Graph compares $\varepsilon_r^{''}$ against frequency for 0%, 0.1%, 1% and 10% chlorobenzene in cyclohexane, respectively. Literature data [8] for $\varepsilon_r^{''}$ accurate to ±1% [8]. Error bars [1] for measured data represent standard error with 95% confidence. Note the data points represent the raw data; the connecting lines are for indication purposes

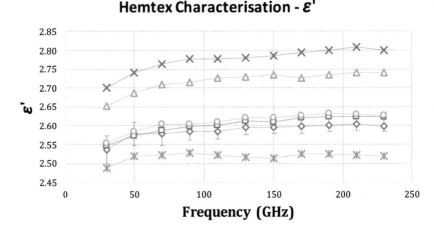

Fig. 11.6 Real permittivity against frequency for six simulant materials. Note the data points represent the raw data and the connecting lines are for indication purposes. For clarity, error bars are only shown for Hemtex009. As stated in the results discussion, the average error across the measured spectrum for ε_r' is ~1%

reported in this work have refractive indexes ranging from 1.58 to 1.68, across the frequency range of 30–230 GHz, which is within the limits of real Semtex implying that Hemtex materials are viable simulants.

On average across the spectrum taken, measured data for ε_r' and ε_r'' had a standard error with 95% confidence of ~1% and ~50%, respectively. The high error in loss measurements particularly varied at lower frequencies, likely due to scattering from voids. The error in ε_r'' for measurements taken between 150 and 230 GHz improves to 35% when excluding batch Hemtex014 which introduced significant uncertainty in loss measurements likely due to void sizes and batch variation.

11.5   Conclusions

The main conclusion of this work is that the Hemtex class of modular simulants reflect fundamental properties of explosives that millimetre wave and submillimetre wave technologies are sensitive to.

We characterised known liquids to test the Teraview CW400 system. We showed that at the frequency range of 30–230 GHz, for higher loss

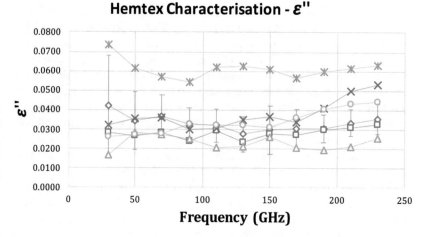

Fig. 11.7 Imaginary permittivity against frequency for six simulant materials. Note the data points represent the raw data and the connecting lines are for indication purposes. For clarity, error bars are only shown for Hemtex009. As stated in the results discussion, the average error across the measured spectrum for ε_r'' is ~50%

materials, the results are well within the acceptable range for characterisation, given that threat simulant materials vary between batches produced. These reference material measurements give confidence in the accuracy of the complex permittivity measurements we have recorded for simulants.

At the lower end of the spectrum ~30 GHz and for low loss materials is still a significant error of 20% in ε_r''. This is due to the system limitation in the characterisation of loss for non-absorbing materials, due to the transmission setup used and diffraction effects having a prominent effect on results.

Simulant characterisation provides good evidence of the variability in the refractive index that we would expect between different batches of energetic materials. Absorption of Hemtex samples posed more problems as grain size and quality of material have a visible effect at the millimetre wavelengths. Ideally, particle size distributions of the grains should be determined and related to absorption. Nevertheless, these characterisation results show that a modular approach potentially enables design of appropriate explosives simulants.

The value of appropriate simulants for millimetre wave and submillimetre wave threat detection technologies is significant. Simulants are safer to handle than real explosives in trials, especially where Person-Borne Improvised Explosives Devices (PBIED) are to be detected. Moreover, they are easier to store and cheaper than real explosives, which allows development times for new instruments to be reduced. It must, however, be noted that any detection instruments should be checked with real materials before submission for regulatory accreditation and certainly before deployment in real situations. If these checks have been conducted, then Hemtex simulants should also be viable for daily or routine verification/calibration of approved millimetre wave detection systems.

The work reported has to be applied to millimetre and submillimetre wave detection systems seeking to find illicit materials hidden, especially on people, at non-contact distances ranging ideally from metres to tens of metres. To accelerate research and development of such systems, often referred to as 'stand-off' detection, in a safer and cost-effective manner instead of using real materials with attendant hazard is an aspiration. Future work could include further characterisation of chlorobenzene in cyclohexane to understand the uncertainties with the CW400. A solid reference material of low, medium and high loss will also need to be selected to test the true capability of this system.

Acknowledgements Funding for the work was gratefully received from Home Office CAST's 'Innovation' initiative.

REFERENCES

1. Kirby, B. J. (2010). *Micro- and nanoscale fluid mechanics: Transport in microfluidic devices.* Cambridge: Cambridge University Press.
2. Agilent Technologies. (2006). *Agilant basics of measuring the dielectric properties of materials.*
3. Afsar, M. N., & Button, K. J. (1983). Precise millimeter-wave measurements of complex refractive index, complex dielectric permittivity and loss tangent of GaAs, Si, SiO_2, Al_2O_3, BeO, macor, and glass. *IEEE Transactions on Instrumentation and Measurement, MTT-31*, 217–223.
4. Haken, H., & Wolf, H. C. (2005). *The physics of atoms and quanta – Introduction to experiments and theory* (7th ed., pp. 5–26). Heidelberg: Springer.
5. LOT-Oriel Group Europe. *Terahertz CW400 Spectra 400 system.*

6. Wilk, R. (2010). Continuous wave terahertz spectrometer based on two DFB laser diodes. *Optica Applicata, XL*(1), 119–127.
7. Tani, M., Morikawa, O., Matsuura, S., & Hangyo, M. (2005). Generation of terahertz radiation by photomixing with dual- and multiple-mode lasers. *Semiconductor Science and Technology, 20*, 151–163.
8. Chantry, G. (1980). *High frequency dielectric reference materials BCR project 43. Final report of phase 1.* Luxembourg: Commission of the European Communities.
9. Greenall, N., Valavanis, A., Desai, H. J., et al. (2017). The development of a Semtex-H simulant for terahertz spectroscopy. *Journal of Infrared, Millimeter, and Terahertz Waves, 38*, 325. https://doi.org/10.1007/s10762-016-0336-z.
10. Saenz, E., Rolo, L., Paquay, M., Gerini, G., & de Maagt, P. (2011, April). Sub-millimetre wave material characterization. In *5th European Conference on Antennas and Propagation (EUCAP)*, Rome, Italy, pp. 3183–317.

CHAPTER 12

Law Enforcement Priorities in the Era of New Digital Tools

Georgios Kioumourtzis, Patrick Padding,
Natalie van de Waarsenburg, Zale Johnson,
Shaun Mallinson, Pepe Jose Lopez, Rashel Talukder,
Jarmo Puustinen, and Panagiotis Papanikolaou

12.1   Introduction

I-LEAD project is funded by the EU Commission Horizon 2020 (H2020), 2016–2017 Work Programme under the heading of "Secure Societies – Protecting the Freedom and Security of Europe and its Citizens" [1]. The I-LEAD project commenced in September 2017 with a duration of 5 years, coordinated by the Dutch National Police (NPN). Consortium

G. Kioumourtzis (✉)
European University Cyprus, Nicosia, Cyprus
e-mail: g.kioumourtzis@i-lead.eu

P. Padding · N. van de Waarsenburg
National Police of Netherlands, Central Unit,
Driebergen-Rijsenburg, The Netherlands
e-mail: patrick.padding@politie.nl; natalie.van.de.waarsenburg@politie.nl

© The Author(s), under exclusive license to Springer Nature   197
Switzerland AG 2021
B. Akhgar et al. (eds.), *Technology Development for Security
Practitioners*, Security Informatics and Law Enforcement,
https://doi.org/10.1007/978-3-030-69460-9_12

members include 12 law enforcement agencies (LEAs) and 7 research institutions who represent 13 EU Member States.

I-LEAD builds upon the work of the European Network of Law Enforcement Technology Services (ENLETS) [2] that brings EU law enforcement together to share best practices, activate co-creation and stimulate research for operational purposes.

I-LEAD's key focus is to contribute to the development of new and existing crime fighting capabilities of LEAs across Europe. I-LEAD will achieve this objective by firstly, engaging and consulting with operational police officers and supporting staff from all Member States. This will be undertaken via a series of bespoke practitioner workshops that have been exclusively designed to identify and define end-user priorities in 25 subject-specific areas of law enforcement. The findings from the workshops will be used to identify "fit for purpose" solutions within the marketplace and/or direct research and innovation within academia, with small and medium-sized enterprises (SMEs) and/or those organisations within the security industry environment. The overall results of this work will effectively con-tribute to the development and enhancement, where relevant and required, of the existing policing "crime fighting tool kit".

I-LEAD will also examine and recommend opportunities for standardi-sation and joint procurement, concentrating on the technical, human, organisational and regulatory elements. This work aims to produce a posi-tive impact that will improve connectivity and enhance interoperability between European LEAs.

Z. Johnson · S. Mallinson
Home Office, London, United Kingdom
e-mail: zale.johnson@homeoffice.gov.uk; Shaun.Mallinson@homeoffice.gov.uk

P. J. Lopez
Spanish National Police, Madrid, Spain
e-mail: josef.lopez@policia.es

R. Talukder
Polish Platform for Homeland Security, Warsaw, Poland
e-mail: rashel.talukder@ppbw.pl

J. Puustinen
Police of Finland, Helsinki, Finland
e-mail: jarmo.puustinen@poliisi.fi

P. Papanikolaou
Center for Security Studies, Athens, Greece
e-mail: p.papanikolaou@kemea-research.gr

I-LEAD recognises that science and innovation have a leading role to play in the future of policing. However, history has taught us that it is the policing community that should be giving direction to this work. It is this very principle that lies at the heart of the I-LEAD project and embraces the "triple-helix" concept, that being promoting a proactive collaboration between law enforcement, the scientific community and industry. Having this conviction will ensure that maintaining public safety and security for all citizens across Europe is optimised.

The need to identify, develop and implement technologies and methodologies within the security arena to support policing across Europe has never been as vital to society as it is at this present time. Therefore, I-LEAD will be committed to undertake an active role in the future of policing by contributing to the momentum of partnerships between policing and research and innovation.

12.2   EUROPEAN LAW ENFORCEMENT NETWORKS

I-LEAD will establish a sustainable and permanent cooperation framework (preferably with a legal entity endorsement) that will represent European LEAs. This will be based on the insights and mechanisms developed within the I-LEAD project. Its programme of work includes the common end-user priorities:

- Monitor research and innovation
- Standardisation and procurement recommendations
- Improved collaboration with industry and research
- Capacity building and knowledge exchange

In 2018, I-LEAD was successful in bringing together 78 experts from operational law enforcement from across 21 Member States via their first 5 subject-specific practitioner workshops. These facilitated events took place in the UK, the Netherlands, Spain, Romania and Belgium and provided a conducive environment for participants to collaborate as a community, discuss end-user requirements and ultimately identify an agreed set of priorities in the following topics:

- Open Source Intelligence (OSINT)
- Mobility for Officers
- People Trafficking

- Intelligence Analysis
- Technologies in DNA

The results of these workshops are documented in the following sections.

12.3   OPEN SOURCE INTELLIGENCE (OSINT)

Within the policing environment, Open Source Intelligence (OSINT) is data that is gathered from publicly available sources and used within an intelligence context. In relation to law enforcement and security, this intelligence is utilised in the prediction, prevention, investigation and prosecution of all types of crime, including acts of terrorism. This form of data gathering by law enforcement agencies across the world has been exploited for decades, and sources include the Internet, public agencies and the private sector, with one of the major contributors being that of social media. OSINT is not only a strategic enabler for decision and policy makers; increasingly it is being used by criminals to the detriment of law-abiding citizens. The rapid development of technology in this area has given rise to a more sophisticated and "tech savvy" criminal that is able to undertake a wide variety of unlawful criminal activities across borders and jurisdictions. This is a major challenge for LEAs and one that is growing exponentially. Therefore, it is essential that innovative research is carried out within this discipline so that police officers not only maintain the highest standards but, additionally, further develop and improve on current capabilities that will enable them to carry out their work successfully.

12.3.1   Priorities of the OSINT Community of Practitioners

Through the work undertaken at the I-LEAD OSINT practitioner workshop, it was found that amongst the fundamental requirements of the OSINT practitioners was a need to have access to tools that can effectively monitor, gather and analyse data, differentiate between useful and non-useful data and manage large amounts of datasets. Additionally, being able to validate data and understand the correlation between separate pieces of information is extremely important while ensuring against data overload and maintaining control of the gathered intelligence. Presently practitioners are utilising several different tools for different purposes. This technology is either freely available online, commercially procured or developed in-house. Commercial companies all want to sell their products and therefore promote their product as "the best", yet often, these tools do not adequately fulfil end-user requirements. Furthermore, the practitioners are called upon to "fine-tune" or update the acquired technology to suit their needs. Another issue is for those tools that are readily available online, which are then removed from the market, leave the OSINT practitioner with no adequate replacement capability.

12.3.2 Opportunities for Development Within OSINT

Practitioners agree that at present the landscape of OSINT technology needs to be less fragmented and more harmonised and coordinated and the discipline would greatly benefit from technological development in the following areas:

- *Increased automation of capabilities*

The OSINT practitioner is still required to input data and search assignments manually, which is labour and time-intensive, and the results of this work are highly dependent on the competency of the OSINT officer, e.g. experience, skill, knowledge and ability. As yet there is no artificial intelligence or self-learning computer system that can help with this activity.

- *Improved interoperability between systems*

Those working within the OSINT discipline need to utilise many types of tooling and software and have to work between different systems. As there is no link between the systems, open source to open source as well as open source to closed source, this is very problematic, especially in the case of open source to closed source, as they have to deal with differences in legislation and jurisdiction.

- *Enhanced management of information*

Due to the large amount of data generated, OSINT would benefit from a developed management information system with a centralised intelligence repository.

- *Advanced methods of monitoring intelligence*

To have the capability to be able to monitor intelligence in real-time would be of great benefit. An example of this would be the ability to identify early signs of radicalisation or potential new modus operandi which has the potential to prevent criminal activity before it occurs.

12.4   Mobility for Officers

Police presence within any community is vital, as it plays a major role in creating partnerships with citizens, preventing crime and building trust. It has been found that fighting crime and bringing criminals to justice can be optimised if everyone takes on the responsibility. Therefore, the mobility of police officers provides reassurance to a community due to the visibility of the officers. This then provides the means for improved interaction with the general public and an effective way to promote a safe and secure society. The mobile police officer does in fact fulfil many of the principles of law enforcement laid down by Sir John Peel in 1829, especially that which is covered by Principle 7 [3]. This directs the police to maintain a relationship with the public and whereby the police and the public are one entity in which both are responsible for the welfare and existence of the community. Furthermore, it is seen that the trust built up between themselves and the community will encourage enhanced dialogue and contribute toward the concept of intelligence-led policing in fighting organised crime and terrorism. The mobility of officers is therefore extremely vital, and any technology that can contribute effectively to this would be extremely beneficial.

12.4.1   Priorities of the Mobility for Officers Community of Practitioners

The I-LEAD practitioner workshop found that LEAs across the EU are strategically signed up to the "concept" of officer mobility however; take-up of new technologies, methodologies and processes is often not embarked upon due to expenditure and the inability to evidence tangible cost savings. In reality, change always has a price, and there are little or not easily identifiable and calculable monetary advantages in mobilisation. The real driver for this is the obligation of law enforcement to meet the expectations of communities by investing in its business and work force in order to keep citizens safe.

For those working within the officer mobility arena, there is a requirement for the harnessing of innovative research and development that not only fulfils end-user requirements but is fit for purpose and future proof. Furthermore, any new technology should enhance the interaction between the general public and the officer. It is vital that the technology is accepted by society as it is this that will enable the real and successful "front line policing" of the future.

12.4.2 Opportunities for Development for the Mobile Police Officer

Those practitioners across Europe representing the Mobility for Officers Community at the I-LEAD workshop stated that there are three main areas of work which would contribute to the operational work of the mobile police officer. These being:

- *The police vehicle*

One of the main priorities put forward by the Mobility for Officers community was that in relation to police vehicles. As the amount of wearable technology for police officers increases, the police vehicles themselves become increasingly incompatible with the officer and their operational duties. The physiological effects on the police officer wearing the technology are becoming more apparent due to the cumbersome nature of the various devices and the difficulty of movement when getting into and out of the vehicle.

- *Drones*

Another of the priorities for the mobile officer was the requirement to be able to use drones especially when police helicopters were not available. Drone capability would be very beneficial in firearms operations, patrolling streets, surveillance, crime scenes and accident recording. Additionally, practitioners stated that they could be used in mass operations such as football games, festivals and public order situations.

- *Mobile facial recognition systems*

The next generation of facial recognition technology should be that which can be used by the mobile officer on a handheld device. Mobile access to facial recognition technology will allow speedy on the spot identification of persons and also ensure the safety of the officer and the public if a person is quickly identified as dangerous.

12.5   PEOPLE TRAFFICKING

In the last century, cases of human trafficking were considered an isolated phenomenon. However, in recent years the numbers of such cases have grown exponentially, and in response to this rise, in 2007 the United Nations implemented three new protocols named the Palermo Protocols [4]. These were introduced to supplement the 2000 Convention against transnational organised crime. The investigation of people trafficking by LEAs across Europe has traditionally been conducted using similar technologies to that used in the investigation of transnational organised crime, e.g. border control, documents falsification, maritime surveillance, intelligence (routes and organisations), exchange of information amongst LEAs, electronic transnational surveillances (video, audio and tracking) or even detection of hidden persons technologies at borders. However, in the investigation of people trafficking, many other special issues have to be considered by police officers such as the societal and psychological impact and the approach to, and protection of, victims as a way to obtain intelligence and information.

12.5.1   *Priorities of the People Trafficking Community of Practitioners*

It is recognised that across the EU, the present situation with regard to people trafficking and related technology is an area that needs to be investigated and developed. Some of these key areas include exchange of information, intelligence, detection technologies (detection of hidden people in transports), document falsification technologies, cross-border surveillances, technologies for identification and technologies helping to avoid victimisation. Therefore, any new and emerging technologies must be fit for purpose to take into account the requirement for the different approaches in tackling these crimes.

12.5.2   Opportunities for Development Within People Trafficking

During the practitioner workshop, those working in this field identified three major priorities, these were:

- *Language translation tool*

One of the priorities for practitioners working to mitigate the criminal activity of people trafficking was that of a "real-time" translation tool that could help overcome the language barrier between police officers and witnesses, victims and suspects. An improved capability in this area would ensure better efficiency of police resources, as seeking out competent interpreters took time, but money spent on interpretation services was costly. Additionally, practitioners stated that on many occasions, they were not convinced that the translators converted the language correctly.

- *Management of big data*

Practitioners in this field generate large amounts of data, and being able to determine and discriminate between useful and non-useful data is difficult and time-consuming. The capability to be able to extract information from a dataset and transform it into a logical and understandable form would be of great benefit. The practitioner is aware that the technology is out there; however, a tailor-made system would definitely be an advantage in fighting the crime of people trafficking.

- *Data analytical tool*

It was recognised by the practitioners that other sectors such as banking and the retail industry utilised data analytics in a much more intelligent and constructive way. For example, supermarkets are using data-driven science to predict their customers buying/spending habits, which is based on studies of present and past data. It is believed that this type of technology could be adapted to be used in the fight against those criminals working within the people trafficking industry.

12.6   Intelligence Analysis

Intelligence Analysis (I.A.) is a concept which emerged from within the military and intelligence services field, which was then utilised by law enforcement agencies (LEAs). Early adoption of its use was in the USA in 1981 in which a group of professionals formed the International Association of Law Enforcement Intelligence Analysts, Inc. (IALEIA) and also in the UK [5].

In the early 1990s, the concept of intelligence-led policing (ILP) emerged, with I.A. products being developed that benefitted managers and operative police personnel alike. In less than a decade, areas of expertise arose, such as operational, tactic and strategic assessments. The products (deliverables) of an intelligence analyst received various names: case analysis, comparative case analysis, problem and subject profile, etc. I.A. also gained momentum due to the expansion of the I.T. sector moving from instruments like ANACAPA (visual matrix often designed manually) to dedicated software for processing, analysis and visualisation of data. The ability to analyse and visualise intelligence created an "information hungry" environment which, for most LEAs, translated into a quest for building applications with the ability to store information. It would not be too far from the truth to say that in the early 2000s, the motto for the intelligence analyst was "one problem one app". This resulted in LEAs having a number of separate systems for storing data, for example, information relating to theft of cars, theft of documents, theft of jewellery, etc. The benefits of technological development in this area would be multifold with emphasis on reducing cost, time and effort and improving detection rates, interoperability and information sharing.

12.6.1   Priorities of the Intelligence Analysis Community of Practitioners

Practitioners attending the workshop agreed that a refresh of the discipline and improved communications amongst European analysts would have a wide-reaching crime fighting benefit and would better contribute to the safety and security of all European communities. The lead in this area of work defined their future priority by saying that all intelligence analysts should be working towards "one single aim, all using the same tools and technologies". Presently, LEAs find themselves in an "island like" situation with numerous data bases, having little perspective of connecting the dots due to high costs, proprietary data formats, lack of specialised assistance, legal boundaries and poor inter- and intra-communication between various stakeholders.

12.6.2   Opportunities for Development Within the Intelligence Analysis

The Intelligence Analysis practitioner workshop proved extremely productive and those attending fully contributed to putting forward their priorities and underlining their priorities as a single community. These being:

- *Platform for the exchange of information within and amongst the community*

A platform to facilitate exchange good practices, increased awareness of the work of the intelligence analyst, Q&As between analysts, better understanding of each countries' legislation and data sharing protocols, discussion forums and reviews of analytical tools used by LEAs to inform of best tools (these tools could then be reviewed by security accreditation). This platform should not be used to house sensitive data.

- *Mobile platform for real-time data sharing: "actionable information"*

The concept of a "real-time" sharing of information capability has many benefits to policing. For example, the provision of risk indicators for police officers, i.e. to let them know if they are in a dangerous situation and real-time maps to inform police officers of people of interest in their area – with photos. However, the security and connectivity of this type of device must be at the forefront of any development, and to optimise the "mobile" capability of the device, it must also be compatible to the mobile officer's needs and requirements.

- *European information system*

Another priority of the intelligence analyst was that of a European Data Lake (data warehouse) for the upload of all reported crimes from all European countries and to have the functionality of indexing data for matching information within the system. For this system to work across all

Member States, there is a need for standardised fields for data input with an agreed structure to what data is appropriate and required for each country. For example, place of birth is important for crime investigation in Poland, whereas this is not so for the UK; and where a National Insurance Number is important in the UK, it is the National Identity Card that is used in many other countries.

- *Specific tools for the intelligence analyst*

There were two priorities that emerged from the workshops relating to intelligence tools, that being optical character recognition (OCR) solutions to enable the identification of printed characters using photoelectric devices and computer software and an application that can capture an image with its attributed metadata, such as date, time and place. At present this work is carried out manually and is prone to error.

12.7   EMERGING TECHNOLOGIES IN DNA

Since the mid-1980s, DNA profiling has been used by law enforcement agencies across the world to identify crime scene traces to convict the guilty and exonerate the innocent. Yet, despite the success of DNA and its use in genetic identification within the forensics arena, several issues remain, and new challenges continue to emerge. DNA profiling has become a victim of its own success with investigations relying more and more on the technology and hence causing a back log of work. Efficiency and cost-effectiveness of systems are aspects that are often raised by those working within the DNA environment, stating that there is still so much more that can be achieved to assist police investigations.

12.7.1   Priorities of the Emerging Technologies Community of Practitioners

During the practitioner workshop, all areas of the DNA profiling process were considered and discussed including laboratory techniques and procedures, crime scene protocols, contamination issues and the integrity of forensic evidence.

12.7.2   Opportunities for Development Within DNA Technologies

The community the practitioners decided that their priorities were

- *Rapid DNA: faster results*

Across Europe there are expectations within the DNA discipline to achieve faster results of DNA profiling, and these expectations are increasing all the time with investigations becoming more reliant on DNA evidence. However, it was emphasised by the workshop practitioners that this requirement must be balanced against that of maintaining the integrity of the DNA evidence. At present some LEAs within Europe are testing various types of "Rapid DNA" systems, where others are not able to fund the technology. Additionally, there is concern that as the chemistry becomes more sensitive, Rapid DNA technology used within the field is more prone to contamination and does not have the same amount of assurance as that of laboratory analysis.

- *Body fluids: automating the stain search*

During the workshop, it emerged that locating body fluids on items within the laboratory and at the scene was a very time-consuming task and that automation of the process would be of great value. Presently this is carried out manually using light sources; however automation of the process could be via a type of non-destructive scanner that could pass over the item of interest and give an indication when the body fluid was detected. At the scene a handheld device could be used for full spectral analysis of DNA material and perform a quality check regarding the presence of DNA; this could save time sending a sample to the lab when it isn't necessary.

- *Phenotyping: ancestry/age prediction*

Practitioners agreed that having the means to carry out DNA phenotyping analysis would be a great addition to the present "crime fighting toolbox" and a marked step change within the security industry. Having this capability could potentially steer the direction of an investigation with regard to the age and visible characters of a perpetrator. Additionally, this technology could offer the police assistance in those investigations that involved unidentified human remains. In some European countries, phenotyping is used to provide a statistical value to the eye colour of a person. It is not used on a routine basis, but when required physical traits resulting from phenotyping are provided and in some cases an e-fit of a person is created. Most DNA laboratories across Europe are at the same developmental stage, and therefore this would be an optimum moment for all to embark on this work together.

- *I.T. systems within the laboratory*

Practitioners agreed that all were experiencing a lack of capability from their own I.T. systems within the laboratory. It was pointed out that this was a limiting factor to their work, as those working within the DNA laboratory were scientists and did not have the required I.T. "know-how" to be able to deal with issues when required. It was stated that it should be DNA that drives the technology and not vice versa as is the case in most laboratories across Europe.

12.8   Conclusions and Future Work

We presented in this chapter a selection of the results of the five LEA's Workshops that brought together 78 experts from operational law enforcement from across 21 EU Member States. The workshop areas are related to Open Source Intelligence (OSINT), Mobility for Officers, People Trafficking, Intelligence Analysis and Technologies in DNA.

The workshops have proven to be extremely successful and have been evidenced by the identification of a list of key priorities. These priorities were attained through careful and methodical discussions amongst the practitioners and moreover have been found to be new and unique requirements that have not previously emerged from the policing community.

The impact of these findings will contribute to the formation of a set of "grand challenges" for those who seek to find or undertake research to address the issues. Thus, directing those to produce "fit for purpose" solutions and to improve the way LEAs fight crime and keep citizens safe and secure.

In our future work, I-LEAD will continue releasing new results in accordance with the project workplan. We now look forward to the 2019 workshops and are confident these will be as successful as those previous.

The subjects covered are related to financial investigation, drug trafficking, public order, digital forensics and digital investigations.

Acknowledgements

  I-LEAD project is funded by the European Union's Horizon 2020 – Research and Innovation Framework Programme, under grant agreement no. 740685.

BIBLIOGRAPHY

1. I-LEAD project web site. https://i-lead.eu/.
2. European Network of Law Enforcement Technology Services (ENLETS). http://www.enlets.eu/. Last accessed on 9 Dec 2019.
3. Durhan Police UK. https://www.durham.police.uk/About-Us/Documents/Peels_Principles_Of_Law_Enforcement.pdf. Last accessed on 19 Jan 2018.
4. United Nations. *Protocol to prevent, suppress and punish trafficking in persons especially women and children, supplementing the United Nations Convention against Transnational Organized Crime.* https://www.ohchr.org/en/professionalinterest/pages/protocoltraffickinginpersons.aspx. Last accessed on 19 Jan 2018.
5. International Association of Law Enforcement Intelligence Analysts, Inc. (IALEIA). https://www.ialeia.org/. Last accessed on 19 Jan 2018.

PART III

Border Security

CHAPTER 13

Threats and Attack Strategies Used in Past Events: A Review

Konstantinos-Giorgos Thanos, Dimitris M. Kyriazanos, and Stelios C. A. Thomopoulos

13.1 INTRODUCTION

The European Union's strategy for integrated border management across all border modalities (air, land, sea) is based on the four-tier access control model "which covers measures in third countries, measures with neighbouring third countries, border control measures at the external borders, risk analysis and measures within the Schengen Area and return" [1]. TRESSPASS EU research project [2] works on assessing the operational benefits and added value from deployment of risk-based border security management concept across all tiers of the access control and all border modalities. An innovative concept and a paradigm shift from current

K.-G. Thanos (✉) · D. M. Kyriazanos · S. C. A. Thomopoulos
Institute of Informatics and Telecommunications, National Centre for Scientific Research "Demokritos", Agia Paraskevi, Greece
e-mail: giorgos.thanos@iit.demokritos.gr; dkyri@iit.demokritos.gr; scat@iit.demokritos.gr

215

practice, the risk-based approach aims to assist Border Guard Authorities to focus their resources where and when it matters, based on a dynamic and intelligent analysis of risk. The expected impact aims at smarter and more efficient security controls while reducing waiting times and frustration for the increasing number of travellers and passengers across Europe.

Figure 13.1 depicts the TRESSPASS risk-based border security management concept with the TRESSPASS Front End technologies covering Tier 3 Border Control Point (BCP) area, Tier 1 and 2 connected through use of (i) TRESSPASS International Alert System (IAS) and (ii) legacy information systems (e.g. Visa Information System, Schengen Information System, Passenger Name Record, Advance Passenger Information), while Tier 4 is addressed through advanced correlation and analytics, capable of identifying patterns within the Schengen Area that can be connected with higher-risk ranking.

IAS provides a protocol and intelligence sharing component for collaboration with neighbouring and third countries, being critical links in the chain of intelligence. This includes information from all involved authorities, such as border guards, police and customs, about persons of

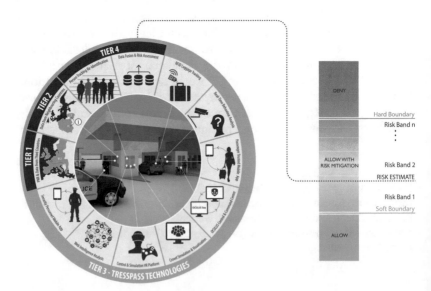

Fig. 13.1 Overview of TRESSPASS concept for border security risk analysis and management

interest or even high-risk warnings for illegal activities. Front End technologies include all the deployed TRESSPASS sensors and components on the field and in the area of the BCP: surveillance cameras and computer vision algorithms, location-sensing beacons for travellers and carry-on luggage, location-based services offered to travellers and assistive mobile applications for enhanced awareness offered to border guards. Finally, simulation and VR can offer valuable training and "what-if" scenario decision support, feeding also pattern recognition and deep learning algorithms with data that are scarce to find in normal everyday operations. All the aforementioned components provide input to the data fusion and risk assessment procedure, which is responsible of providing risk metrics across the four tiers of access control and most importantly to the right place, time, authority and security officer.

Risk assessment procedure is based on the analysis of risk factors of potential risk for malicious incidents jeopardizing critical infrastructures related to border crossing such as airports and harbours. These risk factors are determined by authorized security authorities and refer to indicators of risk of undesired of illegal activity to take place within the infrastructure. The risk factors can either be determined by reports of security personnel, or by the real-time information stemming from surveillance equipment installed on each infrastructure pre-processed by intelligent components that leverage raw data to meaningful high-level information or by infrastructures visitors profile constructed by travel documents, PNR and other available official documentation. Between risk factors and values and high-level information resulting from surveillance intelligent components or profile data, there is a conceptual gap which is covered by the data fusion component. Data fusion role is to aggregate the available input from the available heterogeneous information sources and approach each risk factor. Although these information sources are defined with the aim of providing evidence about the risk factors, the factor approximation cannot be performed without uncertainty. This uncertainty occurs due to the prediction errors that may result from the raw data pre-processing of the surveillance systems and the statistical uncertainty that result from the probabilistic models used to approach risk factors from the heterogeneous sources input. As a result, the data fusion algorithms were designed accordingly in order to be robust in cases of uncertainty or erroneous input. In the following sections, there will be presented briefly a state-of-the-art overview, various techniques along with corresponding pros and cons and finally the determination of the algorithm realized in the DFA component and the several options and decisions that were needed to be taken. The proof of concept and the corresponding evaluation is presented in the last section.

13.2   RELATED LITERATURE

13.2.1   *Information Fusion*

In this regard, Bayes rule (13.1) is exploited in order to link the posterior probability which corresponds to the estimated incident value given the available evidence, with prior beliefs about the expectation of the occurrence and the respective likelihood of the incident.

$$P(z) = \frac{P(z \mid x) \cdot P(x)}{P(z)} \tag{13.1}$$

Although this technique provides more accurate results compared to other method, it is not appropriate for heterogeneous types of sensors and for cases where assignment of probabilities to unknown propositions beforehand is inevitable.

Interval-Based Fusion
Interval-based methods mainly address a crucial weakness of Bayesian methods having to do with uncertainty management. In this approach uncertainty is represented as an interval between upper and lower parameter limits (e.g. $x \in [a, b]$) where no any probabilistic distribution of x over the interval is implied.

Interval-based fusion method has the benefit of providing a good measure of uncertainty in case of lacking probabilistic information. However, these algorithms cannot guarantee convergence. Moreover, these methods are not appropriate for encoding dependencies between variables.

Fuzzy Logic-Based Fusion
Fuzzy logic is a generalization of rule-based reasoning by extending each rule outcome and related fact values from binary (true or false) to real number ranging from 0 to 1. Fuzzy logic inference follows the process below:

1. Fuzzification of input values: Map input variables to membership function.
2. Apply fuzzy rules and compute the output membership functions.
3. Defuzzification of output memberships to specific values, which corresponds to the outcome estimation.

Fuzzy logic inference demands expert knowledge to be provided and is characterized by high complexity in the learning phase of membership functions.

Evidence-Based Fusion
Evidence-based fusion is distinguished from the other probabilistic methods by treating uncertainty not with the strict sense of probability but in a more generalized context where Kolmogorov axioms are not verified. In these methods mass functions are assigned to elements, sets and subsets of elements. Particularly, each incident is represented by a basic probability in the interval [0, 1], which expresses the amount of relevant evidence which is available for supporting this incident. Moreover two uncertainty measures are defined, the belief function *Bel()* and the plausibility function *Pls()*, both ranging from zero to one, which correspond to an upper and lower bound, respectively, of the incident uncertainty.

Evidence-based fusion methods generalize Bayesian inference and additionally have the capability of managing heterogeneous types of information sources, and it is efficient in cases of assignment probabilities to unknown propositions beforehand. However, evidence-based fusion methods are less accurate than Bayesian methods, and they are characterized by higher time complexity.

Summary
The above-mentioned methods are summarized below:

Fusion algorithmic approach	Pros	Cons
Bayesian fusion	More accurate compared to other fusion methods	Not appropriate for heterogeneous types of sensors A priori assignment of probabilities to unknown propositions

Fusion algorithmic approach	Pros	Cons
Evidence-based (Dempster-Shafer)	Generalizes Bayesian fusion Capability for heterogeneous sources fusion Assignment of a priori probabilities to unknown propositions is not needed It is closer to human perception and reasoning process	Less accurate than Bayesian approach Higher time complexity
Fuzzy logic-based fusion	Ability of enhancing data quality	Expert knowledge is needed High complexity of membership learning
Interval-based fusion	Intervals provide a good measure of uncertainty in case of lacking probabilistic information	Difficult to get results that converge to specific value Difficult to encode dependencies between variables

13.3   Proposed Solution

The proposed solution regards the intelligent automated real-time surveillance of a critical infrastructure and the corresponding risk detection. In this regard, several sensing components are distributed within the infrastructure which in real time record measurements related to crowd behaviour. These measurements correspond to raw numerical data related to crowd behavioural aspects. To this end, an efficient methodology is needed capable of tolerating the possibility of faulty or missing values, but on the other hand keeping the performance to a satisfying level. Moreover, due to the nature of the sensors input, it is mandatory to the implemented algorithms to be robust with heterogeneous types of sources and have to be flexible with expert knowledge provided in the system. From the operational point of view, the proposed approach consolidates and reconciles usage requirements from all involved security practitioners and parties. This includes requirements, modus operandi and legacy systems within a very fragmented operational ecosystem: border guards, custom authorities, infrastructure and transport operators and neighbouring and third country authorities. Based on these requirements, there is no perfect match by any of the proposed in literature algorithms; thus it is needed an adaptation of some of the proposed approaches which would regularize the weaknesses of one algorithm with the capabilities of another. In this context, a hybrid algorithm is proposed that relies on Dempster-Shafer algorithm and inherits the uncertainty capabilities of the evidence-based algorithms and the robustness in heterogeneous data sources. The system embeds a fuzzy logic rule-based classifier which has the role of regularizing the algorithm's result with any (if any) expert knowledge provided to the system. In this context, the proposed algorithm initially pre-process raw data coming from the sensors in order to distinguish exploitable data from non-exploitable ones and then apply machine learning methods (dimensionality reduction, supervised/unsupervised classification) for extracting high-level meaningful knowledge regarding potential incident factors that impact the risk of undesired behaviours. This information is considered as evidence for potential risk of unforeseen crowd behaviour. To this end this evidence is examined by a fusion system which evaluates the respective information and concludes to potential risk factors related to undesired behaviours and their intense level. Apparently, due the uncontrolled process of measurements acquisition, uncertainty cannot be neglected. As a result, at the information stage, it is proposed a Dempster-Shafer-based fusion algorithm [4, 5], which can tolerate not only uncertainty but lack of a priori knowledge of uncertainty level as well. Dempster-Shafer algorithm however corresponds to a high computational complexity. To this end, evidence sources are cluster based on their relevance to each risk factor. This way each risk factor is estimated taking into account only the most relevant and omits incorporating the ones with least contribution. The fusion process is presented in Fig. 13.2.

The first stage comprises the translation of the high-level information, (a) as results from the pre-processing of the raw measurements by each sensing device and (b) every available profiling knowledge base to linguistic variables each one corresponding to a quantified representation of the concept that the respective high-level information is referring to. These variables are the bases for defining the mass functions of each evidence. To this end these mass functions are defined by expert knowledge given a priori in two ways: (a) either as a direct assignment of an uncertainty value based on the estimated contribution of the respective source to risk evidence (b) or as a rule of combination of various information sources resulting in a specific evidence. Unlike the first case where mass function values are defined directly, in the second case each rule provided by expert knowledge is implemented as a fuzzy rule, and the concluded evidence mass value is calculated based on fuzzy logic inference. The method is

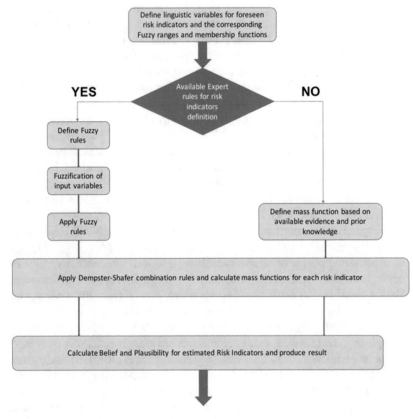

Fig. 13.2 Algorithmic process flow

based on literature proposed approaches [6, 7] where fuzzy logic inference is used for determine the mass function of evidence that is determined by a specific expert rule that combines high-level information
resulting from pre-processing of sensing components data streams. The
next stage corresponds to the solution of potential conflicts of evidence
where Dempster-Shafer combination rule is applied where m_i, m_j are
observations of sensor S_i and S_j, respectively. This rule can be generalized
iteratively where the result of each iteration is fed to the following one as
sensor measurement:

$$\left(m_1 \oplus m_2\right) = \{0; \quad A = 0 \frac{\sum_{B_i \cap B_j = A} m_1\left(B_i\right) \cdot m_2\left(B_j\right)}{1 - \sum_{B_i \cap B_j \neq 0} m_1\left(B_i\right) \cdot m_2\left(B_j\right)}; \quad A \neq 0 \quad (13.2)$$

Finally, based on the mass functions values of the available evidence by determining the upper and lower level, the risk factors uncertainty values are estimated by determining the lower and upper values corresponding to the calculated plausibility (13.3) and Belief (13.4) functions respectively.

$$\mathrm{Pls}\left(A\right) = \sum_{B \cap A \neq \varnothing} m\left(B\right) \tag{13.3}$$

$$\mathrm{Bel}\left(A\right) = \sum_{B \subseteq A} m\left(B\right) \tag{13.4}$$

where A = risk factor and B = evidence related to risk factors.

Finally, the result corresponds to a set of risk factors along with their uncertainty interval.

13.4   Conclusion

Real-time risk estimation exploits various types of heterogeneous sources in order to calculate the risk level of a potential malicious behaviour of persons at border crossing points. Risk estimation is based on the realization of a risk assessment model incorporating high-level factors that may be considered as evidence to potential future malicious actions. These high-level factors are approached via an information fusion model which processes input from heterogeneous sources and calculates each high-level factor along with the respective confidence level. The information fusion algorithm relies on the Dempster-Shafer theory where heterogeneous sources values are considered as evidence for the pre-defined risk factors. Additionally, fuzzy logic is incorporated wherever expert knowledge is applicable, in order to assess evidence mass functions with higher certainty. Finally, the algorithm concludes to risk indicators values that feed the risk estimation model, with the aim of assessing potential risks for malicious or suspicious behaviour. The future work will be directed in two ways: (a) evaluation of the proposed method in an experimental realistic use case scenario, where the system will be tested against several behaviour types, normal and abnormal where some will correspond to malicious intensions and some not. Moreover, (b) there will be research of the capabilities of imposing legal/ethics framework as domain knowledge in the system with the aim of constraining the system's response within pre-defined ethics/legal limits.

Acknowledgement

 Part of the work presented in this chapter is supported by and in the context of EU Research Project TRESSPASS – robusT Risk basEd Screening and alert System for Passengers and luggage – funded by the Horizon H2020 Framework Programme of the European Union for Research and Innovation under Grant Agreement number 787120.

BIBLIOGRAPHY

1. EU Integrated Border Management scheme. Retrieved on January 2020 and available online at: https://ec.europa.eu/home-affairs/content/european-integrated-border-management_en.
2. TRESSPASS project website. Accessed on January 2020 at: https://www.tresspass.eu/.
3. Durrant-Whyte, H., & Henderson, T. C. (2016). Multisensor data fusion. In *Springer handbook of robotics* (pp. 867–896). Berlin, Heidelberg: Springer.
4. Zheng, Y. (2015). Methodologies for cross-domain data fusion: An overview. *IEEE Transactions on Big Data, 1*(1), 16–34.
5. Wu, H., Siegel, M., Stiefelhagen, R., & Yang, J. (2002). Sensor fusion using Dempster-Shafer theory. In *IEEE Instrumentation and Measurement, Technology Conference*, Anchorage, AK, USA, 21–23 May 2002. Author, F. (2010). Contribution title. In *9th International Proceedings on Proceedings* (pp. 1–2). Location: Publisher.
6. Challa, S., & Koks, D. (2004). Bayesian and Dempster-Shafer fusion. *Sadhana, 29*(2), 145–176.
7. Maseleno, A., Hasan, M. M., Tuah, N., Fauzi, & Muslihudin, M. (2015). Fuzzy Logic and Dempster-Shafer belief theory to detect the risk of disease spreading of African Trypanosomiasis. In *Fifth International Conference on Digital Information Processing and Communication (ICDIPC)*, IEEE, 7–9 October 2015.
8. Yen, J. (2018). Generalizing the Dempster-Shafer theory to fuzzy sets. In *Classic works of the Dempster-Shafer theory of belief functions* (pp. 529–554). Berlin, Heidelberg: Springer.

CHAPTER 14

Early Warning for Increased Situational Awareness: A Pre-Operational Validation Process on Developing Innovative Technologies for Land Borders

Dimitrios Myttas, Pantelis Michalis, and Maria Kampa

14.1 Introduction

EWISA project was the result of a call for proposals restricted to a consortium of National Border Authorities from Greece, Finland, Spain, and Romania. The 58-month project's objective was to provide an operational and technical framework that would increase situational awareness and improve the reaction capability of authorities surveying the external land borders of the EU. EWISA provided an innovative system for warning about possible threats for all border control relevant systems, equipment, tools, and processes for the surveillance in selected areas.

D. Myttas (✉) · P. Michalis · M. Kampa
Center for Security Studies, Athens, Greece
e-mail: d.myttas@kemea-research.gr; p.michalis@kemea-research.gr; m.kampa@kemea-research.gr

© The Author(s), under exclusive license to Springer Nature
Switzerland AG 2021
B. Akhgar et al. (eds.), *Technology Development for Security
Practitioners*, Security Informatics and Law Enforcement,
https://doi.org/10.1007/978-3-030-69460-9_14

EWISA promoted further cooperation among public authorities in charge of surveillance of selected parts of the external EU land borders, so as to improve the quality and competence of their services (as related to security), through the Pre-Operational Validation (POV) concept for novel solutions.

In the context of EWISA, for the first time, four EU MoI (Ministry of Interior) Authorities jointly determined defined:

- A vision to improve overall situational awareness
- A common concept – the same core technologies for all areas of implementation
- The validation strategy

Pre-Operational Validation process provides a tangible assessment of the performance levels offered by innovative technologies in a realistic user-defined operational scenario, where a trade-off between efficiency, effectiveness, and cost can be aligned with actual needs.

14.2   EWISA CORE SYSTEM

EWISA concept is based on the development of a flexible, modular surveillance capability which maximizes the use of existing sensor types, including both static and mobile/deployable sensor platforms, following the concept of a unified integrated solution for the external EU borders based on data fusion from heterogeneous sensors, including Video Analytics Technologies generating intelligent analysis reports (Fig. 14.1).

The common core of the project was the development and the validation of the video analytics and data fusion components which were represented as Centralized in National Coordination Center (NCC) level and Decentralized in Command Center/Regional Command Center (LCC/RCC) level. Other sensors or sources at the national or regional level were also integrated within the core system in order to support the proof of concept of EWISA.

The objective was to increase intelligence in surveillance both in a qualitative and quantitative manner. The project provided an innovative system for warning on possible threats, enhancement of effectiveness and efficiency of all land border control relevant systems, equipment, tools, and processes for the surveillance in the selected areas.

The core of the EWISA system is introduced in Fig. 14.2 with the two fundamental components which are:

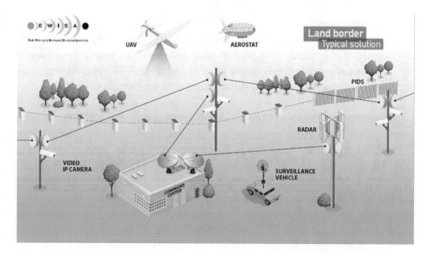

Fig. 14.1 Land border typical solution

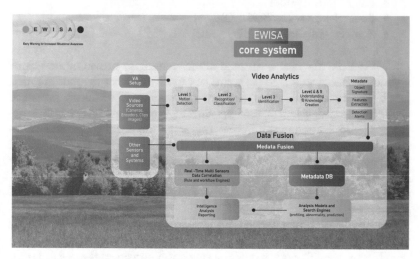

Fig. 14.2 EWISA core system flow

- Video Analytics Component (VAC) fed by video sources
- Data Fusion Component followed by Intelligence Analysis Reporting fed by the VAC information and the input of other Surveillance Supporting sensors/systems such as radars and ESM (electronic support measures) installed on stationary or mobile platforms

In order to achieve a practical implementation of the EWISA core system, other supporting equipment had been deployed along with video analysis components, either to facilitate the provision of coherent inputs to the VA system or just to guarantee adequate performance of the overall surveillance deployment. The supporting surveillance equipment included in the EWISA project consisted of the following:

- Land Vehicle with EO/IR/SWIR/RF/SL/LP
- Low Emission Radars
- ESMs
- Fiber Optics system
- Boat with EO/IR/SWIR/RF/SL
- Aerostat with EO/IR/SWIR/RF/SL

EWISA did not deal with stand-alone technology providing new capabilities. It rather validated (in terms of capacity to meet the requirements set by the public authorities) the integration of novel solutions, proposed by technology developers, into the current/legacy surveillance infrastructure. The realization of the aforementioned setup was through an innovation procurement procedure which concluded with two successful tenderers developing their own technical solution approach.

14.3   EWISA VALIDATION METHODOLOGY

The consolidation of a concrete validation strategy that could be utilized also for other similar testing activities was one of the core activities of the EWISA project. The outputs of the validation process of the offered solution had to ensure that the partners and other stakeholders in land border surveillance framework to:

- Check that the implementation process followed by the contractor had been correct
- Measure the level of compliance of EWISA solutions with the partners' operational objectives
- Compare two different alternatives based on different aspects of interest such as performance, deployability, operational value, etc.

In this sense, the validation process was based on being able to provide answers to the following questions:

- How does the EWISA solutions perform? (answered by testing the technical performance of the solutions)
- Does the EWISA solutions fit to End User's expectations? (answered by measuring the user acceptance of the EWISA solutions over a group of users with responsibility in land border surveillance)

Thus, the EWISA concept of validation, as it is depicted in the following Fig. 14.3, proposes an assessment of the solutions from two complementary perspectives: technical and operational. A third interesting perspective that needs to be taken into account is the cost analysis which is not going to be analyzed in the context of this chapter.

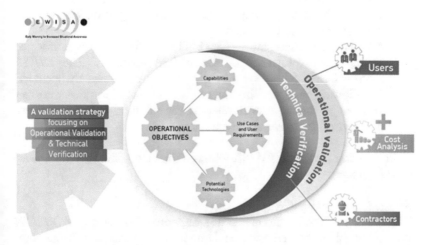

Fig. 14.3 Validation strategy phases

14.3.1   Technical Verification

The technical verification consisted in checking if the contractor has "built the solution right." This was realized through the following means:

- Monitoring the progress in the development of the technical solutions towards providing the R&D service
- Measuring the level of compliance of the EWISA solution with the system requirements and performance levels

This technical verification was ensured throughout the project in the form of continuous monitoring activities and visits on the contractors' premises. In addition, the level of compliance was measured in EWISA project in three different stages:

- LAT (Laboratory Acceptance Tests): This verification is the natural first stage at every development. Each company carried out their own tests prior to step towards the next phase of the deployment.
- FAT (Factory Acceptance Tests): New tests, both modular and as a whole system, were carried on in the companies' facilities by themselves, but under the supervision of the EWISA Consortium this time. Their objective is to check if every requirement is properly fulfilled and the maturity level of the system is high enough to go ahead with the on-site deployment and the integration with the legacy systems of the End Users.
- SAT (On-Site Acceptance Tests): This verification was performed during the deployment of the EWISA solutions. The EWISA Consortium checked that the solutions delivered meet the technical specification by supervising the verification procedures that were carried out by the industry developing the solutions in every scenario, as a part of the scope of the contract.

For the last two tests, EWISA followed the below-mentioned stepwise methodology:

Step 1. Identification of Requirements
The longlist of the requirements was categorized using the MOSCOW method based on their relevance and importance on each test. It was commonly decided that the "Would" will not be checked in the FAT neither the SAT.

Step 2. Identification of Test Scenarios
The test scenarios were initially built during the start of the project; however, a refinement was performed before each testing procedure in order to ensure their alignment with the testing objectives. The reference scenarios were used to experiment variations along the contract execution derived from factors such as changes in modus operandi of different targets, the availability of assets or modifications in the area of interest.

Step 3. Identify Team
For both tests the team was comprised by one technical representative of each partner, with the support of experts.

Step 4. Preparation for Test
This step has been mainly undertaken by each contractor before the start of each testing activity.

Step 5. Run Test and Track Results
The scenarios were executed, and the technical representatives were requested to assign a successful or not verdict to each requirement.

Step 6. Checking Whether the Requirements of the Customer Are Accomplished by Analyzing the Verdicts
Following the test execution, each member of the FAT team assigned a grade of severity (Step 5) to all defects identified on the requirements and the scenarios included into the checklists during the test execution. Based on their classification in Step 1, the requirements with defects were placed on a classification table.

14.3.2 Operational Validation

The operational validation is about answering the question "Did we provide the right service?" In other words, it consists in evaluating whether the service to be delivered meets the End Users expectations. With this aim, an operational evaluation process was set up in EWISA to be executed during the operation stage.

End Users as main beneficiaries of the technologies were responsible for validating the solutions built and tested by the industry. This operational validation was deemed necessary because a correct technical implementation compliant to requirements does not necessarily imply a high

End User satisfaction. In some cases, the operational needs are not translated accurately into the technical specification, and thus the solution built does not provide to the End User the operational added value expected. The same solution may have a different operational value for each End User and scenario, so it is needed that the End Users have a tool that allows them to measure the operational value that the solutions add to the execution of their operational tasks in their real scenario.

The validation procedures established for EWISA project fulfilled the following conditions:

- Flexible enough to be adapted to the needs of EWISA project. Some factors as changes in operational environment derive in changes in the validation needs and though processes should be easily adapted to changes in context. These types of changes could imply to upgrade the type and content of the information to be gathered or the way to gather it.
- End Users must understand what they measure within each indicator in order to obtain added value objective results. In this sense it was important to ensure that the MoEs and the metrics used to evaluate were interpreted equally by the whole community of End Users and no ambiguity existed when providing the measures. Thus, training sessions have been imparted in order to unify End User's criteria and solve doubts.
- Measurements must be effective. The results obtained after analyzing the information gathered should help decision-makers to understand project issues and to evaluate services aspects such as performance, costs, or maturity. Obtaining useful measurements requires the fulfillment of the two previous statements.

The operational validation comprised the following activities:

- Planning validation: comprised all the activities for launching the validation process.
- Information gathering: End Users gathered information during the operation of the EWISA Solutions and evaluated the indicators according to the information gathered.
- Processing the evaluation provided by End Users and generated conclusions.

14.3.2.1 Definition of Validation Concepts

The operational validation process was, therefore, devoted to determining at what measure the EWISA solutions complied with End User's objectives, ensuring that they fulfilled the requirements established in the terms of reference from an operational standpoint. This validation strategy was, therefore, sustained on a validation taxonomy built upon aggregated measures which addressed the effectiveness of the developed solution. This taxonomy was composed of five main concepts that expound on the following sections:

- Operational Obstacles: the main difficulties detected in the different scenarios that complicate the detection and prevention of illegal activities at the border.
- Key Performance Area (KPA): areas defined as most important in determining whether a system has been improved by a new operational measure.
- Key Performance Indicator (KPI): critical subset of performance parameters representing the most critical capabilities and characteristics.
- Measures of Effectiveness (MoEs): These KPIs will be composed of measures of effectiveness (MoE) intended to provide a measure of the expected systems performance in the operational environment according to what the End User expects.
- Metrics: These MoEs, likewise, can be further broken down into metrics when necessary, in order to increase the granularity of the measurement done on the system.

As a preliminary step, prior to the definition of the metrics to evaluate the operative value of the solutions, it was necessary to define the Operational Obstacles faced by the different End Users in their daily work within the framework of border surveillance. The goal is that, once the validation process is finished, it will be possible to determine to what extent the technical solutions of EWISA have contributed to overcome the operational obstacles defined by the End Users. The validation strategy shall allow End Users to measure at what level their expectations have been fulfilled. With this aim, the End Users have translated their expectations into a set of operational objectives which are the indicators to be evaluated using the operational validation process. For the purpose of the project, the operational objectives have been defined using the requirements, capabilities, and the principles of the Concept of Land Border Surveillance established.

The operational objectives have been classified into six key performance areas (KPAs) which are the areas of capability to be reinforced through the solution under validation in order to increase the operational effectiveness of land border surveillance. The KPAs match with the six areas of capability for classifying the system requirements. A number of operational goals were structured into a set of key performance areas. Each area comprised a set of capabilities that led to an improvement of the operational value of the solution.

KPA 1: Command, Control, and Coordination

This area comprised a set of capabilities for improving command, control, and coordination during the operations at different operational levels (tactical, strategic). The main capabilities were related to support planning and decision-making through an enhanced situational awareness and an efficient use of the resources. This Capability Area directly related to the project objective "Achieve a high level of control" had a multiplying effect as it maximized the effect of the rest of the capabilities.

KPA 2: Acquisition

This area comprised capabilities for improving the detection, monitoring, and identification of targets of interest in land borders through the acquisition of more reliable and precise information. The acquisition of information from external sources such as new sensors/platforms, open sources, or external DDBB were also considered. This Capability Area was directly related to the project objective "Detect irregular movement," "In-depth observation/identification."

KPA 3: Exploitation and Analysis

This area comprised a set of capabilities for fuse, correlate, process, and exploit the information acquired from different sources (sensors, platforms, external systems) to generate intelligence from the raw data acquired.

KPA 4: Communications

This area comprised capabilities for the well-dimensioned, robust, and secure transmission of data between the different assets/centers involved in land border surveillance in order to allow the availability of the necessary information at the precise moment and location.

KPA 5: Mobility and Projection
This area comprised capabilities for disposing of the necessary means in order to allow strategic deployment and high mobility of assets and personnel as required by the operations. The objective is to allow the intervention in the area of interest at the required moment.

KPA 6: Sustainability
This area comprised capabilities for guaranteeing the sustainability of the resources in the area of operation during the mission.

Figure 14.4 shows the classification of EWISA operational objectives into areas of capability (or KPAs). The EWISA solutions provided new and/or enhanced functionalities to improve the operational value of the solution perceived by the End User in one or more KPAs, contributing to the consecution of one or more operational objectives. Moreover, KPIs were used to measure the level of improvement provided by the solution on each key performance area (KPA). These measures described how well a solution achieved its objectives. They were the critical subset of operational performance parameters representing the most critical capabilities and characteristics in each particular area, and, of course, they excluded the evaluation of the performances of the legacy systems.

14.4   EWISA Operational Validation Execution

Following the definition of the abovementioned methodology and metrics, both solutions were deployed in four diverse geographical areas of EU external land borders (Figs. 14.5 and 14.6) as agreed by the consortium of EWISA.

In this regard, the validation of the solutions from an operational perspective was performed for an 8-month period in a real environment in surveillance operation. The EWISA concept was tested in a real operational environment, based on well-defined scenarios, representing the EU external borders environment and concept of operation, as follows (Fig. 14.7):

	Mandatory	Critical	High	Low
Must	Extreme	High	Medium	Low
Should	High	Medium	Low	Low
Could	Medium	Low	Low	Low

Fig. 14.4 Classification table

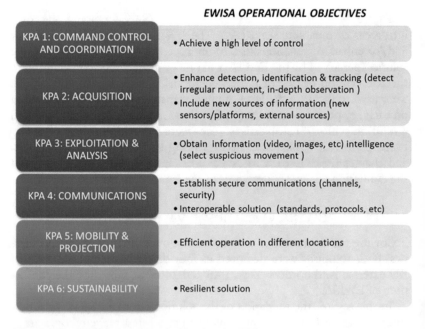

Fig. 14.5 EWISA operational objectives by KPA

- Greece: Surveillance of north area of Evros River in open and semi-open area, on the borders with Turkey.
- Finland: Surveillance of border line and border opening combined with surveillance of border zone boundary in a forest area, uneven, or rough land with Russia.
- Spain: Surveillance of the border line in Melilla area with Morocco.

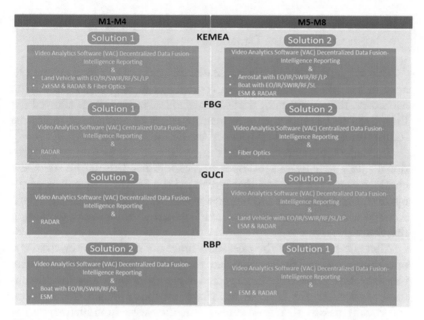

Fig. 14.6 Phase 3 solution testing

Fig. 14.7 EWISA test sites

- Romania: Surveillance of the border line with Serbia at terrestrial border and along the Danube River.

This validation provided a qualitative measure of the level of compliance of the EWISA solution with respect to the End Users' objectives. Moreover, the operation of the solutions in real environment allowed the End Users to measure the abovementioned set of operational indicators (or measures of effectiveness) defined by the EWISA Partners (End Users).

So, the solution developed by the contractor was integrated in additional legacy systems into real operation conditions. During this stage, the contractors ran the solution that integrated the EWISA concept, as developed and tested in previous stages for 4 months at each test site. Each scenario had its own schedule and ran for a certain amount of time during that frame.

After the first half of the operation time, the End Users reviewed the operation status and the intermediate results in order to determine if there were any deviations from the expected deployment by conducting a technical verification, the SAT, as described previously. When the solutions successfully passed all tests meaning that they were identified to run as expected, the solutions continued their operation without interruption to each site.

14.5   EWISA OPERATIONAL VALIDATION RESULTS

In continuation to the above, an online survey was organized for the End Users. The validators have received a link to participate in the validation survey, and they decided which module to evaluate, according to their operation scope in EWISA project. Each evaluator filled a set of MoEs depending on their role and location from which they have been operating.

They had to evaluate the applicable metrics for each MoE, scoring the performance brought by the solution using a 1–5 scale:

- 1: the solution provides very poor performance (it is not implemented or it is not operative due to serious malfunctions).
- 2: the solution provides a slight performance.
- 3: the solution provides acceptable performance.
- 4: the solution provides considerable performance.
- 5: the solution provides great performance.

Besides, for each metric, the validators were also able to say how it has been the EWISA experience compared with "outside EWISA" systems (e.g., existing legacy systems), if any. This was very important also for the evaluation because this way the Consortium could assess not only if the systems performs well, but also if it has provided added value to the End Users. In this case, the evaluator had to check a higher, equivalent, or lower performance of EWISA with respect to outside EWISA systems. The N/A option was also available.

Based on the input provided by the evaluators, a summary of the results per solution has been created, as presented below (Figs. 14.8 and 14.9):

Both solution's overall results have been positive. The validation process has been globally performed as it was planned in each prescribed phase. An interesting comparison between the delivered solution and the current used system has been performed, demonstrating its added value with the respect to the state of the art. Results have been assessed and discussed, showing an adequate satisfaction of user needs in both solutions.

14.6   Conclusions

Considering that the core element of the EWISA project is the POV of technological solutions in an operational real scenario, End Users were heavily involved in all procedures. Despite the inherent complexity of the EWISA project, the operations of both solutions were implemented and experimented in real environmental scenarios. In general, the developed technologies delivered outstanding added value in the scope of the border surveillance.

More specifically, the LE radar was considered as an important asset to the technical surveillance capacity. For the radar sensor, target detection, tracking, and separation performance was sound for all tests, with highly accurate, near real-time results. Radar coverage was continuous with high quality and reliability. However, it is true that in some cases, the detection of people by radar is difficult due to the proximity (less than 50 m) of roads and houses with a massive influx of people on the surveillance area.

Demonstration offered also a good opportunity to test the video analysis features. In this case, the detection of movements of groups through video analysis resulted more effectively in the short range. Additionally, it was considered that the new sensor could work also as an important deterring element helping the End Users to perform their daily activities.

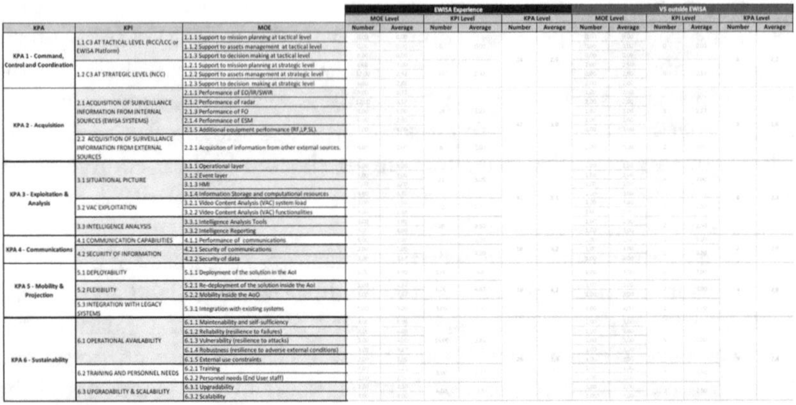

Fig. 14.8　Solution 1: General statistics

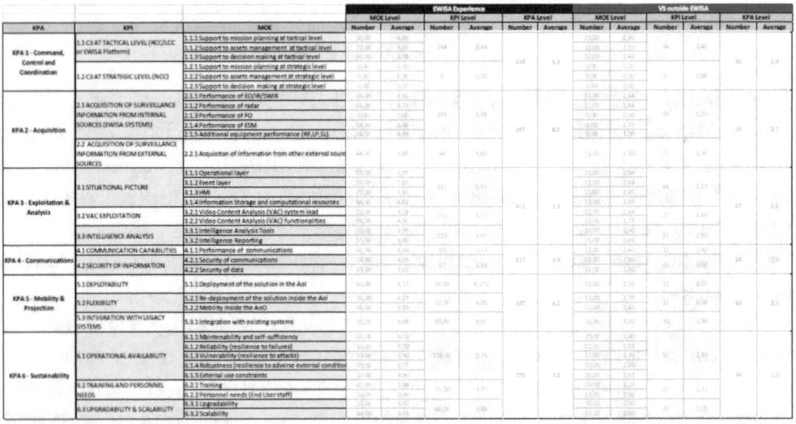

Fig. 14.9　Solution 2: General statistics

Regarding the ESM, it was considered of high value for the surveillance activities. The FO sensor was also considered interesting new technology which could enhance the performance of the technical border surveillance. The sensor was able to detect single targets, distinguish between simultaneous targets, as well as split groups to individual targets, performing within the required range and accuracy specifications.

From the operational point of view, the integration of terrestrial radar systems and different sensors provided new capabilities for border early warning.

The implementation of this project introduced some innovative features such as (i) the diverse environments where all demonstrations took place, covering many different environmental setups, (ii) long-lasting demonstrations of 8 months, and (iii) frequent presence of End Users throughout the 8-month period for validating the solutions and infusing insights and feedback valuable for delivering accurate results.

Acknowledgments

 This project has received funding from the European Union's Seventh Framework Programme for research, technological development and demonstration under Grant Agreement No 608174. The content of this chapter does not reflect the official opinion of the European Union. Responsibility for the information and views expressed in the chapter lies entirely with the author(s)

BIBLIOGRAPHY

1. EUROSUR Regulation. https://eur-lex.europa.eu/legalcontent/EN/TXT/?qid=1418993536491&uri=CELEX:32013R1052.
2. EWISA project. http://www.ewisa-project.eu/.
3. Frontex. (2019). *Migratory routes.* Eastern Borders Route. https://frontex.europa.eu/along-eu-borders/migratory-routes/eastern-borders-route/.
4. European Commission. *Pre-Operational Validation: Examples of Public Procurement of R&D services within EU funded Security Research actions,* Paolo Salieri 1/2/2017. http://www.seren-project.eu/images/Documents/Presenttions/PCP_PPI/3_Paolo_Salieri_EC_DG_HOME_POV.pdf.
5. Schengen Borders Code. (2016). https://eur-lex.europa.eu/legal-content/EN/TXT/?uri=celex%3A32016R0399.

CHAPTER 15

Border Surveillance Using Computer Vision-Enabled Robotic Swarms for Semantically Enriched Situational Awareness

Georgios Orfanidis, Savvas Apostolidis, Georgios Prountzos,
Marina Riga, Athanasios Kapoutsis,
Konstantinos Ioannidis, Elias Kosmatopoulos,
Stefanos Vrochidis, and Ioannis Kompatsiaris

15.1 Introduction

Political instabilities, war conflicts, economic crises and the maximization of personal profit comprise few of the main causalities that result in increased illegal events at border territories. Cross-border crime is referred to any serious crime with a cross-border dimension committed at or along the external borders [1]. Towards maximizing the overall profit, such activities involve in many cases the utilization of recent technological

G. Orfanidis (✉) · S. Apostolidis · G. Prountzos · M. Riga · A. Kapoutsis
K. Ioannidis · E. Kosmatopoulos · S. Vrochidis · I. Kompatsiaris
Centre for Research and Technology Hellas, Information Technologies Institute,
Thessaloniki, Greece
e-mail: g.orfanidis@iti.gr; sapostol@iti.gr; gprountzos@iti.gr; mriga@iti.gr;
athakapo@iti.gr; kioannid@iti.gr; kosmatop@iti.gr; stefanos@iti.gr; ikom@iti.gr

243

B. Akhgar et al. (eds.), *Technology Development for Security
Practitioners*, Security Informatics and Law Enforcement,
https://doi.org/10.1007/978-3-030-69460-9_15

advances such as innovative sensory systems and specialized equipment. Such technological tools facilitate the activities of criminals which eventually might lead even to human casualties as, for example, drug trafficking using unmanned aerial vehicles.

The effective control and identification of transnational crime activities are essential for ensuring peace and stability and for promoting pertinent political and socio-economic activities. At tactical level, European Border Surveillance System (EUROSUR) is a common example for such initiatives. EUROSUR [2] establishes a common framework for the exchange of information and cooperation between EU member states and Frontex to improve situation awareness and reaction capabilities at the external EU borders confronting cross-border crime and protecting lives of migrants. At operational level, considering also the diversity and the increased number of operational aspects, border authorities and relevant practitioners face important challenges in patrolling and protecting areas under their jurisdiction. The heterogeneity of the threats, the wideness of the surveyed areas, the complexity of the operational environments and the adverse weather conditions are some characteristic subjects under consideration from border practitioners. Thus, it is considered imperative in many cases for the operational personnel to be equipped with advanced surveillance systems in order to effectively complete their objectives.

Such systems mostly involve video and thermal cameras; dedicated sensors for motion, pressure, etc.; RFID tags; radars; and satellite images. Despite their sufficient effectiveness, each system displays either environmental restrictions or limited capacities due to spatial heterogeneity. In addition, the majority of these sensory systems are static resulting in restricted monitored areas strictly depending on their technical specifications. As a result, border authorities currently exploit novel technologies posing existing infrastructure as legacy systems. Unmanned vehicles (UxV) provide such cutting-edge technologies that can be utilized as either independent or complement of existing border surveillance equipment. In this book chapter, we introduce and analyse relevant robotic technologies combined with swarm intelligence for a completely autonomous border surveillance system. In addition, pioneer visual detection approaches are presented for increased efficiency, while semantic data representation models upgrade the overall capacities for optimum situation awareness.

The rest of the chapter is organized as follows. Section 15.2 introduces swarm intelligence as an autonomous navigation scheme, while Sect. 15.3 presents enhanced visual detection models. The following section describes semantic enrichment models towards increased situation awareness, while Sect. 15.5 concludes the chapter by highlighting the benefits of such technologies.

15.2 SWARM INTELLIGENCE FOR AUTONOMOUS NAVIGATION

The utilization of different UxVs acquires much popularity in missions that demand immediate situation awareness or are considered as hazardous for the integrity of human lives. Due to these technologies, data acquisition from the operational areas of interest is obtained currently safer, faster and more affordable as higher objectives can be accomplished without the need of specialized sensors. However, despite the convenience that a UxV can offer, such systems prerequisite a specialized operator in order to command and manipulate the assets. The complexity of the process is increased in missions where multiple UxVs are commanded to complete one major objective. In such cases, not only the total operator number is increased accordingly, but also the personnel must be in continuous communication to achieve the overall mission.

An autonomous, yet safe and secure, navigation system for operating UxVs has been proven to be essential in numerous application fields. Introducing autonomy for navigation objectives decreases the operator's interference in the overall operations since his involvement from a low-level operator is converted into a manipulator of higher-level objectives for the defined missions, without the requirement of a priori knowledge of utilizing multiple and heterogeneous UxVs. After the identification of high-level objectives, the navigation system will commence to design robot trajectories in order to successfully complete the overall goal of the defined mission. During the execution of the defined mission, the operator acts only as a supervisor nonetheless, for safety reasons; the system is responsive to any interference at any moment. Thus, the process is more effective since the operator can utilize multiple UxVs, without any special expertise and training, while simultaneously, the efficiency of the mission is increased, and the operational time is reduced.

The presented autonomous navigation system, developed specifically for border security operations, supports three different types of missions. More specifically:

- Strictly user-defined paths to be executed separately from UxVs
- Complete coverage of a polygon region of interest (ROI) over a map, utilizing multiple UxVs
- Continuous surveillance of an unknown, dynamically changed ROI utilizing multiple UxVs

For the first and most simple mission type, the operator/practitioner identifies a set of waypoints for a UxV over a map corresponding to the area of interest. The module provides high-level controls for the UxVs without the need of special training courses or awareness of technical limitations.

Moreover, operating multiple UxVs simultaneously is simplified, while the requirement of using multiple operators is no longer valid. This mission type is considered appropriate for objectives when specific locations must be monitored continuously.

The second type of mission provides the feature of commanding a swarm of UxVs to completely scan a user-defined ROI. Thus, the module is appropriate in covering wide, arbitrary-defined territories benefiting from the number of UxVs in order to significantly limit the overall execution time of the mission and constrain human interference. In addition, it is suitable for different types of UxVs, requiring just minor adjustments on the mission's parameters according to the UxVs' specifications. The overall mission is reduced to a multi-robot Coverage Path Planning (CPP) [3] problem. Receiving as input a polygon for ROI, the number of UxVs and a scanning density (distance between two sequential trajectories), the polygon is represented on an optimized grid for the specific problem, obtaining values that correspond to free space or an obstacle. The entire region is divided into exclusive subregions for every UxV with DARP algorithm [4]. For every subregion, an independent Spanning Tree Coverage (STC) [5] problem is solved. A Minimum Spanning Tree (MST) [6] is constructed, and a circumnavigating path is outlined. These paths incorporate energy aware features, posing them as resource efficient (Fig. 15.1).

Finally, the third mission type provides the capability to the operator to select a region over the map and continuously calculates the optimal monitoring position for every UxV, in order to provide complete situation awareness of the region. The morphology of the region may be completely unknown and dynamically changed, while the number of UxVs may similarly modified even during the mission. The autonomous navigator will

Fig. 15.1 Multi-robot coverage paths in polygon ROI

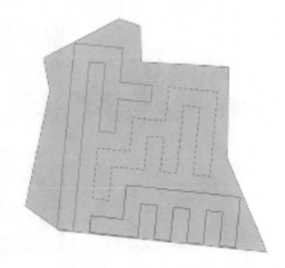

reallocate the available resources to provide the best possible result and fulfil the overall objective.

A relevant module as reported above was implemented according to a distributed, plug-n-play algorithm for multi-robot applications with a priori non-computable objective functions [7]. This algorithm extracts a sub-cost function individually for each UxV and achieves the overall objective of the swarm by optimizing them combined. Towards this objective, a distributed methodology according to the cognitive-based adaptive optimization (CAO) algorithm [8] is implemented that approximates the evolution of each robot's cost function and adequately optimize its decision variables. The entire training procedure is performed online focusing only on problem-specific characteristics that affect the completion of mission objectives. The fast convergence of the algorithm can ensure fast adaptation of the swarm to the mission, not only during the first stage but also during modifications of the ROI or the swarm itself (Fig. 15.2). As a result, border personnel acting as operators can leverage such systems without requiring specialized training courses, while operational effort is retained at low levels as the feature of autonomy is inherently integrated.

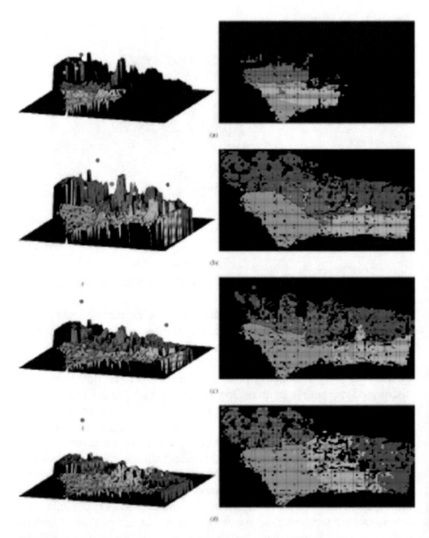

Fig. 15.2 Swarm adaptation to unknown ROI for surveillance during subsequent time steps (**a**–**d**)

15.3   Visual Detection Capabilities

Similarly, due to the heterogeneity of the identified threats, systems utilized by border practitioners should be equipped with enhanced capabilities in identifying specific objects of interest. Considering also that a deployed surveillance system relies on robotic technologies, navigation systems are strictly related to object detection capacities for completeness in the context of autonomous functionalities. In principle, an object detection model corresponds to a schema for simultaneous recognition and localization over the projection plane of objects of interest within a visual representation.

Therefore, the real objective of object detection is to scan the acquired images for identifying any appearance of objects of interest and localizing the detected instances in the processed images. The localization result corresponds to a bounding box surrounding each object of interest, which can be provided in various formats, for example, in upper left and lower right coordinates, centre coordinates, width and height of the bounding box, etc. There are two main categories for visual object detectors: two-step and single-step approaches. The former perform an additional initial step for deciding the "objectiveness" of the area included in a bounding box to determine the best candidates for objects included in the image. The latter category performs both area selection and label assignment (classification) in the same step. The predominant method belonging to the first category is Faster RCNN [9] and typical examples of the second category are Single-Shot Detector (SSD) [10] and You Only Look Once Detector (YOLO) [11] with the latter having several improved versions. The object detector output involves a list of bounding boxes along with their corresponding class labels and their confidence scores. The latter roughly represents the estimation of how confident is the model for the assigned to this bounding box label. Object detection as a capacity is considered overall precise nonetheless, depending on the level of some limitations, inefficient. Thus, a typical approach is to combine this functionality with a tracking module in order to monitor the identified objects. A tracker comprises a module which is provided with an initial bounding box for each detected object and attempts to estimate its motion from a sequence of images or video streams. In most cases, the application of an object tracker is computationally more effective rather than feeding continuously an object detector with sequential frames in systems that require visually identification of specific objects. A typical, yet efficient and fast, tracker relies on the Kernelized Correlation Filters (KCF) [12].

Towards identifying the most efficient object detection model for border surveillance applications, multiple relevant models were deployed and properly evaluated considering both accuracy and execution time. After extensive experiments and evaluations, Faster RCNN [9] resulted in the most sufficient outcomes for the objects of interests as typically, the objects to be identified display small sizes (due to the height and angle of perception) and the model is reported as the most efficient for this objective.

Towards decreasing the overall execution time of the visual identification system, a KCF tracker [12] is applied between two subsequent frames. At every key-frame, an object id is assigned to each distinct object in order to uniquely identify its presence. During the tracking frames, which are typically larger in number than the key-frames, the object ID remains unchanged. At the next key-frames, an Intersection-Over-Union comparison against a fixed threshold of the two bounding boxes is applied. The two bounding boxes, deriving from the object detector and the tracker respectively, are utilized to estimate if the same object is depicted within the bounding boxes' boundaries. The entire scheme is depicted in Fig. 15.3.

For the evaluation process in order to identify the adequacy of the module, the PascalVoc evaluation metric was exploited [13]. The resulted object detection accuracy values are provided in Table 15.1 where 11 classes of objects of interest were used. The presented work emphasized mostly on identifying maritime vehicles leading to identifying four relevant classes: ships, speedboats, inflatable and regular boats. The latter class corresponds to vehicles that could not be categorized in the other classes; nonetheless, the object corresponds to a boat instance. This fact reveals the high importance of maritime border surveillance since the measures

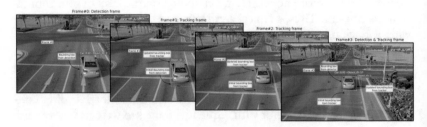

Fig. 15.3 Pipeline for an object tracker in surveillance application

Table 15.1 Accuracy results of Faster-CNN with PascalVoc metric

mAP 0. 6651									
UAV	Boat	Bus	Truck	Car	Helicopter	Inflatable	Person	Motorcycle	Ship	Speedboat
0.68169	0.56543	0.70576	0.64993	0.75568	0.67105	0.41560	0.84015	0.76698	0.73174	0.53311

that should be considered for each maritime vehicle are diverse; thus, it is imperative to be able to classify such type of vehicles. On the contrary, the performance for some classes suffers since the distinction between these classes is occasionally vague. A typical example of such case would be a light speedboat compared with an inflatable boat with a powerful engine. Figure 15.4 depicts some characteristic examples of visual results acquired with the application of the Faster-RCNN model.

The integration of cognition functionalities comprises a real multilevel asset of the system as object of interests can be identified accurately via processing visual data. Following a hierarchical data flow, the outcomes can be enriched with additional information, while the feature of autonomy can be significantly extended for various operational scenarios.

Therefore, a detailed, yet comprehensive, operational overview can be presented to the operator decreasing the required commanding effort and focusing more on operational goals.

Fig. 15.4 Visual results of Faster-RCNN

15.4 SEMANTIC ENRICHMENT FOR INCREASED SITUATION AWARENESS

Such surveillance systems display an increased complexity at operational level from the practitioners' perspective as usually, they are not familiarized with such technologies. Noncomprehensive sensor readings and detection outcomes might result in an obsolete system, and eventually, practitioners exploit traditional methods of monitoring the areas of their jurisdiction. In order to facilitate the operational activities of border practitioners and increase their situation awareness, relevant systems integrate technologies at a higher level of implementation to obtain the desired objectives. Such technologies involve the utilization of semantics which refer to the linguistic study of meaning in language coherent to the operator. Therefore, semantic enrichment provides a knowledge framework built upon the acquired data and the detection outcomes so that the operator could be comprehensively be informed.

More specific, ontologies are a means for specifying a vocabulary for conceptualizing and representing a shared domain of discourse [14] in a formal, structured and semantically enriched way. Knowledge in ontologies is modelled via the knowledge graphs by defining common components, like classes (objects, concepts and other entities existing in a domain of interest), properties (attributes, relationships that hold between them), axioms (expressed in a logical form) and rules (if-then statements for logical inferences). With the use of semantic reasoners such as FACT++ [15], Pellet [16] and HermiT [17], logical consequences and new assertions (facts) that are not explicitly expressed in an ontology can be derived.

Ontologies play a key role in facilitating the understanding, sharing and reuse of knowledge between different components within complex systems such as swarm robotics. They have been widely used for situation awareness [18] and decision- making [19] and in IoT infrastructures [20], natural language processing [21] and many more. They demonstrate multiple benefits and capabilities in improved searching, data integration, interoperability, multilinguality and dynamic content generation in an extensive range of areas such as security, healthcare [22], telecommunications, archive portals and law [23].

In the current work, we focus on the semantic representation and enrichment of sensor-based data sourced from different surveillance components (additional sensors, etc.), for extracting potential threats and alerts in the surveillance area, enhancing the representation of the derived detections and improving the situation awareness of the end-users. Eventually, the displayed information to the operator is formatted according to common representation models that are widely utilized in their operational activities at a daily basis.

Therefore, the corresponding service of the increased situation awareness is strictly dependent with the application and the described operational scenarios. More specifically, an ontology was developed for the representation and semantic integration of heterogeneous data generated and exchanged across the cooperative surveillance systems. The proposed semantic model is compliant and extends the EUCISE2020 data model [24], a CISE (Common Information Sharing Environment)-based collaborative initiative for promoting automated information sharing between maritime monitoring authorities. In a nutshell, the CISE data model identifies seven core data entities (Agent, Object, Location, Document, Event, Risk and Period) and eleven auxiliary (Vessel, Cargo, Operational Asset, Person, Organization, Movement, Incident, Anomaly, Action, Unique Identifier and Metadata). An illustration of our ontology-based serialization of the EUCISE2020 model is presented in Fig. 15.5.

The proposed extension of the EUCISE2020 model is related to the following types: (i) further specialization of objects and vehicles and (ii) addition of classes and properties representing the detection of incidents, objects and persons. For demonstration purposes, we consider one rather common scenario in maritime surveillance that involves the detection of an oil spill over sea surface. Whenever an oil spill is detected, an instance of PollutionIncident class (Fig. 15.6) is created, which involves an incident of OilSpill and is associated with respective PollutionType and

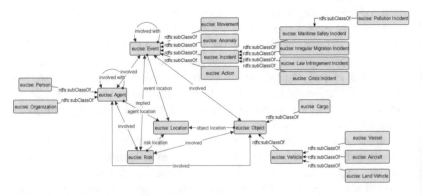

Fig. 15.5 Core classes of our ontology-based serialization of the EUCISE2020 model, along with their main interrelationships

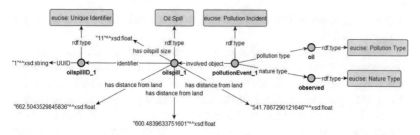

Fig. 15.6 An instance of oil spill associated with a pollution event of specific pollution and nature type

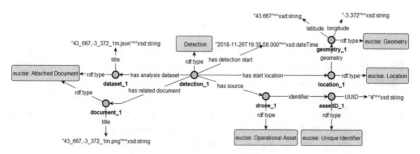

Fig. 15.7 An instance of Detection type associated with an operational asset, a document of reporting and the location of interest

NatureType instances. Also, an instance of Detection is created (Fig. 15.7), which is associated with all relevant information populated in the AttachedDocument, Geometry and the OperationalAsset classes that made the detection via the appropriate data and object properties including hasAnalysisDataset, hasStartLocation and hasSource.

On the basis of the implemented ontology, semantic reasoning techniques (SPARQL rules and constraints) might be additionally adopted to aggregate data from various sources and to achieve both low-level fusion from external resources (such as geospatial services) and high-level fusion by combining information from geographically dispersed and heterogeneous sensors. This approach facilitates the automatic detection and inference of complex events of interest like threats, abnormal activities and illegal border trespassing. In general, SPARQL is a highly expressive RDF query language that allows querying the linked data, by matching one more or patterns against the relationships of the knowledge base while supporting features like aggregation, negation, filtering, constraints and property paths.

Overall, such technologies target eventually to present the system's outcomes within a common representation framework. The displayed alerts and information follow a widely utilized template which was derived from the operational needs of the corresponding experienced personnel. Thus, the system interacts with the operator using one common basis for which the results are comprehensive and intentionally simplified in order for the operators to increase their situation awareness and focus on operational tasks.

15.5   Conclusions

Recent technology advancements are considered to be sufficiently mature for integration in many systems and applications. Even in very complex operational scenarios like border surveillance, cutting-edge technologies can perform adequately well. The relevant practitioners can benefit of such systems towards improving their operational capabilities. As the challenges that they have to confront display significant diversities, the utilized surveillance systems must integrate specialized capacities.

Towards this objective, swarm robotics can broaden the solutions that are provided to the border practitioners. Such systems enhanced with additional features can be used effectively to monitor distant territories. In this chapter, three different pillars of services in different levels of implantation were presented towards describing a fully autonomous and operational surveillance systems. More specific, an optimizer for autonomous navigation of a swarm was presented. The service provides high-level commands to the practitioner to mitigate the complexity of operating such systems while retaining, nonetheless, their effectiveness in monitoring tasks. In addition, visual recognition of object of interests can increase the detection capabilities of the overall system leading to a truly autonomous surveillance framework. Finally, the integration of semantics improve the practitioners' perception for the identifying events increasing the level of the current situation awareness. These three types of technology have been proven particularly efficient in monitoring tasks since they have been extensively deployed in relevant systems as independent features.

Therefore, their integration along with their combination comprises a significant added value for an autonomous surveillance system since each additional feature increases its main operational objective.

Acknowledgements This work was supported by the ROBORDER project funded by the European Commission under Grant Agreement No 740593.

REFERENCES

1. Passas, N. (2003). Cross-border crime and the interface between legal and illegal actors. *Security Journal, 16*(1), 1 37.
2. Eurosur homepage. https://ec.europa.eu/home-affairs/what-we-do/policies/borders-and-visas/border-crossing/eurosur_en. Last accessed: 2 Nov 2019.
3. Cabreira, T., Brisolara, L. R., & Ferreira, P. (2019). Survey on coverage path planning with unmanned aerial vehicles. *Drones, 3*(1), 4.
4. Kapoutsis, A. C., Chatzichristofis, S. A., & Kosmatopoulos, E. B. (2017). Darp: Divide areas algorithm for optimal multi-robot coverage path planning. *Journal of Intelligent & Robotic Systems, 86*(3-4), 663–6 0.
5. Gabriely, Y., & Rimon, E. (2001). Spanning-tree based coverage of continuous areas by a mobile robot. *Annals of Mathematics and Artificial Intelligence, 31*(1-4), 77.
6. Gower, J. C., & Ross, G. J. (1969). Minimum spanning trees and single linkage cluster analysis. *Journal of the Royal Statistical Society: Series C (Applied Statistics), 18*(1), 54–64.
7. Kapoutsis, A. C., Chatzichristofis, S. A., & Kosmatopoulos, E. B. (2019). A distributed, plug- n-play algorithm for multi-robot applications with a priori non-computable objective functions. *The International Journal of Robotics Research, 38*(7), 13–32.
8. Kosmatopoulos, E. B. (2009). An adaptive optimization scheme with satisfactory transient performance. *Automatica, 45*(3), 716–723.
9. Ren, S., He, K., Girshick, R., & Sun, J. (2015). Faster r-cnn: Towards realtime object detection with region proposal networks. In *Advances in neural information processing systems* (p. 1).
10. Liu, W., Anguelov, D., Erhan, D., Szegedy, C., Reed, S., Fu, C. Y., & Berg, A. C. (2016). Ssd: Single shot multibox detector. In *European conference on computer vision* (pp. 21–37). Springer.
11. Redmon, J., Divvala, S., Girshick, R., & Farhadi, A. (2016). You only look once: Unified, real-time object detection. In *Proceedings of the IEEE conference on computer vision and pattern recognition* (pp. 77–77).
12. Henriques, J. F., Caseiro, R., Martins, P., & Batista, J. (2014). High-speed tracking with kernelized correlation filters. *IEEE Transactions on Pattern Analysis and Machine Intelligence, 37*(3), 5 3–5 6.
13. Everingham, M., Van Gool, L., Williams, C. K., Winn, J., & Zisserman, A. (2010). The pascal visual object classes (voc) challenge. *International Journal of Computer Vision, 88*(2), 303–333.

14. Guber, T. (1993). A translational approach to portable ontologies. *Knowledge Acquisition, 5*(2), 1–22.
15. Tsarkov, D., & Horrocks, I. (2006). Fact++ description logic reasoner: System description. In *International joint conference on automated reasoning* (pp. 2 2–2 7). Springer.
16. Sirin, E., Parsia, B., Grau, B. C., Kalyanpur, A., & Katz, Y. (2007). Pellet: A practical owl- dl reasoner. *Web Semantics: Science, Services and Agents on the World Wide Web, 5*(2), 51–53.
17. Motik, B., Shearer, R., & Horrocks, I. (2007). Optimized reasoning in description logics using hypertableaux. In *International conference on automated deduction* (p. 67–3). Springer.
18. Frank, A. U. (2001). Tiers of ontology and consistency constraints in geographical information systems. *International Journal of Geographical Information Science, 15*(7), 667–667.
19. Hogenboom, A., Hogenboom, F., Frasincar, F., Schouten, K., & Van Der Meer, O. (2013). Semantics-based information extraction for detecting economic events. *Multimedia Tools and Applications, 64*(1), 27–52.
20. Bimschas, D., Hasemann, H., Hauswirth, M., Karnstedt, M., Kleine, O., Kroller, A., Leggieri, M., Mietz, R., Passant, A., Pfisterer, D., et al. (2011). Semantic-service pro- visioning for the internet of things. Electronic Communications of the EASST 37.
21. Hellmann, S., Lehmann, J., Auer, S., & Brümmer, M. (2013). Integrating nlp using linked data. In *International semantic web conference* (p. 113). Springer.
22. Liyanage, H., Krause, P., & de Lusignan, S. (2015). Using ontologies to improve semantic interoperability in health data. *BMJ Health & Care Informatics, 22*(2), 30315.
23. Benjamins, V. R., Casanovas, P., Breuker, J., & Gangemi, A. (2005). *Law and the semantic web: Legal ontologies, methodologies, legal information retrieval, and applications* (Vol. 336). Berlin/Heidelberg/New York: Springer.
24. Eucise2020 fp7 project. http://www.eucise2020.eu/. Last accessed: 06 Dec 2019.

CHAPTER 16

FOLDOUT: A Through Foliage Surveillance System for Border Security

Christos Bolakis, Vasiliki Mantzana, Pantelis Michalis,
Aggelos Vassileiou, Roman Pflugfelder,
Martin Litzenberger, Michael Hubner, Gaetano Pastore,
Domenico Oricchio, Marie Desplas, Marie Ansart,
Maria Rosaria Santovito, Giulia Pica, Luis Patino,
James Ferryman, and Andreas Kriechbaum-Zabini

16.1   Introduction

In the border control context, as defined by the Schengen Border Code, border surveillance is defined as "the surveillance of borders between border crossing points and the surveillance of border crossing points outside the fixed opening hours, in order to prevent persons from circumventing border checks" [1]. Border surveillance shall be to prevent unauthorised border

---

The original version of this chapter was revised. The correction to this chapter is available at https://doi.org/10.1007/978-3-030-69460-9_31

C. Bolakis (✉) · V. Mantzana · P. Michalis · A. Vassileiou
Center for Security Studies, Athens, Greece
e-mail: c.bolakis@kemea-research.gr; v.mantzana@kemea-research.gr;
p.michalis@kemea-research.gr; a.vassileiou@kemea-research.gr

© The Author(s) 2021, Corrected Publication 2022
B. Akhgar et al. (eds.), *Technology Development for Security Practitioners*, Security Informatics and Law Enforcement, https://doi.org/10.1007/978-3-030-69460-9_16

259

crossings, to counter cross-border criminality and to take measures against persons who have crossed the border illegally. A person who has crossed a border illegally and who has no right to stay on the territory of the member state concerned shall be apprehended and made subject to expulsion.

To achieve an effective and efficient border management, there should be used technologies and personnel that (a) supervise border sections between border crossing points, (b) supervise border crossing points (border gates) outside opening hours and (c) control movement in order to prevent persons from circumventing border checks.

In achieving this and with regard to the surveillance environment, border guards work with shifts on a 24/7 basis that take place at central offices as well as different places along the border. In general, border control units are well equipped with state-of-the-art surveillance equipment. Border guards use systems that produce alarms (a special graphical frame and/or a sound alarm) each time, either a target has been detected without filtering and needed to be clarified through and high-definition (HD)/ thermal camera (fixed or mobile) or through the motion of an object/ human/animal. When a C2 operator performs the programming of the sensor, it is often being done either manually or partially assisted. However, this requires a twofold skill from the operator: knowledge of sensors

R. Pflugfelder · M. Litzenberger · M. Hubner · A. Kriechbaum-Zabini
AIT Austrian Institute of Technology GmbH, Seibersdorf, Austria
e-mail: Roman.Pflugfelder@ait.ac.at; martin.litzenberger@ait.ac.at;
Michael.Hubner@ait.ac.at; Andreas.Kriechbaum-Zabini@ait.ac.at

G. Pastore · D. Oricchio
Thales Alenia Space Italia, Rome, Italy
e-mail: Gaetano.Pastore@thalesaleniaspace.com;
Domenico.Oricchio@thalesaleniaspace.com

M. Desplas · M. Ansart
Thales Alenia Space, Cannes, France
e-mail: marie.desplas@thalesaleniaspace.com; marie.ansart@thalesaleniaspace.com

M. R. Santovito · G. Pica
CORISTA Consortium of Research on Advanced Remote Sensing Syst,
Naples, Italy
e-mail: mariarosaria.santovito@corista.eu; giulia.pica@corista.eu

L. Patino · J. Ferryman
UoR University of Reading, Computational Vision Group, Reading, UK
e-mail: j.l.patinovilchis@reading.ac.uk; j.m.ferryman@reading.ac.uk

needed to perform the mission and knowledge of programming for all sensors involved in the mission. These activities greatly increase the operator's workload without an effective gain. In addition, manual programming makes it harder to distinguish between bad configurations and false negatives (especially an issue for RADAR-like systems).

Border authorities are disadvantaged in preventing illegal border activities in areas where objects to be detected, like people and vehicles, are concealed by foliage. Such environments are extremely challenging due to people and vehicles being hidden behind opaque layers as well as under the cover of darkness and/or under reduced visibility. For example, if a patrol finds people moving into forests or other harsh and unstructured environments, they are not able to follow them. No border fence or surveillance system can protect the border by itself. Rapid Intervention Troops and/or Border Police Teams are required to be deployed at the scene and being thoroughly informed as soon as possible. This would be a key to an effective border security.

The technologies currently available to border guards do not match many of their needs. Large areas require monitoring, and with modern technology (such as smart phones for communicating new routes and exchanging information about the activities of border control units) aiding the smugglers and traffickers, it is necessary to improve the detection capabilities of the border guards. To effectively monitor border areas, it is necessary not only to have the ability to scan a specific area but also to predict where the next illegal crossing will take place.

In particular, solutions are needed that do not only detect persons and vehicles crossing land borders illegally but also being able to do under harsh and unstructured environments, such as a canopy of foliage. Solutions are needed to be able to provide border guards with improved situational awareness of border regions including robust detection of people and vehicles, groups, recognition of abnormal behaviours and predicting routes of individuals and small groups. Technologies customized for foliage penetration should be integrated into a quick decision system that consists of autonomous ground-based sensors, with high mobility even on challenging and harsh terrain (able to perform 24 hours), as well as aerial and/or space-based systems for pre-warning of ground-based sensors and quick interventions.

In this paper, an overview of an EU-funded programme related to border security called FOLDOUT is presented [2]. The main goal of FOLDOUT is to develop, test and demonstrate a system and solution to detect and locate people and vehicles operating in illegal cross-border activities under the coverage of trees and other foliage over large areas. The planned improvements through FOLDOUT to the current situation of border surveillance will be evaluated on a threefold basis through the development of required mechanisms for effective detection of (a) irregular border crossings (illegal migrants + vehicles) in forest terrain, (b) persons and vehicles in a search and rescue situation in forest terrain and (c) illegal transport and entry of goods (i.e. human trafficking and goldmines) in temperate broadleaf forest and mixed terrain. Overall, in order to achieve FOLDOUT's main goal, a multi-sensorial platform will be designed and developed. This platform shall incorporate end-users' requirements by integrating, ground, air, space and in situ sensor systems.

16.2   FOLDOUT User Requirements

Based on interviews with FOLDOUT's end-users' (practitioners from border authorities from Greece, Bulgaria, Poland, French Guiana, Finland and Lithuania), system requirements were identified and defined. Interviewees mentioned that formation of the surveillance ground area varied significantly including landfill and smooth, plains and hills/mountains, rocky ground, big altitude differences, bogs, moraine and uneven woodland. They also stated that they were more interested in detecting people as well as detecting, recognizing and tracking vehicles. The minimum detection distance before crossing the border, necessary to be able to react and intercept, was reported to be a few kilometres from the border line for a vehicle and several meters for people. Regarding response time from detection to tracking, sensor fusion and situation awareness, due to the complexity of the analysis algorithms that are employed, the respondents indicated that the maximum delay should be up to a few seconds. For satellite systems, the delay was defined to almost a day between the event/alarm and data availability. Regarding the importance for the FOLDOUT solution to be integrated with existing systems at borders, it was determined to be a key requirement. In a related context, interviewees indicated that they mostly operate Synthetic Aperture Radar (SAR), RADAR and LIDAR sensors.

In the following Table 16.1, an overview of end-user requirements is presented.

Table 16.1 FOLDOUT end-user requirements

User requirement	Description
Surveillance area: ground	Landfill and smooth; plains and hills/mountains; rocky ground; big altitude differences; bogs; moraine; uneven woodland
Detection needs	Detecting people; detecting, recognizing and tracking vehicles
Minimum detection distance	A few kilometres from the border line for a vehicle; several metres for people
Response time from detection to tracking and sensor fusion	Maximum delay should be a few seconds; for satellite systems, it was defined to almost a day between the event/alarm and data availability

16.3 FOLDOUT Architecture Design

To design FOLDOUT system, we used service-oriented architecture (SOA), which is more flexible and suitable for large and complex systems. In this term, we did not have to describe each single component of the SoS at structural level but just to define a set of services (e.g. command and control (C2) service, data fusion service, sensors service), the interfaces among them and how they collaborate to provide the final service to end-users. In this way, the different system components had been described like services. As a matter of example, each sensor was not seen with respect to its structure, but as an object providing some functions/services to other objects (i.e. C2 service, fusion service, etc.) through well-defined interfaces.

Overall, in order to achieve FOLDOUT's main goal, a multi-sensor platform was designed and will be developed. This platform shall incorporate end-users' requirements by integrating, ground, air, space and in situ sensorial techniques. More specifically, FOLDOUT's architecture design focus is on detecting and tracking activities in foliated areas, in the inner and outermost regions of the EU. FOLDOUT will build a system that combines various sensors and technologies and intelligently fuses these into an effective and robust intelligent detection platform, as illustrated in Fig. 16.1. To support detection and tracking activities of border guards in foliated areas, the FOLDOUT system consists of the following main subsystems:

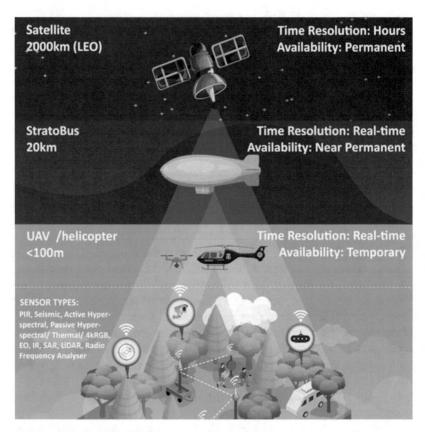

Fig. 16.1   FOLDOUT platform and architecture

(a) Sensors layer that will receive information from registered visual and non-visual sensors. This concept for border surveillance includes mobile platforms equipped with or without wireless connection to ground sensors (radio spectrum, RADAR, LIDAR, EOS, RGB, visible and thermal cameras, acoustic sensors). These platforms are fully autarkical, providing also computational resources for the processing and automatic analysis of the sensor data. Further miniaturization of specific sensors (camera, acoustic) will facilitate deployment of resource limited lightweight smart ground sensors, which are used temporarily and complementarily, in dense forests. StratobusTM is finally introduced to border surveillance as a quasi-static platform able to operate over longer timespans at altitudes above 20 km by that filling a gap between satellites and UAV.

(b) Fusion platform that is a high-level processing component responsible for performing data fusion algorithms based on machine learning and providing sensors' fused detections, tracking and alarms to the C2 platform.

(c) C2 subsystem that combines the information received from the sensors layer and the fusion platform with external data sources (such as weather conditions and maps) and provides alarms and relative information to C2 operators through a GIS-based real-time web platform. The subsystem includes modern command and control tools and provides a live action map with terrain and environment information continuously updated with real-time information. Moreover, through this subsystem, border guards can also (a) register and manage (when possible) sensors and (b) plan interception of targets by utilizing assets from the C2 system.

It reinforces the decision-making process and provides operation dispatching capabilities thus allowing end-users to set and monitor activities, send and receive event-related messages but also to include ad hoc information from sensors or sensor networks. In achieving this paper's aim, in the following paragraphs, ground sensors, StratobusTM and satellite sensor technologies that had been used in the design of FOLDOUT will be further analysed.

16.3.1   Ground Sensors

Ground sensors considered in FOLDOUT include radio spectrum, LIDAR, EOS, RGB, visible and thermal cameras and acoustic sensors. While all sensors can complement to detect an object, foliage detection introduces challenges to state-of-the-art camera-based systems [3–5] which basically breaks down in the case of fragmented occlusion [6]. Fragmented occlusion occurs very frequently in forests as tree and bush leaves occlude target objects irregularly. This is in contrast to simpler cases of occlusion such as partial occlusion where parts of the object are still visible and recognisable. In FOLDOUT, a new kind of ground sensor will be developed: SMARTSENSE (Fig. 16.2). This new sensor technology offers

Fig. 16.2 The SMARTSENSE platform (battery driven and mountable on trees or masts)

Table 16.2 Sensors of the SMARTSENSE platform

Sensor	Property	Function
4K	High resolution	Gives detailed, rich information about target and
RGB	Color information	background appearances
LWIR	Thermal information	Gives information about target location; works day/night and for large distances
IMU	Pose information	Gives information about the pose of the platform

through foliage detection of persons and vehicles by combining high-resolution thermal (LWIR) and visual (4K RGB) sensor modalities and by fully exploiting the available information in the sensory data with innovative analysis techniques based on temporal and spatiotemporal processing, neural networks and deep learning. The software is inbuilt to form a smart sensor running on Nvidia's Jetson Xavier embedded board for efficient computation and data transfer between the inbuilt optical sensors, the internal memory and the FOLDOUT environment. The use of LWIR and RGB data combines complementary properties of the data, for example, high-resolution colour information with contrast in LWIR at farther distances. Furthermore, thermal images are not influenced by the illumination variations and shadows, and objects can be distinguished from the background as the background is normally colder. In addition, thermal infrared tracking can be used in total darkness, where visual cameras have no signal (Table 16.2).

The idea of SMARTSENSE is to generalise single images to video and to spatiotemporally analyse the data. It has been shown that video captures important additional information to solve the problem of fragmented occlusion [7]. We follow an approach where we pre-process the raw video data and extract temporal information by learning online a model of static background. This model is then used to estimate foreground pixels in the video frame which potentially form the appearance of occluded target objects. In a second step, these foreground masks are used as region proposals to refine and improve classical two-stage neural object detectors [3, 5] to better cope with fragmented occlusion.

16.3.2 Sensor Mounted on a StratobusTM

StratobusTM is a high-altitude platform station (HAPS) of between 100 and 140 meters long and 30 meters in diameter that fills the operational gap between satellites and unmanned aerial systems (UAS). This airship-based platform can operate above airplanes at 20 km, in the low layers of the stratosphere. From this operational point, it provides multimission capability with powerful payloads of about 250 kg and 5 kW.

It offers real-time, stationary satellite-like capabilities over wide areas of more than 100,000 km² for missions up to 1 year. Exclusively powered by solar energy, it flies autonomously, storing during daytime the energy needed for the night. Thanks to high-density rechargeable batteries, enough solar energy is stored to maintain its position at any time of the year and for wind speeds of up to 25 m/s (90 km/h). Its unique feature of envelope rotation to permanently face the sun allows maximum energy collection all year long. The hull, filled with helium, is made of an advanced high strength and very light material and UV-resistant and with very low permeability.

StratobusTM offers flexibility in missions: it provides permanent surveillance, telecommunication and monitoring services for both defence, institutional and civil applications. Thales Alenia Space is planning to finalize necessary adaptations to provide solutions by 2023.

Thanks to the power and mass available for payloads on the StratobusTM, various remote sensors are conceivable to perform a permanent surveillance above a specific area or above the border. StratobusTM could be easily connected to satellites and drones or interconnected to other StratobusTM, via RF or laser link for combination of multiple sensors over different areas to achieve global missions. StratobusTM is an unmanned

Fig. 16.3 FOLDOUT StratobusTM system

platform, piloted from the ground to perform its mission. It requires an annual ground preventive maintenance of few days and a major overhaul after 5 years of operation. Maintenance is also an opportunity to switch payloads for a different mission or to embark newer payloads to remain at the cutting edge of the technology. Transfer from the take-off site to its operational station-keeping site is easy within few days by using its electrical propellers (Fig. 16.3).

The concept airborne RADAR system of CORISTA (Consortium of Research on Advanced Remote Sensing Systems) represents an excellent candidate to be mounted on board StratobusTM. The mentioned radar concept is a low-frequency radar developed by CORISTA and funded by ASI (Italian Space Agency). It is a multi-mode and multi-frequency radar, which has been designed with the aim of transportability and as easy installation. The instrument is completely stand-alone, with the power supply connector being the only electrical interface [8].

Currently the system works in sounder mode and in the SAR mode at two different P-Band carrier frequencies. Radar sounding is a powerful technique for detecting, localizing and identifying dielectric interfaces underneath planet's surface. The transmitted radar pulse is capable of penetrating below the surface and is reflected by dielectric discontinuities. SAR is a technique that allows to get high-resolution radar images from data collected by side-looking radar instruments carried by aircraft or spacecraft. The entire CORISTA Radar System is quite compact: its dimensions are 50 cm × 50 cm × 60 cm, for a weight of about 35 kg; it can be easily mounted on board relatively small airplanes or helicopters [9].

During the FOLDOUT programme, the feasibility of embarking the CORISTA concept P-band radar on board stratospheric platform as well as the suitable and necessary modifications (mechanical and electronic) will be verified. The possibility of having a radar system with foliage penetration functionality on a HAPS like StratobusTM would give the possibility of continuously monitoring (without delay of relevant data) forested areas of interest any time (day/night) and regardless of weather conditions, with a direct access link to the instrument.

16.3.3   Sensor Mounted on a Satellite

The satellite system studied in the frame of the FOLDOUT project aims to provide geo-located images (2D) and derived products (target detection metadata) with the use of Synthetic Aperture Radar (SAR). The system is based on constellation of LEO orbit (around 600 Km) satellites. The satellite SAR works at low frequency, 435 MHz, which permits foliage penetration capabilities and the detection of metallic objects (e.g. trucks, infrastructures) with a footprint up to 100×100 Km^2 (swath-width) covered by vegetation. The layout of the system architecture and interfaces is depicted in the following Fig. 16.4:

The *Ground Segment (G/S)* aims to perform the main functions/operations, at ground level, needed to manage the FOLDOUT mission, in terms of both satellite control and data management:

- *Satellite Control System (SCS)*: performs routine activities on the satellite and execution of planned payload operations (mainly, instrument data acquisitions and transmission to the ground).
- *Mission Control System (MCS)*: is mainly devoted to the development of planning activities. The activities include the preparation of the mission plans, solving the possible conflicts on the spacecraft, the commands to be uplinked (safety on board, attitude and orbit maintenance, sensor operative mode setting, on-board S/W patches to be uplinked).
- *Data Processing System*: is the core element which is charge of the processing of the satellite raw data with the aim to provide Level 1

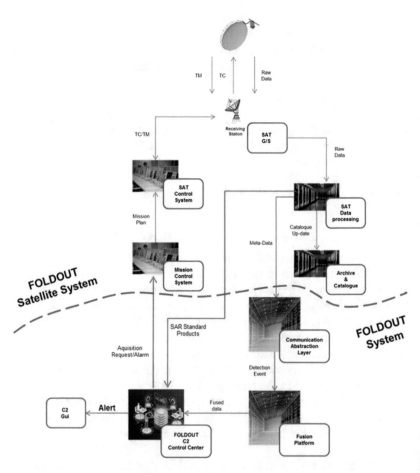

Fig. 16.4 E2E Earth observation satellite subsystem architecture layout

(images) and Level 2 (metadata) products to the FOLDOUT interface. Images and metadata are stored in the Archive and
Catalogue system.

Centralized vs Distributed Ground Segment configurations are envisaged and will be ranked according to evaluation criteria which take into
account performances (i.e. system response time), costs (CAPEX and
OPEX) and complexities:

- *Option 1 – Centralized Architecture:* envisages the use of a single data receiving ground station located near the polar region in order to maximize the contact time per day. The data acquisition ground station will be located in Svalbard (or Kiruna). The Data Processing and Data Distribution are located in the same centre and are common for all countries/border authorities.
- *Option 2 – Distributed Architecture:* envisages to use of dedicated Data Processing and Distribution Centre per each country in order to allow an autonomous data processing and storage.

The *Space Segment* is constituted of a constellation of LEO satellites which preliminary orbit data are described in Table 16.3. The orbit has been selected according to the following criteria and assumptions:

- Access to all FOLDOUT border area of interest (AOI) including French Guiana (global access)
- SAR instrument access area better than 20° to 45° in term of incidence angle (that corresponds at about 300 km on ground)
- Provision of the right operational conditions for the SAR instrument

Table 16.3 SAR performance summary

Parameter	Value
Antenna area	99.5 m^2
Antenna dimensions (range × azimuth)	6.7 × 14.8 m^2
Antenna range aperture	6.2°
Antenna azimuth aperture	2.8°
Antenna gain	32.3 dB
Range resolution	35–73 m
Azimuth resolution	35 m (5 looks)
Access area	20–45°
Antenna pointing	21.3–37.1°
Transmitted peak power	1200
Azimuth resolution (single-look)	7 m
PRF	>1570 Hz
Tx duty cycle	19–34%
Swath	40–100 km
NESZ	-47/-41 dB

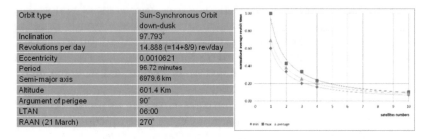

Orbit type	Sun-Synchronous Orbit down-dusk
Inclination	97.793°
Revolutions per day	14.888 (=14+8/9) rev/day
Eccentricity	0.0010621
Period	96.72 minutes
Semi-major axis	6979.6 km
Altitude	601.4 Km
Argument of perigee	90°
LTAN	06:00
RAAN (21 March)	270°

Fig. 16.5 E2E Earth observation satellite subsystem orbit characteristics

A single satellite with a SAR instrument allows a 100% access to the target AOI in less than 9 days. In particular, the AOIs with small surface, such as Greece and French Guyana, are covered in about 3 days (always inside of the cycle of 9 days). Countries like Finland with a very large surface require many passages for complete coverage (about 7 days). In order to reduce the gap between the acquisitions, the number of satellites on the orbit needs to be increased. In particular, according to Fig. 16.5 with two or three satellite, it is possible to reduce of two or three times the revisit time.

The preliminary *satellite SAR* performances, according to system requirements and analyses, to achieve the FOLDOUT target detection objectives are provided in Table 16.3.

In order to validate and refine the satellite SAR performances, the acquisition of airborne datasets which are representative of the satellite system will take place. The campaign will be based on an airborne P-band SAR that has been developed by CORISTA in the framework of the Italian Space Agency (ASI) technological project (contract ref. I/062/10/0 and ref. 2015-029-I.0). The campaign is expected to be performed at Bulgarian border area. The Fig. 16.6 provides a snapshot of the layout of the CORISTA system (kindly granted for publication by Italian Space Agency).

16.3.4   *Fusion of Ground Sensors, StratobusTM and Satellite Data*

FOLDOUT's core functionality is the combination of various sensors and technologies and intelligently fusing these into an effective and robust intelligent detection platform. The clear advantage is that fusing several sensor signals increases the effectiveness of detection. Furthermore,

Fig. 16.6 CORISTA SAR demonstrator Layout

combining information from various sensors allows for a better interpretation of the current situation in the surveyed area (situational awareness) and inference of possible threats (alarming).

Fusion can only be performed on registered data. In the FOLDOUT system, a registration component will work explicitly to convert local coordinate measurements from heterogeneous sensors into a common reference frame (i.e. WGS84). If, for instance, SMARTSENSE, StratobusTM and/or satellite sensors achieve the detection of a target, the positions of the target will be converted from the sensor frame of reference to the chosen common reference frame (i.e. WGS84). The fusion engine will then associate the observations from different sensors so that targets which are detected on several sensors will be unified into a single entity. The fusion module will work on detection variability likely to exist among sensors; that is, depending on sensor characteristics, weather conditions, etc., some sensors will detect the object with high confidence, some sensors may have only partial detection, and some sensors may not detect the object at all. For instance, specifically in relation to the StratobusTM and satellite sensors, data from the latter provides large area coverage but refreshed only on hourly basis, while data from the former should be near to permanently available and in real-time resolution but on a smaller coverage. SMATSENSE detections may correspond to a smaller selected area where cameras are pointed out. The fusion module will thus aggregate evidence from all sensors into individual "heat maps", which can be interpreted as detection probability maps with a time delay. The FOLDOUT Fusion system can also be perfectly used with current technology. Current border surveillance includes the use of visible, thermal cameras, PIR sensors, seismic, RADAR and LIDAR sensors as among the most employed types of sensors. While current solutions provide stand-alone systems to end-users, FOLDOUT will fuse all information and give a unified picture of activity to end-users. Fusion will as well solve inconsistencies that may exist on object classification (person/vehicle) and through synchronisation of heterogeneous data; by this, it is meant that data itself is heterogeneous as input to fusion which may include location of detected objects, object classification, type of material, etc.

The last layer of analysis on fused objects is situational awareness and alarming. This component reasons on the behaviour of detected targets and current situation to decide whether to issue an alarm [10–12]. The current situation will be partly asserted from modelling threat situations incorporated from end-users' requirements and from real-time querying C2 system for real-time external/contextual data (including map information such as nearby routes to border, type of terrain, elevation, etc.). Such contextual information will allow probabilistic inference of next movements of the targets to allow border guards to take appropriate action and prove them sufficient response time.

16.4   SCENARIOS DESCRIPTION

The architecture will lead to the development of FOLDOUT platform that will be tuned and tested with data collected in four different European land borders under realistic and harsh conditions in Bulgaria, Finland, French Guiana and Greece. In doing this, the main difficulty when developing such systems, which is the lack of representative data for tuning and testing, will be overcome. As a minimum, the following use cases will be considered: (i) detection of irregular border crossings (illegal migrants and vehicles) in forest terrain, border surveillance (Bulgarian and Greek scenarios); (ii) detection of illegal transport and entry of goods (trafficking) in temperate broad leaf forest and mixed terrain, border surveillance (French Guiana Scenario); and (iii) detection of persons and vehicles in a search and rescue operation in forest terrain. For each scenario, practitioners will present relevant use cases (e.g. use of an unmanned vehicle to confirm and verify detection made by ground mobile and fixed ground surveillance assets or re-planning and re-tasking on a mission that shifts from illegal immigrants tracking to search and rescue).

In addition, to optimize data analytics, reference data is needed fitting two major constraints:

(a) Representativeness: the reference data must correspond to real-life scenarios, including events/actions as those encountered by border guards in their daily practice; the quality of data must correspond to the data used for investigation (e.g. video surveillance camera); the data must be sufficiently represent different variations in the environment and the events.

(b) Availability of the "ground truth" annotation corresponding to the data: the events/persons/objects to be detected must be known and precisely documented in order to measure the performances. To take an example, for the measurement of person tracking performances, the ground truth must document the actual track of the various persons. In doing this, two purposes will be served: on one hand, the scenarios will form a body of preliminary work to be used as a baseline for end-user expectations. On the other hand, the use cases will be extremely important for technical adjustments in particular concerning necessary sensors, interfaces and input/output.

16.5   CURRENT RESULTS

In order to give a demonstration of the importance of the core capabilities of the FOLDOUT system, its main components (sensor layout, fusion, command and control) have been deployed and connected to some of the most commonly employed sensors in current border surveillance technology, namely, visible, thermal cameras and PIR sensors. Figure 16.7 shows the layout of sensors deployed for this demonstration in a simulated border between two countries (Fig. 16.8).

Employing the FOLDOUT system, border guards have the benefit of exploring terrain activity on a global map instead of being obliged to look at separate stand-alone systems. The system will not only unify detections from different sensor types but will also analyse the confidence of individual detections to decide if one sensor might be firing incorrectly and thus decrease the false-positive ratio overall. Indeed, an important aspect

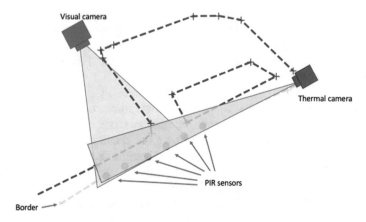

Fig. 16.7 Sensor deployment in a FOLDOUT demonstrator system. Three different types of sensors are deployed: visible, thermal cameras and PIR sensors. Cameras are mounted in observation towers for better view through foliage. The border between the two countries

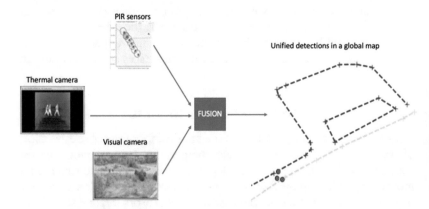

Fig. 16.8 FOLDOUT fusion demonstrator system takes detections from three independent sensor types and unifies detections to eliminate inconsistencies and allow for a global view of activity in the border area

of through foliage detection is the frequent occurrence of false-positive detections due to movement of vegetation, of branches and leaves, in wind and rain. Many sensor technologies employed in border security, such as RADAR and passive thermal IR detectors, are affected by this, leading to additional work load on the operators to manually verify the situation in visual or thermal IR camera images. Tests of the FOLDOUT demonstrator in strong windy conditions confirmed the appearance of false-positive detections, at the level of individual sensors, which could successfully be filtered out with the FOLOUT system. The proposed multi-sensor fusion approach thus potentially allows to reduce the number of false positives by filtering out detections that are not coinciding in the spatiotemporal domain. Through foliage detection, challenges are facilitated in FOLDOUT in part by including a fusion system in its core functionalities.

16.6   CONCLUSION

The ambitions of this concept for border surveillance with respect to air and space are threefold:

(i) To improve situational awareness through fusion of advanced aerial and space-based sensor platforms into one surveillance solution. Currently no high-rising, fixed sensor platform offering an optimal, unobstructed field-of-view exists for border surveillance. StratobusTM as a quasi-static high-rising sensor platform promises to fill this gap.

(ii) To exploit low frequency SAR in the P and L bands which in principle allows to penetrate foliage. Tomography, interferometry techniques with change detection are developed, to extract suitable data from the SAR sensor.

(iii) To fuse aerial, space as well as ground-based sensor data by registering multiple events into a common geographical map. Ground sensors play a vital role in border surveillance, hence ideas of mobile, autarkical, smart sensors allowing a better coverage of dense forest. Ground and airborne sensors will together provide and enhanced coverage of the area, particularly important in FOLDOUT through foliage challenge. Abnormality detection is performed by behavioural analysis within the common geographical map by also taking contextual information into account. It is believed that the developed concept of border surveillance is also fruitful for studies of the environment where instead of vehicles and persons, larger gradual spatiotemporal changes as well as specific local patterns are of interest.

FOLDOUT has received funding from the European Union's Horizon 2020 research and innovation programme under grant agreement no. 787021. The opinions expressed in this paper reflect only the author's view and reflects in no way the European Commission's opinions. The European Commission is not responsible for any use that may be made of the information it contains.

Acknowledgement

 This article is based on research, inter alia, undertaken in the context of the EU-funded FOLDOUT project, which has received funding from the European Union's Horizon 2020 research and innovation programme under grant agreement no. 787021. The views expressed in this article are those of the authors alone and are in no way intended to reflect those of the European Commission.

BIBLIOGRAPHY

1. Regulation (EC) No 562/2006 of the European Parliament and of the Council of 15 March 2006, establishing a Community Code on the rules governing the movement of persons across borders, OJ 2006, L 105, p.1.
2. Grand Agreement, "FOLDOUT" Number 787021, European Commission H2020, April 2018. (www.foldout.eu).
3. Ren, S., He, K., Girshick, R., & Sun, J. (2017). Faster R-CNN: Towards real-time object detection with region proposal networks. *IEEE Transactions on Pattern Analysis and Machine Intelligence, 39*(6), 1137–1149. https://doi.org/10.1109/TPAMI.2016.2577031.
4. Liu, W., et al. (2016). SSD: Single shot multibox detector. In B. Leibe, J. Matas, N. Sebe, & M. Welling (Eds.), *Computer vision – ECCV 2016. ECCV 2016. Lecture notes in computer science* (Vol. 9905). Cham: Springer.
5. He, K., Gkioxari, G., Dollár, P., & Girshick, R. (2020). Mask R-CNN. *IEEE Transactions on Pattern Analysis and Machine Intelligence, 42*(2), 386–397. https://doi.org/10.1109/TPAMI.2018.2844175.
6. Pegoraro, J., & Pflugfelder, R. (2020). *The problem of fragmented occlusion in object detection*, to appear at the Austrian Joint Computer Vision and Robotics Workshop, Sept 2020.
7. Black, M. J., & Anandan, P. (1996). The robust estimation of multiple motions: Parametric and piecewise-smooth flow fields. *Journal of Computer Vision and Image Understanding, 63*(1), 75–104.

8. Papa, C., et al. (2014). Design and validation of a multimode multifrequency VHF/UHF airborne radar. *IEEE Geoscience and Remote Sensing Letters, 11*(7), 1260–1264.
9. Perna, S., et al. (2019). *The ASI P-Band Helicopter-Borne integrated sounder-SAR system: Preliminary results of the 2018 Morocco desert campaign,* submitted to International geoscience and Remote Sensing Symposium (IGARSS), Yokohama, Japan.
10. Patino, L., & Ferryman, J. (2016). *Semantic modelling for behavior characterization and threat detection,* IEEE conference on computer vision and pattern recognition (CVPR) workshops.
11. Patino, L., Ferryman, J., & Beleznai, C. (2015). *Abnormal behaviour detection on queue analysis from stereo cameras,* AVSS 2015 - 12th IEEE international conference on advanced video and signal based surveillance.
12. Patino, L., & Ferryman, J. (2014). Multiresolution semantic activity characterization and abnormality discovery in videos. *Applied Soft Computing, 25,* 485–495.

CHAPTER 17

Identifying and Prioritising Security Capabilities for the Mediterranean and Black Sea Regions

George Kokkinis and Georgios Eftychidis

17.1 INTRODUCTION

The countries located at the Southern and Southeastern Europe are neighbouring to some of the most unstable regions in the world. The Mediterranean and Black Sea (M&BS) region – a crossroad between Europe, Africa and the Middle East – is experiencing the last decades an unprecedented regional security crisis. In the vicinity of the South and Eastern EU, external borders, humanitarian, security and climate change challenges affect the life of a large number of vulnerable populations, which leads to regional instability. As a result, the number of migration and asylum seekers to Europe is continuously increasing with a significant impact to border controls and security resilience. Moreover, the M&BS region experiences also increased organised crime activity and terrorism issues attributed to the return and relocation of foreign terrorist fighters

G. Kokkinis (✉) · G. Eftychidis
KEMEA – Center for Security Studies, Athens, Greece
e-mail: g.kokkinis@kemea-research.gr; g.eftychidis@kemea-research.gr

© The Author(s), under exclusive license to Springer Nature   281
Switzerland AG 2021
B. Akhgar et al. (eds.), *Technology Development for Security
Practitioners*, Security Informatics and Law Enforcement,
https://doi.org/10.1007/978-3-030-69460-9_17

to EU countries. From another viewpoint, the rapid social and economic development during the last decades have influenced rural and urban land-scapes in the Mediterranean. Wildland-Urban Interface fires occur more and more often, while their severity and consequences are adding new dimensions to the current risk profile of these regions. As a result of the above, the security practitioners operating in the Mediterranean and Black Sea countries are facing more threats than ever before. There is a strong demand for new capabilities in fighting against organised crime, in responding to natural disasters and technological accidents, in protecting large energy infrastructures of European interest, in managing a continuously increasing number of migrant and asylum seekers and in supervising the external European borders in a politically fragile and religiously rigid region. Scenarios of deliberate migration flow, which may lead to radicalisation and extremism challenges combined with epidemic or pandemic crisis can't be considered unrealistic. In this context, MEDEA project[1] aims to develop a regional network of security practitioners that will be able, using a scenario-based approach, to identify and analyse the research priorities to develop the desired capabilities from the operational viewpoint. This paper presents the methodology that will assist the project members to achieve their research objectives.

17.2   The MEDEA Network of Practitioners

The MEDEA is a five-year Coordination and Support Action [1] (CSA) under the Horizon 2020 topic of SEC-21-GM-2016-2017 - Pan European Networks of practitioners and other actors in the field of security [2]. The MEDEA's scope is to establish and further develop a regional network of practitioners (NoP) engaged in security operations in the M&BS region to involve them in EU R&D activity. To accommodate the interest and facilitate the co-existence of practitioners from different countries with different operational interest, MEDEA NoP established the following four working groups organised as Thematic Communities of Practitioners (TCP) [3]:

TCP1: Managing of migration flows and asylum seekers
TCP2: Border management and surveillance

---

[1] Mediterranean practitioners' network capacity building for effective response to emerging security challenges

TCP3: Fight against cross-border organised crime and terrorism
TCP4: Natural hazards and technological accidents

The TCPs were stuffed initially with consortium members, based on their operational and professional experience and knowledge in relation to the subject of the TCPs. These groups were extended then including first responders, border guards, firemen, police officers, civil protection and emergency teams, humanitarian/social workers, army officers, policy makers and advisers participating as experts and peers in the project activity. The aim is to identify, in a professional community setting, existing barriers and capability gaps that prevent security practitioners to respond effectively to a series of regional and common security threats. The findings and conclusions from all four TCPs will be analysed in context of MEDEA activity using a four-dimensional analysis known as THOR (Technology, Human, Organisational, Regulatory) as explained later in this paper.

MEDEA network has a layered structure with the project partners (consortium) at the core, the practitioners (TCPs) in the middle, the associated experts and R&D providers as third layer and the decision and policy makers who are placed at the external layer. All these groups form the MEDEA regional network. The challenge of MEDEA consortium is to trace the capability radius that link these layers starting from the centre of the structure to reach the external layer.

The external layer is where the findings of the TCPs and the results of the THOR analysis will be reported, by publishing them in the Mediterranean and Black Sea Security Research and Innovation Agenda (MSRIA).

17.3   MEDEA METHODOLOGY BUILDING BLOCKS

There are three main building blocks defined in MEDEA methodology, which are used to define challenges, identify capability gaps and elaborate potential solutions in relation to each of the four TCPs. These are (1) a scenario development block where security practitioners from the M&BS region will first develop at organisational level and then jointly elaborate with other stakeholders from their TCP operational scenarios that will be used to identify their needs for enhanced or new operational capabilities and (2) an impact analysis block for the needed capabilities. A four-dimensional analysis will examine the impact of the requested capabilities with respect to the Technology, the Human, the Organisational and the

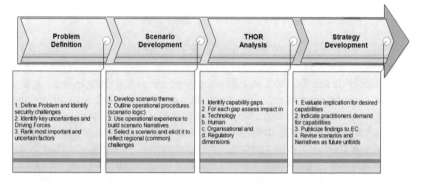

Fig. 17.1 MEDEA scenario development, impact analysis and recommendations process

Regulatory dimension. Based on the findings from the impact analysis, (3) the practitioners will arrange their expectations in three horizons, while at the same time they will document and objectify their priorities to acquire the respective capabilities in three distinct time horizons: short term, mid-term and long term.

The gap analysis process, which MEDEA adopted, is structured in four consecutive steps from the problem definition to scenario, multidimensional analysis and strategy development (Fig. 17.1).

17.4   PROBLEM DEFINITION
AND SCENARIO-BASED APPROACH

Taking as starting point the scenario definition given by Carrol (1999) where "[...] scenarios are stories about people and their activities" [4], MEDEA defines scenarios as operational cases concerning security practitioners, their activities and their duties. Scenarios are used by many security institutions for a variety of purposes (planning, exercises, operations, etc.). As example, the US Defence Modelling and Simulation Office (DMSO) uses scenarios to establish "An initial set of conditions and timeline of significant events imposed on trainees or systems to achieve exercise objectives". To address the needs of identifying operational gaps in the capabilities of security practitioners, MEDEA adopted the relative approach of Whitworth et al. [5] who consider the scenario as "a representation of the state, and present actions, of a set of animate and/or inanimate objects, so as to permit the exploration of, or reasoning about, their future state and the events that lead to it". This approach is applied to reconstruct past security incidents properly. As such, MEDEA scenarios can assist practitioners, from different security organisations to describe new incidents related to similar threats that may occur modified somehow. The approach used is aligned to the works of Rigland and Schwart (1998) [6] and Van der Heijden (2002) [7] who demonstrated how scenarios can be used as a planning instrument for strategic foresight. In MEDEA, practitioners use scenarios as a mean to describe potential future situations that are attributed to present circumstances. A repository of scenarios corresponding to interesting and important security cases related to the topics of the four TCPs of MEDEA has been created. The developed scenarios can be used to predict a probable situation evolvement without using a straightforward projection of the current conditions. On the contrary, scenarios might have as a starting point a present situation, and they can be evolved and transformed using the practitioners tacit knowledge to foresee possible outcomes very different and irrelevant to the present situations, based on operational experience and expertise.

17.4.1 A Systems Approach to Scenarios

In order to determine the descriptive fields that should be considered in a specific scenario, information from a number of real cases or relative scenarios developed by practitioners in context of past EU-funded projects is analysed. This material is organised in a list of distinct "scenario instances", which take the form of individual sections in the scenario design. The basic principles in scenario structure are the following: Each scenario has an introductory section where contextual information is provided regarding an envisaged event type. A setting of the socio-economic and political context using eventually references from past cases. Information concerning the threat or risk and the specific challenge considered, the likelihood of the scenario (or its specific instance), the expected impact, the purpose of the analysis and the interested countries are determined in the introductory section. Succeeding sections accommodate specific information concerning instances of the scenario. This information includes entries for the initial conditions, the place, site, incident type, stakeholders involved and the scenario storyline (sequence, duration/pace, facts, actions/injects). The sections may refer to any stage of crisis or disaster management, i.e.

prevention, mitigation, preparedness and early warning, response and recovery [8].

Thus, in MEDEA, scenarios are stories developed by practitioners, highlighting challenges to meet their operational objectives against virtual operational cases related to current or new security threats. Using scenario as an analysis tool, the practitioners coopérate with the R&D community to identify missing capabilities to address relevant artificial incidents. This approach allows the MEDEA network to become a research-practice interface, able to provide focused and documented recommendations to relevant decision and policy makers for specific research topics.

Apart from the structure, the scenarios in MEDEA have characteristic elements. They include or presuppose a setting (typically a location in M&BS region) and additional setting elements like the number of irrespective organisational objectives and operational mandates. Each practitioner organisation typically has its own goals or objectives. These are the operational mandates that the practitioner should achieve under challenging circumstances of the setting. Every scenario involves at least one practitioner organisation and at least one operational objective. The elaborated scenarios will most likely include more than one practitioner organisation or country.

Scenarios have also a plot; they include sequences of facts/injects and actions. These refer to things that practitioners involved in the scenario can do, things that happen to them, changes in the circumstances of the setting and so forth. Particular actions and events can facilitate, obstruct or be irrelevant to given goals. Notably, actions and events can often change the goals – even the defining goal – of a scenario. Scenarios and the elements of scenario-based design rationale can be generalised and abstracted using theories of human behaviour, enabling the cumulation and development of knowledge attained in the course of design.

17.4.2   Scenario Planning and Horizon Scanning

In [9] Rowe et al. quoting the works of [10] and [6] define scenario planning (SP) as a "collaborative process to envision alternative future environments, articulate their implications, test the logic of long term plans, strategies and policies". In this approach, a single scenario gives a deterministic view of the future – whereas multiple scenarios depict a number of prospects and deepen the focus, expression and understanding of possible changes and developments. As such, by considering multiple possible scenarios, recognition is given to the indeterminate and emergent nature of the future, in contrast to forecasting-based approaches of the future, which often simply extrapolate on the basis of present situation and past trends.

The first French white paper on Defence and National Security issued after the end of Cold War cites that "there is a need to develop a 'horizon-scanning' approach by the Government, in universities and in defence and security circles, in order to anticipate emerging risks and threats, opportunities for French and European interests, and to guide preventive policies and assets in a timely fashion"[11].

The use of horizon scanning (HS) approach in MEDEA is intended to develop a practitioner's organisation capability for identifying subtle security changes, allowing relevant organisations to cultivate a high awareness and understanding of their needs for future capabilities, leading to a quick and effective response to contextual changes and new threats (unexpected events) as per the work of Miles and Saritas [12]. Practitioners claim the true value of SP and HS lies in enhancing the "cognitive agility" of planners by extending long-term thinking and exploring future developments.

17.5   The Origins and the Evolution of THOR Methodology

The MEDEA project has adopted a comprehensive approach, which is widely known as THOR (Technology-Human-Organisational-Regulatory) to analyse the capability gaps and the relative solutions that the network of practitioners considers. The THOR methodology [13] had been originally introduced by the FP7 CAMINO[2] project [14], a project aiming to provide a realistic roadmap for improving resilience against cybercrime and cyberterrorism. Four dimensions of analysis have been defined, the combination of which can efficiently enhance resilience against cybercrime. These dimensions concern Technical and Human issues inter-related with Organisational and Regulatory aspects. The work performed in CAMINO resulted to a number of resilience topics. Each topic was related to high priority (core) activities that were addressed in the CAMINO Roadmap. Each one of these topics was assigned to one of the four (THOR) dimensions, and each topic was further divided into

---

[2] https://cordis.europa.eu/project/id/607406/pl

objectives. For every objective short-term, mid-term and long-term goals "milestones" were defined.

The THOR methodology was further elaborated in two recent EU H2020-funded projects (INSPEC²T [15] and TRILLION [16]), both within the FCT-14-2014[3] – Topic 2: Enhancing cooperation between law enforcement agencies and citizens. The work performed in INSPEC²T project revealed that a topic might be related to more than one dimension of analysis [17]. That is, an objective or solution, which is a subset of a topic, can be assigned to more than one dimension. For instance, a specific technical solution might require a legal amendment (Regulatory dimension), while most likely its introduction might necessitate acquisitions of specific skills (Human dimension).

The concept of "objective", used in the aforementioned two FCT-14 projects, is perceived in MEDEA as "identified gaps". The objectives are classified in short-, mid- and long-term time horizons, while every objective is prioritised over the others [18]. For instance, a technological solution nowadays can be preferred over the others (short time frame), while in the mid- and long term, its impact might be less severe; therefore other solutions might be prioritised.

Following the relative outcome of CAMINO, INSPEC²T and TRILLION projects, the THOR methodology is considered as a concrete framework to analyse the missing capabilities that practitioners need to prevent, mitigate and respond to various security-related challenges.

17.5.1   The MEDEA Approach to Identify Missing Capabilities

A key objective of the MEDEA project is to identify the gaps in the capabilities of the security practitioners in the Mediterranean and Black Sea region to address current challenges and emerging threats. These findings will be organised to define a Mediterranean Security Research and Innovation Agenda (MSRIA), which may feed relevant future security programmes or policies in EU. The use of THOR methodology has been adopted for the needs of MEDEA project and the MSRIA scope in order to formulate a "footprint" of needed practitioners' capabilities. The methodology will be applied to a number of selected scenarios.

---

[3] FCT-14-2014 - Ethical/Societal Dimension Topic 2: Enhancing cooperation between law enforcement agencies and citizens - Community policing

Initially the topics of interest are defined by the core of the MEDEA network, which is formed by the project consortium partners and organised in four focused, thematic working groups related to (a) migration and asylum, (b) border security, (c) organised crime and terrorism and (d) natural and technological hazards. These groups are open and ensure the communication of the project with the diverse communities of practitioners while supporting the growth of the MEDEA network with new members. In the MEDEA jargon, the groups are named Thematic Communities of Practitioners (TCPs). Scenarios related to the topics of interest are first developed by practitioners and submitted to the TCPs, using a structured template. A number of these basic scenarios are selected then for further processing, based on community decisions. There are two iterations in elaborating the scenarios. In the first, the practitioners co-create a set of scenario cases considering short-term needs (threats currently experienced or likely to occur within the next three years). In the second iteration, a new set of scenarios will be developed concerning the same topic of interest for mid-term security threats (that most likely will appear within the next decade).

In the course of the MEDEA project, the network members and associated expert practitioners will engage into a scenario-based assessment of capability gaps, to address present (0–3 years) and emerging (3–10 years) threats. This will allow to be timely prepared to mitigate these threats and to respond effectively, acquiring new and improved capabilities. Having identified and documented the present and the emerging gaps in the capabilities of the practitioners, the MEDEA core group will engage in discussion with industry, academia and research communities to share their findings and identify potential solutions and prioritise future research needs in MSRIA.

To develop the short-term scenarios, the practitioners use their operational experience to outline their needs, using the scenario structured template, attributed to current and potential challenges that might become security threats within the next 3 years. For the mid-term horizon, the practitioners shall use both their operational experience and their tacit knowledge to outline situations that might evolve to security challenges in the near future. Based on practitioners' experience and security concerns, worst-case scenarios for the considered time period (next 3–10 years) are also considered. An example related to the migration issue is the case of deliberate flow of migrants into Europe through the south and eastern borders due to geopolitical conflicts. In such case security practitioners in the M&BS region will need to cope with an unprecedented situation, which can eventually be combined with a pandemic.

MEDEA aims to gather the direct feedback from practitioners to define security risks and the need to develop the appropriate solutions, using the European R&D capacity. However, considering the need to identify challenges and threats in the long term, which is 10 or more years from now, the contribution of the practitioners might be limited. To this purpose, HS techniques will be utilised to elaborate relative foresight scenarios. Interaction with academia, technology pioneers and policy makers will be used by the core group to elaborate relevant security capacity building approaches and define new capabilities that may mitigate such risks.

17.5.2   Application of MEDEA SP and HS Approaches

When a scenario is created, it is stored in the MEDEA's collaboration workspace, which is accommodated in a secure platform, maintained by the project. Invitations are promptly sent to members of the respective TCP of the network to inform them on the new scenario and ask them to elaborate its content and review it online. At the same time, a virtual presentation to TCP members is scheduled by the scenario creator. The members of the network can ask information and clarifications on the context of the scenario or highlight solutions, already in place, that may fulfil relative mission objectives, propose modifications and recommend variations or additions. At a later stage, the interested practitioners may co-define an operational scene, which will host a number of different scenarios. By doing this, a number of different and divert scenario cases are developed using a formalised theatre of operations where all threats will be studied and analysed during an interactive workshop session.

The scenario templates host a special section where the maximum impact is defined for the societal, economic, reputation or environment aspect. However, the use of taxonomies for the scenario scene and the categorisation of threats might also be convenient for a more comprehensive examination of practitioner needs. It is up to the practitioner communities (TCPs) to select the scenarios of their interest. Scenarios developed for the needs of other EU-funded security projects are considered as well, further developed and possibly adapted to match the peculiarities of the M&BS security topics. From the developed scenarios, the most comprehensive and complementary ones are selected to be further examined in proper practitioner's capability analysis workshops.

Following the scenario selection, physical interactions, preferable in TCP Capability Gaps Workshops, are taking place. A preferred audience for the workshop, a part of the MEDEA members and invited subject matter security experts, are members (practitioners) from other practitioners' network and/or members from other EU-funded projects with profound interest in the workshop's objectives. During the Capability Gaps Workshops, audio and video material is used to visualise the scenario and to trigger the interest and feedback of the participants. The practitioners are confronted to respond to a number of security issues and challenges relevant to the envisaged scenario. Initially, the practitioners outline the currently existing capabilities, which are able to respond to the challenges under analysis, and then they are indicating (to the best of their knowledge and operational experience) the needed capabilities that will enable their organisations to become more effective. At the final workshop stage, the practitioners will confirm and rank the identified capability gaps and outlined a timeline for acquiring these capabilities and incorporate them in their organisations. This is illustrated in Fig. 17.2.

The preliminary outcomes of the workshops are identification and prioritisation by practitioners of capability gaps on existing systems and procedures and requirements for additional functionalities. Specific emphasis is placed on the interoperability of information systems and the use of existing EU databases. The output of the scenario analysis, following the Capability Gaps Workshops, takes the form of practitioners' related "user

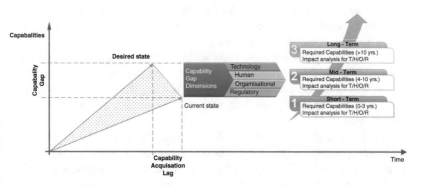

Fig. 17.2 Practitioner's capability gap identification and analysis process

requirements", which are further processed and organised to be conveyed to academia and industry at the next stage.

17.5.3   THOR Methodology in Context of MEDEA

The outcomes of the Capability Gaps Workshops (a priority list of missing capabilities) are further processed by the core group of MEDEA (project consortium). The capability gap findings are pre-processed according to the four THOR dimensions by the core group of MEDEA. For each capability gap, the THOR methodology is applied based on the following.

For the *Technology* dimension, practitioners examine a variety of already developed solutions as well as developing ones and assess their adequacy to address the envisaged gap. The MEDEA practitioner's network aims to short list and prioritise the missing technical capabilities, which are (1) currently needed or (2) desired in the mid- and long term by security practitioners.

The *Human* dimension corresponds to the analysis of the practitioner's capabilities (both current and desired), in regard to possible new skills and training required to suppress new and emerging threats. Eventual social implications, which may follow the introduction of new capabilities, will be examined under this dimension as well.

The *Organisational* dimension studies the reorganisation of procedures and operational activity that can improve response to current and resilience to emerging and future threats. Further to an organisation's reform, the organisational analysis covers the impact collaboration and corporate culture may have to the operational performance and suggest standardisation of procedures between different practitioners' entities or across neighbouring member states of EU.

Lastly the *Regulatory* dimension seeks to identify gaps in institutional, policy and legal frameworks that may influence the capability to respond to current and new security challenges. Besides, there is in many cases a need for common policies and adoption of unified regulations across all EU MS.

After the conclusion of THOR analysis, there can be cases where addressing capability gaps might require more than one dimensional component (attribute). The impact of each dimension to an envisaged capability is described by a relative attribute. Solutions to address an identified gap can be related to more dimensions. As an example, a capability gap might be attributed to two technological and one regulatory dimension,

Table 17.1 MEDEA attributes database for each THOR dimension

Attributes class and number	T-dimension	H-dimension	O-dimension	R-dimension
Interoperability	3	2	2	2
Dual use and misuse	1	1	2	2
Migration and asylum seeking	2	3	2	2
Early warning and situational awareness	2	1	3	3
Cross-border crime and terrorism	2	2	0	2
Natural hazards and technological accidents	1	3	2	0
Soft skills	2	4	7	2
Standardisation issues	4	1	2	2
Uncertainties (local, national, regional)	3	3	2	8
Other security stakeholders	1	1	1	1

i.e. the performance of Electronic Support Measures (ESM) sensors might need to become more stable (technology maturity) and might necessitate interworking with legacy command and control solutions (open/standardised interfaces). However, their usage, for example, might not be regulated in some EU MS despite their proven operational advantage they offer to practitioners.

MEDEA has created a database of attributes (as shown in Table 17.1) where the dimensions of each capability gaps are compared to.

The database (of the THOR attributes) is open and linked to the contribution of MEDEA members participating in THOR analysis workshops for the attributes listed above and will help the researchers and the practitioners to co-develop a list of recommendations for the Mediterranean Security Research and Innovation Agenda (MSRIA). New attributes defined during the THOR analysis workshops are recorded and considered in future analysis cycles.

17.6 PRIORITISATION OF PRACTITIONERS' CAPABILITIES IN THREE HORIZONS

THOR workshops with relevant practitioners' representatives are organised then to analyse and prioritise missing capabilities (gaps). Prioritisation is based on implementation time and impact criteria. The European dimension, beyond the M&BS regional interest, of the identified capability gaps will also play a role in defining the priority and urgency to invest research effort to address them.

Following the identification and documentation of the attributes, every attribute is prioritised against each of the THOR dimensions. This is done by the practitioners using a number of prioritisation techniques, e.g. either a five level Likert scale or the MoSCoW (Must have, Should have, Could have, and Won't have) method. Each prioritisation technique possesses certain advantages. The preferred method is decided as soon as the capability gap attributes are adequately described. Because of the particular nature of the Regulatory dimension, there are attributes that might be related to legal, ethical and regulatory issues. In such case, the 1 to N (Likert scale) quantification approach might not be suitable. Thus, a weight of 0 or 1 corresponding to impact and no impact is utilised instead. The practitioners involved in the TCP workshops are asked to respond to questionnaires that assist MEDEA project to prioritise the need for capabilities in the short, mid- and long term. It is anticipated that the need for developing solutions to fulfil the identified capability gaps will vary over time.

Radar plot will be utilised to highlight the importance (impact) of every attribute including information for the three time-horizons i.e. (1) Short-term (0–3 years), (2) mid-term (3–10 years) and (3) long term (10+ years). The plot chart uses the data of the 1-N Likert/MoSCoW quantification exercise, in order to visualise the priority needs to fulfil the capability gaps.

Apart of the identification and prioritisation of capability gaps, their initial prioritisation to qualify for further analysis using the THOR methodology, the attributes resulted and their prioritisation, the MEDEA partners should also consider the solution development perspective. The desired attributes should be verified if they are available or will be delivered as part of a solution. If the needed capabilities are part of an available solution, the consortium should approach solution providers to arrange for a proof of concept demonstrations for the required capabilities. If the desired capabilities are not covered by existing industry portfolios, a design, development and testing cycle is required, which will delay the introduction of new solutions. As such, the time to deliver and acquire solutions to fulfil the identified capability gaps should be considered in the MEDEA recommendations to the MSRIA.

17.7 CONCLUSIONS

This paper presents the methodology utilised by MEDEA project to identify, prioritise and analyse missing capabilities to address current security challenges and emerging threats in the M&BS region. However, the proposed approach can be applied universally.

Scenarios developed by practitioners combined with a horizon-scanning approach are used to document a list of (a) desired capabilities linked to current/existing threats, which will most likely encountered in the next 3 years (short term); (b) capabilities that will be needed to address emerging threats foreseen to challenge security in the next 3 to 10 years (mid-term), according to the practitioners' operational experience; and (c) capabilities to address improbable future threats that can't be excluded to appear beyond the 10-year horizon (long term). To be successful with the long-term planning, practitioners experienced in the operational and strategic planning should interact with the research community and solution providers to identify capabilities that might be required beyond a decade from nowadays.

MEDEA project utilises Scenario Planning and Horizon scanning approaches as well as Capability Gaps and THOR analysis workshops to identify and analyse the missing and desired capabilities. This is organised in close cooperation between practitioners, experts and researchers to define concrete user requirements, which are conveyed to the R&D community as well as to relevant policy and decision-makers. Interaction with Academia/Research and Industry is planned in the form of open call for ideas, Research Development and Industry days, conferences and proof of concept demonstration activities aimed to bring together the solution provider and the end users. These planned interactions will assist the formed TCPs to proceed with solution selection that will be trialled and demonstrated to practitioners. The Desired, Foreseen and Expected Capabilities, as well as their prioritisation and the results of the THOR analysis, will be documented in the Mediterranean Security Research and Innovation Agenda (MSRIA).

At the end of 2019, in month 18 of the project, MEDEA members have released the first version of the MSRIA. There will be four versions of the MSRIA with the final one scheduled to be delivered in 2023. More interactions in each TCP, cross TCP activities and joint events like the Mediterranean Security Event 2019 are planned to bring together security practitioners across EU MS and enable them to interact with representatives from Academia/Research and Industry, policy makers and other security stakeholders.

Acknowledgements This project has received funding from the European Union's Horizon 2020 research and innovation programme under grant agreement no. 787111. The support is gratefully acknowledged. The views expressed are purely those of the authors and may not in any circumstances be regarded as stating an official position of the European Commission.

REFERENCES

1. [Online]. Available: https://www.medea-project.eu/. Accessed 29 11 2019.
2. [Online]. Available: https://cordis.europa.eu/programme/rcn/701794/en. Accessed 30 11 2019.
3. Wenger, E. (2011). Communities of practice: A brief introduction.
4. Carrol, J. (1999). Five reasons for scenario-based design. In *32nd annual Hawaii international conference on systems sciences.*
5. Whitworth, I. R. e. a. (2006). How do we know that a scenario is 'appropriate. In *11th international command and control technology symposium.* Cambridge, UK.
6. Ringland, G., & Préfacier Schwart, P. (1998). *Scenario planning: Managing for the future.* Chichester: Wiley.
7. Van der Heijden, K., Bradfield, R., & Burt, G. (2002). *The sixth sense: Accelerating organizational learning with scenarios.* Chichester: Wiley.
8. Coppola, D. (2006). *Introduction to international disaster management.* Burlington: Elsevier.
9. Rowe, E., Wright, G., & Derbyshire, J. (2017). Enhancing horizon scanning by utilizing pre-developed scenarios: Analysis of current practice and specification of a process improvement to aid the identification of important 'weak signals. *Technological Forecasting and Social Change, 125,* 224–235.
10. O'Brien, F., Meadows, M., & Murt, M. (2007). *Creating and using scenarios: Exploring alternative possible futures and their impact on strategic decisions. Supporting strategy: Frameworks, methods and models.* Chichester: Wiley.
11. Sarkozy, N. (2008). *The French white paper on defence and national security.* Paris: Odile Jacob Publishing Corp.
12. Miles, I., & Saritas, O. (2012). The depth of the horizon: Searching, scanning and widening horizons. *Foresight, 14*(6), 530–545.
13. Choras, M. e. a. (2015). Comprehensive approach to increase cyber security and resilience. In *10th international conference on availability, reliability and security.*

14. [Online]. Available: https://cordis.europa.eu/project/rcn/185485/factsheet/en. Accessed 03 08 2019.
15. [Online]. Available: https://cordis.europa.eu/project/rcn/194895/factsheet/en. Accessed 25 11 2019.
16. [Online]. Available: https://cordis.europa.eu/project/rcn/194841/factsheet/en. Accessed 10 11 2019.
17. Leventakis, G., & Kokkinis, G. (2018). Developing and assessing next generation community policing social networks with THOR methodology. In *Community-oriented policing and technological innovations*. Cham: Springer.
18. Patrikakis, C., Konstantas, A., Kogias, D., & Chor, M. (2017). TRILLION project approach on scenarios definition for citizen security services. *International Journal of Electronic Governance, 9*, 267.

CHAPTER 18

The Andromeda Galaxy: Legal and Ethical Aspects of Technology-Aided Maritime Border Surveillance Operations

Dimitra Papadaki and Marina Markellou

18.1 Introduction

In the past decades, there has been noticed a consolidation of the trend of 'techno-securitisation', namely, of the deployment of technological means in the security domain, including maritime border surveillance [1]. However, the usage of technology in the name of public and national security raises various ethical and fundamental human rights concerns. This chapter aims to address those concerns in the context of technology aided-maritime surveillance operations with a focus on border control activities. The rationale behind this choice is the chapter's inspiration from the European Union (EU)-funded H2020 research project 'ANDROMEDA', which aims at further developing the EU Common

D. Papadaki (✉) · M. Markellou
Center for Security Studies, Athens, Greece
e-mail: d.papadaki@kemea-research.gr; m.markellou@kemea-research.gr

299

B. Akhgar et al. (eds.), *Technology Development for Security Practitioners*, Security Informatics and Law Enforcement, https://doi.org/10.1007/978-3-030-69460-9_18

Information Sharing Environment (CISE), for Border Command, Control, and Coordination Systems [2, 3].

The research methodology is a combination of doctrinal and theoretical legal research, as the chapter invokes legal provisions which are simultaneously activated by the conduct of surveillance activities and attempts to evaluate and soothe the potential tension between them, by recourse to case law applied by analogy. The research design is also empirical, as it is based upon ANDROMEDA's research project, which provides for the conduct of pilots.

The chapter will firstly introduce the reader to the general legal framework for technology-supported maritime border surveillance, including EUROSUR, the European Border and Coast Guard (EBCG) Regulation (FRONTEX Regulation), and the new merger Regulation [4–6]. Then, it will mention the key tools deployed for maritime border surveillance; by critically evaluating the legal provisions authorising their use, it will assess their main ethical and fundamental human rights implications, especially concerning the right to privacy, the protection of personal data, and asylum. Lastly, it is going to refer to judgements of the Court of Justice of the European Union (CJEU) and the European Court of Human Rights (ECtHR), to conclude on the possibility of balancing conflicting or complementary interests.

18.2   Overview of Technology-Aided Maritime Border Surveillance Operations Framework

The European Union Maritime Security Strategy (EUMSS) Action Plan coordinates efforts towards various challenges stemming from the global marine environment, such as environmental degradation, illegal fishing, human and goods trafficking and threats to defence and navigations [7].

One of its elements, integrated maritime surveillance, is aspired to be revolutionalised via CISE, by integrating existing surveillance systems and making them interoperable, in order to facilitate the exchange of information between EU and national authorities responsible for different aspects of maritime surveillance [3].

EU maritime border surveillance activities are regulated under the new FRONTEX Regulation repealing EUROSUR and FRONTEX Regulation [4–6]. The key element of EUROSUR and FRONTEX system is information exchange at the national and European level. EUROSUR had introduced a cooperation mechanism to enable the exchange of information within and between member states for the purposes of reinforcing border surveillance, preventing and tackling cross-border crime but also saving lives of refugees and migrants trying to reach EU shores by sea [8]. To ensure the effective implementation of European integrated border management, repealed FRONTEX Regulation had established 'the Agency' which shares responsibilities with the national authorities competent for border management, including Coast Guards to the extent that they carry out border control tasks [5].

National authorities cooperate via National Coordination Centres (NCCs) and exchange information via national situational pictures (NSPs). At the European level, the NCCs exchange information with each other and the EBCG Agency via the EUROSUR communication network (ECN), with unlimited access to the European situational picture (ESP) and the common pre-frontier intelligence picture (CPIP) [4]. The Agency also cooperates with other EU bodies, such as Europol, Eurojust, the European Maritime Safety Agency, and the EU Satellite Centre to make the best use of the information, capabilities, and systems available [9].

To further support EU's framework on external border control, the new FRONTEX Regulation, merging repealed EUROSUR and FRONTEX Regulations has been adopted [6]. It incorporates the EUROSUR system into the Agency's framework and strengthens the latter's role in border control, returns, and cooperation with third countries [10]. The text of the Regulation authorises surveillance and communication based on state-of-the-art technology, including large-scale information systems [6]. Also, it states that the Agency shall support national authorities carrying out Coast Guard functions, by technological means, including space-based, ground infrastructure, and sensors mounted on any kind of platform [6]. Last but not least, it refers to a fundamental rights strategy, seeking to ensure, among others, full respect for the right to life, the prohibition of torture, the right to liberty and security, the right to the protection of personal data, the right to asylum and to protection against removal and expulsion, non-refoulement, and non-discrimination [10].

18.3   Legal and Ethical Challenges for Technology-Aided Maritime Border Surveillance Operations

18.3.1   *Technological Means for Maritime Border Surveillance*

The needs of maritime surveillance operations call for the integration of data collected from land-based, naval, and aerial equipment [11]. These technologies, known as SOST (surveillance-oriented security technologies), may be rather sophisticated and indicatively refer to satellites, UAVs (Unmanned Aerial Vehicles) or RPAS (Remotely Piloted Aircraft Systems) or drones, radars, scanners (seismic, magnetic), CCTV, cameras (smart, infrared, high-resolution optical), the fusion of data collected from those sources, and the performance of data analytics (such as predictive analytics) [12, 13].

18.3.2   *Assessment of Technology-Aided Maritime Border Surveillance Operations from an Ethics and Fundamental Human Rights Perspective*

Maritime surveillance activities via technical means as depicted in the EUROSUR and FRONTEX framework raise, among others, the fundamental human rights and ethics issues of the intrusion into the citizens' private life, the inappropriate use of personal data and that of sharing information with third countries which may interfere with the principle of non-refoulement, the prohibition of torture and ill-treatment, and the right not to be discriminated against [11, 14].

The Right to Privacy
Interference with the right to privacy, as the latter is guaranteed under Article 7 of the Charter of Fundamental Rights of the European Union (the Charter) and Article 8 of the European Convention on Human Rights (ECHR), is observed when people are monitored by surveillance technologies in a systematic way or via intrusive means [15, 16].

Therefore, the deployment of technical means for maritime surveillance may lead to ethical implications interrelated with privacy and democracy. The most significant of those include the 'dehumanisation of the surveilled', when UAV pilots are remoted from the field of operations leading to the minimisation of empathy towards the surveilled; the constant feeling of being watched either due to warnings issued by the conventional aerial surveillance means (helicopters) or due to the uncertainty caused by the lack of them when it comes to small drones; the collection of information without the person's knowledge; the lack of transparency and visibility as regards the operation of a surveillance system and the purposes of such operation; the 'function creep', namely, the situation when the technological system is used for purposes which conflict with the ones that it was originally designed for; and the threat to body privacy, privacy of location and space, and privacy of association [17].

Specifically regarding drones, it has been supported that, with reference to the ECtHR's reasonable expectation of privacy argument, their use in a public space may, in principle, cause interference with the right to privacy: (i) when drone operators monitor and record data systematically and permanently, irrespectively of whether the surveillance is covert or overt; (ii) when they reveal previously collected images of someone; and (iii) when they do not record images, but monitor a public space through 'sophisticated' means [17].

Another example is that of Very High Resolution (VHR) satellite imagery which could interfere with location privacy. For example, such imagery can identify a vessel, which can subsequently lead to the indirect identification of its registered captain who is highly possible to be on board, even if the sensor's resolution does not allow for the identification of the specific individual [18]. Tracking the boat can lead to tracking the captain's and potentially the crew's location. If these images are combined with images regarding their sequential activity, sensitive information about them, regarding their social life, beliefs, and health, may be revealed depending on the locations they visit which may relate to certain people and institutions, interfering with their right to privacy [18].

The ECtHR has indeed found in *Uzun* case that the surveillance of the applicant via GPS in the case and the systematic collection and storage of their data determining his whereabouts and movements in the public domain amounted, in principle, to an interference with his private life, as protected by Article 8 § 1 [18].

The Right to the Protection of Personal Data

The potential for direct or indirect identification of an individual during maritime surveillance activities raises the concern of the sufficient protection of their personal data, as safeguarded under Article 8 of the Charter and of the ECHR [15, 16]. The processing of personal data constitutes an exception in the EUROSUR framework and shall be in accordance with EU and national data protection rules [4–6, 10, 19–21].

Although maritime border surveillance systems do not particularly aim at depicting persons, they have the potential to. For example, drones are mobile and may carry a combination of surveillance devices, such as sensors, including sophisticated cameras, GPS, etc. [1]. Drones' capacity for wider areas surveillance and collection of a mass amount of data raises the possibility of the collection of a higher amount of personal data. In addition to that, cameras and GPS devices themselves can collect images, sound, videos, and location, and depending on their properties, they may have the capacity for the collection of biometric data. The use of algorithms to analyse the collected data, in order to classify behaviours from 'raising' to 'non-raising alerts', is common. Summing up, it is more likely that high amounts of data, including special categories of personal data, may be collected by technology-aided surveillance systems, and that data may be further analysed [12].

The risk of unintentional transfer of personal data among the maritime surveillance actors increases, as cooperation between them is encouraged [4, 11]. In the context of EU member states' reporting information on 'unidentified' and 'suspect' individuals and vessels close to the borders, photos and videos are allowed to be attached to an event uploaded in the system. The photos and videos attached to an incident and posted within EUROSUR system may allow for the identification of individuals, either directly or via software or via a combination of data, depending on factors as the weather, the distance, the light, and the potential of the deployed technologies, such as the resolution of an image [11].

Safeguards shall be established, to ensure that personal data are not exchanged in the ESP and the CPIP [11]. This presupposes that a validation mechanism, ensuring that no personal data – except for ship and aircraft identification numbers or necessary for EUROSUR scopes – are forwarded from the NSP to the ESP/CPIP, must be enforced at the national level. This could be achieved by appropriate tagging of data as personal or non-personal by the NCC and taking the respective action, such as anonymisation of the relevant for ESP/CPIP data [9].

One of the dangers lurking within the sphere of personal data processing that could be enhanced by contemporary technological developments, due to the capability for higher quality of the material collected and the complex techniques of analysing it, is the possibility to compare them with national, European, or international databases [11]. It is arguable if running a check of an asylum seeker's photo against EUROSUR records in the absence of existing fingerprints in Eurodac database would comply with the rule of law, as it is not prescribed by Eurodac Regulation, in case of lack of fingerprints of quality ensuring appropriate comparison [11]. Although Article 87 and 90 (a) of the merger Regulation authorise exchanges of personal data between the Agency and EU Agencies, such as Europol, Eurojust, and national law enforcement agencies (LEAs), it is disputable whether they are necessary, which are the purposes for their processing by the latter and if they can be a priori assumed as compatible with or falling under the purposes of their initial collection as defined under Article 18 [6, 10, 22]. Is it justifiable to invoke the umbrella term of 'public interest' to cover all cases? In a similar fashion, the ECtHR, ruled in *Shimovolos* judgement, that the collection and storage of data about the movements of the applicant via train and air in a surveillance database under an unclear and unforeseeable scope and manner was in violation of A.8 of the Convention [18].

However, the above concerns shall be also viewed from the perspective of the obligation for the exchange of information between NCCs and search and rescue, law enforcement, asylum, and immigration authorities, at the national level, when a situation regarding the safety of a person occurs in the context of maritime surveillance operations [4, 23].

Special attention shall be paid to the new provision in the merger Regulation which characterises the 'return' of third-country nationals who do not meet the requirements for their entry, stay, or residence in the EU, as 'an important issue of substantial public interest' [10]. Considering that GDPR provides for the derogation of transferring personal data to third countries in the absence of an adequacy decision, due to reasons of 'substantial public interest', the new provision could lead to the exploitation of return as a justification for sharing a large and indiscriminate amount of information with third countries and consequently to the circumvention of personal data rules [20]. As the term 'derogation' indicates, resorting to the public interest clause should be the exception, not the rule [22]. Moreover, according to GDPR, 'reasons of substantial public interest' may be alleged as a legal basis for the processing of special categories of personal data, namely, data revealing racial or ethnic origin, political opinions, religious beliefs, etc., which is in principle prohibited [20]. This means that 'return' could be used as a means to abuse two crucial personal data protection provisions, that of processing of special categories of personal data and that of data transferring to third countries.

As the data necessary for returns may be limited to the name, date of birth, travel document details, and potentially fingerprints, sharing of special categories of data irrelevant and unnecessary for the identification and documentation of returnees could interfere with the right to data protection because of not complying with the 'proportionality test' [22]. Therefore, it is suggested that an impact assessment is conducted before sharing special categories of personal data, due to reasons of 'substantial public interest' [22].

Adding up the vulnerability of the data subject, who could be an applicant for international protection, sharing of their personal data before the request is decided in the final instance could put in danger the safety and bodily integrity of the applicant and their family, affect the outcome of the application, and unlawfully interfere with the principle of non-refoulement [22].

The Principle of Non-refoulement

The principle of non-refoulement, namely, the obligation not to send a refugee or asylum seeker to a country where she or he may be at risk of persecution or harm, on account of their race, religion, nationality, and membership of a particular social group or political opinion, stems from the right to asylum as recognised under Article 18 of the Charter, the right to life and the prohibition of torture, established under Articles 2 and 3 of the ECHR and applies equally on land, on the maritime territory of a State and on the high seas [26, 27].

The repealed EUROSUR and EBCG Regulations provided for the essential safeguard of prohibiting the transfer of information to a third country that could be used to identify persons who have requested access to international protection, and it is under examination, or they are under a serious risk of violations of their fundamental rights [4]. According to the merger Regulation, the scope of the prohibition which was referring to the exchange of 'information' is currently regulated in the personal data section and limited to the exchange of 'data' [10]. This becomes a threat for the principle of non-refoulement, considering the contemporary enhanced surveillance means' potential for the collection and exchange of a large amount of information in general, not limited to personal data [5, 22].

One of the procedural consequences of the principle of non-refoulement is the prohibition of collective expulsion, meaning that international protection seekers shall have access to a fair and effective refugee status determination by which their cases are evaluated individually [16, 28, 29]. Therefore, the collective disembarkation of persons arriving in the EU borders by sea, even when they are intercepted on the high seas, without any assessment of their individual needs for international protection, by the competent authorities, constitutes a breach of the human right to asylum [29].

Considering that neither the technological tools can distinguish between those entitled to international protection and those who are not, nor border authorities are competent for deciding on the issue, the material scope of the EUROSUR shall find limited application, as ruled, indicatively, in *Hirsi Jamaa* case [29].

To elaborate on that, data collected from sensors for maritime surveillance can be analysed, to classify movements or behaviour of persons or vessels, as 'suspicious', and take further actions, such as activating the 'detection mode' of smart sensors or informing the Coast Guard for further patrolling and automatically issue alarms [31]. The criteria for the classification of alerts, their interpretation, and the actions that are taken by the authorities under the issued alerts must comply with the prohibition of collective expulsion. This means that the content and consequence of issuing an 'alarm' or 'threat' can only be limited. It does not authorise border control authorities to disembark people, and it does not justify 'push-back' operations on the high seas [29]. On the contrary, people arriving by sea must be informed of their rights, taken to the competent authorities for international protection claims, and their cases must be examined individually by them, under the law-prescribed guarantees. The issuing of 'warnings' does not affect the legal status of a person as a beneficiary of international protection, and the reaction capability to such a warning must be narrowed down to what applicable law, including International Refugee Law, prescribes for [29, 32, 33].

18.4   Equilibrium Between Privacy, Personal Data Protection, the Principle of Non-refoulement, and Security

Privacy and personal data protection could be considered as conflicting with security, leading to sacrificing the one for the sake of the other [34]. Contemporary literature, legislation, and case law move away from this notion and suggest a privacy-by-design approach [20, 34, 35]. This can be translated as the application of the proportionality test on decision-making when introducing technology-aided surveillance measures that may infringe on the right to privacy and data protection [34].

Such a balancing is based on the admission that the rights to respect for private life and to personal data protection are not absolute, but they may be limited, if necessary, for an objective of general interest or to protect the rights and freedoms of others. However, the core European human rights instruments provide for specific and strict conditions for lawful interference with those rights [15, 16].

The interference with the right to privacy and the right to personal data protection can be justified when the following criteria are fulfilled cumulatively: 'it is in accordance with the ("domestic") law', it pursues a 'legitimate aim' or an 'objective of general interest recognised by the EU or the need to protect the rights and freedoms of others', 'it respects the essence of the fundamental rights and freedoms', and it is 'necessary in a democratic society' or 'necessary subject to the principle of proportionality', according to the Charter and the ECHR, respectively [15, 16].

As regards the identification of a legitimate aim of the restrictive to the rights measures, the probability or possibility of the establishment of surveillance for public security purposes may be enough to satisfy the Court without reference to the reasons and the ways the intervention serves the aim of security [36]. Nevertheless, the absence of evidence to support that a measure meets the pursued aim suffers from vagueness, and either the measure shall be sufficiently supported or it shall not be proposed. In any case, the ECtHR tends to require that such a measure is subject to adversarial by judicial or independent supervisory bodies; therefore, the assumption that national security is threatened shall be challengeable [37–39].

The ECtHR finds that the Contracting States do not enjoy unlimited discretion on assessing which surveillance measures they deploy [40, 41]. Indeed, taking into account the various technologies involved in maritime surveillance, their particularities shall be examined in the light of strict necessity and proportionality [38, 42, 43]. Therefore, intrusive surveillance technology can be characterised as necessary on condition that similarly available, alternative, and less intrusive means of surveillance were considered comparably ineffective [34]. Moreover, the integration of technology with other technologies and means of surveillance, the level of invasiveness of the technologies deployed, the sensitivity of the information and/or personal data collected, the complexity and extent of the processing operations, the duration of the operations, the number of people they affect, the protective measures taken to limit any negative effects on the rights of the individual, the security safeguards implemented including technical and organisational measures, the storage and retention period, the institutions or people who have access to the information – including personal data – and the purpose(s) for the processing shall be considered to ascertain whether the stricto sensu proportionality test is fulfilled [20, 34]. Such a test to the technologies specificities could be that of the Privacy Impact Assessment (PIA) or Data Protection Impact Assessment (DPIA), to be conducted before and throughout the processing operations [20].

As for the principle of non-refoulement in the context of maritime surveillance, the prohibition of collective expulsion is applicable. Surveillance technologies deployed by the Coast Guard cannot determine whether a vessel is safe to return on the basis of the port of departure without violating the state's International Human Rights obligations, because the border authorities are not competent on deciding on asylum claims, while the usage of surveillance information cannot substitute either amount to a full and meaningful individual assessment of each person's case which is normally conducted via in-person interviews by experts on the field [32, 39]. Indeed, the competent authorities for international protection claims must carry out an individualised analysis of the situation of each applicant for international protection, including on the jurisdiction for such an application, as decided in *Sharifi* ruling [44]. Therefore, the role of maritime surveillance technology on 'controlling migration flows' is rather auxiliary than decisive.

18.5 CONCLUSION

The analysis has identified several conflicts between the legal framework for maritime surveillance aided by technology and International Human Rights Law, ethical principles, and International Refugee Law, especially concerning the right to privacy, personal data protection, and the principle of non-refoulement. It has also assessed the provisions of the merger Regulation for border surveillance which are more relevant to the usage of technological developments in maritime operations. Within this attempt, it has pinpointed the potential interference of the technology-aided maritime surveillance operations with the aforementioned rights.

It has referred to selected case law of the CJEU and the ECtHR and has critically applied certain of their findings by analogy to propose the balancing of the interests at stake, departing from but not necessarily rejecting the notion of the trade-off between them. It has concluded that a counterpoise could be accomplished, by assessing the necessity and proportionality of the technological particularities in the design phase, before and throughout the intended information including data processing, to ensure the legality of a proposed maritime surveillance measure. Lastly, according to the prohibition of collective expulsion, it suggested that the role of maritime surveillance technologies can only be limited and the procedures of individually evaluating each case of arrival by the competent authorities must be followed as prescribed by International Refugee Law and International Human Rights Law.

Acknowledgement

 This project has received funding from the European Union's Horizon 2020 research and innovation programme under grant agreement no. 833881. This research reflects only the authors' views, and the Research Executive Agency (REA) is not responsible for any use that may be made of the information it contains.

BIBLIOGRAPHY

1. Marin, L. (2017). The deployment of drone technology in border surveillance, between techno-securitisation and challenges to privacy and data protection. In M. Friedewald, J. P. Burgess, J. Cas, R. Bellanova, & W. Peissl (Eds.), *Surveillance privacy and security* (pp. 107–114). London, New York: Routledge.
2. H2020 EU-funded research programme "ANDROMEDA" GA No 833881.
3. European Fisheries Control Agency-CISE. https://www.efca.europa.eu/en/content/common-information-sharing-environment-cise
4. Regulation (EU) No 1052/2013 of the EP and EUCO of 22 October 2013 establishing the European Border Surveillance System. Article 13, Recital (13), Article 20 (4); 3(b); 20(5); 2;4.

5. Regulation (EU) 2016/1624 of the EP and EUCO of 14 September 2016 on the European Border and Coast Guard and amending Regulation (EU) 2016/399 of the EP and EUCO and repealing Regulation (EC) No 863/20. Recital (56), (58); Article 14, 34, 54.
6. Proposal for a Regulation of the EP and EUCO on the European Border and Coast Guard and repealing Council Joint Action No 98/700/JHA; Regulation (EU) No 2019/1896 of the EP and EUCO of 13 November 2019 repealing Regulations (EU) No 2013/1052 and 2016/1624.
7. European Commission. *Maritime security strategy* [Internet]. Available from: https://ec.europa.eu/maritimeaffairs/policy/maritime-security_en
8. European Commission. (2013, June 19). *EUROSUR: New tools to save migrants' lives at sea and fight cross-border crime.* Brussels.
9. European Commission. (2015). *Annex to the commission recommendation adopting the practical handbook for implementing and managing the European Border Surveillance System.*
10. Position of the European Parliament adopted at first reading on 17 April 2019 with a view to the adoption of Regulation (EU) 2019/... of the EP and EUCO on the European Border and Coast Guard and repealing Regulations (EU) No 1052/2013 and 2016/1624. Recital (7), (87), Article 70 (b), 16(d) Recital (26), Article 29, 87–91 (g), Recital 84, Article 90; 88 (1) (c), 89 (2) (a) and 89 (3); 18; Recital (67); Article 90(4).
11. European Union Agency for Fundamental Rights. (2013). *Fundamental rights at Europe's southern sea borders.*
12. Centre for the Study of Global Ethics. (2010). *Ethics of border security* [Internet]. http://eapmigrationpanel.org/files/research/en/ethics_of_border_security_report-1.pdf
13. Friedewald, M., Peter Burgess, J., Čas, J., Bellanova, R., & Peissl, W. (Eds.). (2017). *Surveillance, privacy and security citizens' perspectives* (1st ed.). London, New York: Routledge.
14. European Commission. *European Union Maritime Security Strategy, responding together to global challenges. A guide for stakeholders.*
15. Charter of Fundamental Rights of the European Union, 26 October 2012. Article 7, 8; 52(1).
16. Convention for the Protection of Human Rights and Fundamental Freedoms Rome, 4.XI.1950.
17. European Commission, Trilateral Research & Consulting VU. (2014). *Study on privacy, data protection and ethical risks in civil remotely piloted aircraft systems operations summary for industry.*
18. Aloisio, G. (2018). Privacy and data protection issues of the European Union Copernicus Border Surveillance Service; see also ECtHR, Uzun v. Germany Application No. 35623/05. 2010; ECtHR Shimovolos v. Russia Application no. 30194/09.2010.

19. Directive 95/46/EC of the EP and the EUCO of 24 October 1995 on the protection of individuals with regard to the processing of personal data and on the free movement of such data.
20. Regulation (EU) 2016/679 2016/1624 of 27 April 2016 on the protection of natural persons with regard to the processing of personal data and on the free movement of such data, and repealing Directive 95/46/EC. Article 45,49; 49 (1d); 9(1), 9(2g); 44, 45, 46; 49 (1d);25;5;35.
21. Directive (EU) 2016/680 of the EP and EUCO of 27 April 2016 on the protection of natural persons with regard to the processing of personal data by competent authorities for the purposes of the prevention, investigation, or prosecution of criminal offences.
22. European Union Agency for Fundamental Rights. (2018). *The revised European Border and Coast Guard Regulation and its fundamental rights implications Opinion of the European Union Agency for Fundamental Rights*. Vienna.
23. The United Nations Convention on the Law of the Sea. Article 98, 1982.
24. IMO. (1974). International Convention for the Safety of Life at Sea (SOLAS). Chapter V.
25. IMO. (1979). International Convention on Maritime Search and Rescue (SAR).
26. UNHCR. United Nations Convention Relating to the Status of Refugees, 1951, Article 33(1).
27. UN High Commissioner for Refugees Advisory Opinion on the Extraterritorial Application of Non-Refoulement Obligations under the 1951 Convention relating to the Status of Refugees and its 1967 Protocol.
28. Council of Europe. Protocol No. 4 to the Convention for the Protection of Human Rights and Fundamental Freedoms, securing certain rights and freedoms other than those already included in the Convention and in the first Protocol thereto, 1963, Article 4.
29. ECtHR – Hirsi Jamaa and Others v Italy [GC], Application No. 27765/09. [9–14], [27], [91] 2012; Contra ECtHR-N.D. and N.T. v. Spain, Applications Nos. 8675/15 and 8697/15.2020 and criticism in Pichl M., Schmalz M. "Unlawful" may not mean rightless, Center for Global Constitutionalism WZB, 2020.
30. Council of the European Union. (2018, June 26). Council conclusions on the revision of the European Union Maritime Security Strategy (EUMSS) Action Plan.
31. FP7 Project "IPATCH" [Internet]. Available from: http://www.ipatchproject.eu/about.aspx
32. Jumbert, M. G. (2018). Control or rescue at sea? Aims and limits of border surveillance technologies in the Mediterranean Sea. *Disasters*, 683. https://onlinelibrary.wiley.com/doi/epdf/10.1111/disa.12286.

33. United Nations Office on Drugs and Crime. (2009). *Frequently asked questions on international law aspects of countering terrorism.*
34. Somody, B., & Máté Dániel Szabó, I. S. (2017). Moving away from the security–privacy trade-off: The use of the test of proportionality in decision support. In M. Friedewald, J. Peter Burgess, J. Čas, R. Bellanova, & W. Peissl (Eds.), *Surveillance, privacy and security citizens' perspectives* (1st ed., pp. 166–172). London, New York: Routledge.
35. Strauß, S. (2017). A game of hide and seek? Unscrambling the trade-off between privacy and security. In M. Friedewald, J. Peter Burgess, J. Čas, R. Bellanova, & W. Peissl (Eds.), *Surveillance, privacy and security citizens' perspectives* (pp. 255–272). London, New York: Routledge.
36. ECtHR – Mubilanzila Mayeka and Kaniki Mitunga v Belgium, Application No. 13178/03. [79] 2006.
37. ECtHR, Kennedy v. the United Kingdom, Application No. 26839/05. [124] 2010.
38. ECtHR, Klass and others v. Germany, Application No. 5029/71. [36] 1978.
39. Contra ECtHR, Big Brother Watch and others v. the United Kingdom, Applications Nos. 58170/13, 62322/14 and 24960/15.
40. ECtHR, Weber and Saravia v. Germany, Application No. 54934/00. [106] 2006.
41. ECtHR. Guide on Article 8 of the European Convention on Human Rights Right to respect for private and family life, home and correspondence. 2019.
42. ECtHR, Szabo and Vissy v. Hungary, Application No. 37138/14. [73] 2016.
43. European Data Protection Supervisor. (2017). *Assessing the necessity of measures that limit the fundamental right to the protection of personal data: A toolkit.*
44. ECtHR, Sharifi and Others v. Italy and Greece, Application No. 16643/09. 2014.

PART IV

Protection of Critical Infrastructures (CI)

CHAPTER 19

Security and Resilience in Critical Infrastructures

Maria Belesioti, Rodoula Makri, Panos Karaivazoglou,
Evangelos Sfakianakis, Ioannis Chochliouros,
and Alexandros Kyritsis

19.1 THREATS AND RESILIENCE IN CRITICAL INFRASTRUCTURES

19.1.1 Introduction

Critical infrastructures (CIs) are valuable assets for a well-organized state and for a structured society. Within the global digital reality, CIs have become pure "enablers" for growth and development at a multiplicity of

M. Belesioti (✉) · E. Sfakianakis · I. Chochliouros
Hellenic Telecommunications Organization (OTE) S.A.,
Fixed Network R&D Programs Section, Athens, Greece
e-mail: mbelesioti@oteresearch.gr; esfak@oteresearch.gr;
ichochliouros@oteresearch.gr

R. Makri · P. Karaivazoglou · A. Kyritsis
Microwaves and Fiber Optics Lab, Institute of Communication and Computer
Systems (ICCS) of the National Technical University of Athens, Athens, Greece
e-mail: rodia@esd.ece.ntua.gr; pkaraiv@esd.ece.ntua.gr

© The Author(s), under exclusive license to Springer Nature
Switzerland AG 2021
B. Akhgar et al. (eds.), *Technology Development for Security
Practitioners*, Security Informatics and Law Enforcement,
https://doi.org/10.1007/978-3-030-69460-9_19

levels and everyday life. However, CIs are "prime" targets for man-made threats, operation disruption, and organized terrorist attacks but can also be affected by extreme weather events and natural disasters.

A vast literature exists concerning the threats and vulnerabilities in CIs yielding a large discussion concerning their classification. Although various schemes are defined, it is seen that all aim to initiate suitable implementation of corresponding measures with an ultimate goal to enhance the infrastructure's resilience. A solid base to start with is the various standards and models foreseeing different classes of threats and resilience quantities, depending on the point of view and the hierarchy of security principles, such as the NFPA 1600, ANSI/ASIS SPC.1-2009, and ISO 22301 Standards [1–3].

The issue is very important nowadays since, due to the technology advancements, the CIs are dealt as cyber-physical (CPS) systems. In this context, the telecom CIs can be regarded as fundamental, considering the large impact that wireless and data networks have on the CIs, especially in light of the emerging 5G revolution and the Internet of Things (IoT) world approaching fast. Since almost all CIs, such as energy or transport, greatly depend on telecommunications and data networks, it is clear that this interdependence will surely become more evident and profound in the near future.

The threats and resilience in telecom CIs have also been largely dealt in the literature, while specific bodies like ENISA have already made relevant classification [4]. The literature attempts are to identify and classify existing security and resilience metrics and evaluation criteria; however, the descriptions are often vague, and the evaluative factors are rather provided at a conceptual level or through a theoretical formulation. Thus, the focus of resilience metrics remains more on summative indicators rather than meaningful, risk-based ones to determine the effectiveness of a strategy [5].

Various operational organization schemes have been proposed towards that target. Nevertheless, in the majority of the literature, they are seem to be tailored to the specific infrastructure, networks, or systems needs or characteristics that the metrics are applied instead of a more generally employed approaches. And this is reasonable since each network (power grid, gas, or telecom) has its own features, procedures, and technical subsystems. It is thus recognized that crucial gaps are identified regarding the existence and evaluation of (statistically or other) sound metrics in a general manner [6].

Addressing these gaps and challenges, the ongoing RESISTO project introduces a holistic approach for the security and resilience enhancement that could be potentially implemented to serve all types of CIs. In the present work, the relevant proof of concept will be held for the telecom CIs, the significance of which is emphasized previously; these will act as the case study for the RESISTO implementation focus, paving the way for its future expansion and adaptation to other CIs such as energy and ports. RESISTO also encompasses to examine the interconnections and/or the dependencies with other critical infrastructures, presupposing their location in the vicinity of the telecom ones.

19.1.2 Security Threats and Resilience Challenges within RESISTO

Since RESISTO aims to prove a holistic approach that can be employed for all kinds of CIs, a more generic classification of threats is adopted, as following:

- Physical threats that affect physical systems, buildings, and infrastructure
- Cyber threats that exploit vulnerabilities causing possible harm in the digital realm
- And "cyber-physical" threats (combined ones) where exploited physical vulnerabilities can enable security issues in the cyber space and vice versa

The most common impression when discussing about physical security is that of dealing mainly with the protection of building sites and internal equipment from theft, vandalism, natural disasters (i.e., floods, earthquakes, fire), man-made catastrophes, and accidental damage or unintentionally destructive acts. Thus, it requires suitable emergency preparedness and appropriate safeguarding from intruders [7]. Cyber threats on the other side affect the whole operation as a software system and service, basically involving cyber intrusions, cybercrime, and deliberate malware in the CI operator's firmware, causing broader impact to the services and customers' personal data.

Physical security is often thought as only controlling personnel entrance and preventing attackers from gaining access and causing damages. However, its relation to endangering information systems is more than crucial, and it is often overlooked since most organizations focus on countermeasures to prevent hacking attacks [8, 9]. As new technologies such as biometrics and remote security become widely available, the challenges of implementing physical security are much more important now than in previous decades. Traditional card and guard security is being supplanted by identification and tracking systems in and around the facility [10]. Although cyber threats are reasonably given major attention, since data security is a primary factor, the physical ones are not evenly regarded, and physical security is often a second thought [8].

Nowadays the malicious attacks turn to be more sophisticated and aware of new technologies, imposing equally sophisticated countermeasures. Physical threats include intrusion (i.e., unauthorized access causing damages or terrorism actions), airborne and land threats (explosions, bombing by aircrafts or land vehicles, hostile drones, and unmanned aerial vehicles – UAVs – bearing weaponry), and deliberate jamming, apart from the natural hazards, affecting also the vicinity of the CI. In this context, cyber threats, apart from relative direct actions, can be also seen as a result of physical security breaches (cyber-physical threats). Organizations often focus on technical and administrative controls, and as a result, security breaches may not be discovered right away.

Cyber-physical threats include disruptions to information systems, which directly affect physical infrastructure services or intrusions to the physical domain that can cause possible harm in the CI's cyber domain. The main point when addressing and confronting cyber-physical threats is their early detection and correlation between the events. It is assumed that the cyber or physical threat events can be detected independently from the operator's corresponding security systems. However, the timely correlation between the two events is what would need a more concentrated focus to be able to detect early enough if an, i.e., physical intrusion, that would either way be detected in any case, could enable, i.e., a dormant software to the CI's cyber domain. Thus, early detection to provide alerts and intrusion events but also timely correlation between the two types of events are needed as well to improve resilience and security against sophisticated cyber-physical threats. The aim is to contribute to the overall protection concept as well as to the risk and resilience assessment of the whole infrastructure.

Resilience is the system's ability to both absorb the impact and recover rapidly from a disruption and return back to its original service levels [11]. In the context of a CI, resilience is defined as the ability of a facility or asset to anticipate, resist, absorb, respond to, adapt to, and recover from a disturbance [12]. In order to evaluate the infrastructure resilience and the effectiveness of related strategies, metrics are needed to assess and allow the decision makers to check various threat scenarios.

It is generally admitted though that it is difficult to measure resilience since it not directly observable per se but must be placed in relation to a given outcome [13]. Metrics should be specific to the contexts of the considered system, and this precludes generic indicators and thresholds; therefore benchmarks are rather difficult. A spectrum of resilience factors (for the specific system) is more meaningful, due to the dynamic and multi-dimensional nature of resilience and the fact that it is not always easy to obtain reliable, objective, and comprehensive data [13]. Attempts to derive a resilience measurement index specifically for the CIs were carried out in the USA especially after devastating natural disasters (Hurricane Katrina, 2005, and Superstorm Sandy, 2012) [14]. The methodology is based on multi-attribute utility theory, decision analysis, value patterns, and weights. However, it is noted that a relative measure is represented, while the limitations related to the subjective interpretation or use of the collected, through survey tools, data, and associated indices due to the human intervention, need to be considered.

Especially for the telecom infrastructures, apart from certain systematic approaches in recent works [15], ENISA in [4] fully recognizes similar gaps in the whole process including the lack of a standardized framework, common for all telecom providers. Although ENISA's Resilience measurement framework is meant for the existing commercial telecom networks, it addresses the issue by bringing together different taxonomies in an overall, unified, and flexible classification model, successfully addressing the weaknesses found in literature; the model includes a two-dimensional approach, incident- and domain-/discipline-based, with relevant grouping of the various metrics.

19.1.3   The RESISTO Resilience Framework

In light of the above, the ambition of RESISTO is to provide a more holistic approach. The RESISTO integrated risk and resilience management process is based on the ISO-31000 standard for risk management and formulates the Long-Term Control Loop of the RESISTO system, as it will be seen later. The complete process is being analyzed and described in detail in [16, 17] and uses suitable metrics to assess system performance and decides upon mitigation options based on the obtained information; it first identifies the system functions and derives proper resilience-related quantities that will provide the required information to facilitate the decision-making process.

Nine steps are involved, while several input tables are gathered from an extended threat list (step 1). Dedicated information for four main process steps is collected:

- System components → Step 2: System analysis
- System functions → Step 3: System performance function identification
- Threats → Step 4: Disruptions identification
- Mitigation options → Step 8: Selection of options for modifying resilience

The system performance functions (SFs) identified in step 2 constitute resilience quantities that need to be monitored, computed, or generated. In particular, they are necessary input for the pre-assessment of critical combinations of system functions and disruptions (step 5), the resilience quantification (step 6), and its final cost evaluation (step 7). For each one from the list of all SFs, obtained by the telecom CIs operators, several input fields are contained in the relevant template. An important feature of the template is the linkage between the tables, in this case the identification of system components needed for the SF to perform properly (Linked Components). This enables monitoring the propagation of the malfunctioning of a specific device (System Component) due to a disruption (Threat) to the performance loss of a specific service (System Function).

As soon as the mitigation options are finally selected (step 8), their implementation is to be held (step 9) closing the whole tool cycle. The resilience quantification is based on a computation of the performance loss due to the disruption by means of network simulations and exemplary resilience curves. Detailed results can be obtained by using the identified system performance functions, e.g., a certain failure might only affect specific services, while other functions are still working. A schematic representation of the joint risk and resilience process is shown in Fig. 19.1.

In the framework of the RESISTO project, this resilience and risk assessment management tool formulates the Long-Term Control Loop of the overall RESISTO platform that is the subject of the next section.

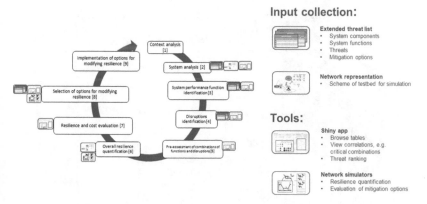

Fig. 19.1 Risk and resilience management process with supporting inputs and tools for the RESISTO project

The concept is proven through implementation in telecom CIs; however it can also adequately be applied to other kinds of CIs.

19.2   The RESISTO Solution

19.2.1   Concept and Approach

The RESISTO concept aims to develop a cooperation platform framework that allows different parts of the overall CI security personnel to exchange data and signals, to recognize complex attack patterns from different sources, and, based on real-time simulation of attack propagation within the CI and across interconnected CIs, to select and implement the best response and the optimal mitigation strategy. RESISTO aims to advance the infrastructure's security and resilience by developing an "entity," encompassing a holistic ecosystem of technology innovations and operational models. In fact, RESISTO takes up this challenge by fostering integrated risk-resilience assessment, faster detection of threats, better informed decision-making, and holistic understanding of a situation across the cyber and physical domain and interlinked CIs, allowing for better reaction and more efficient selection of countermeasure and mitigation actions. The logical architecture of the RESISTO platform is modular and adaptable to interfacing the existing infrastructures addressing the following five core functions, as in Fig. 19.2.

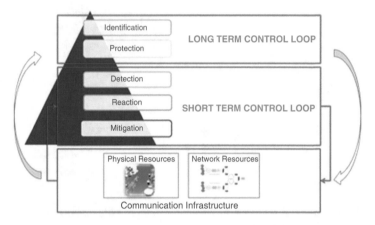

Fig. 19.2   The RESISTO logical architecture

Identification   For defining and maintaining a knowledge base on physical and cyber security risks to systems, assets, data, and capabilities characterizing Telecom CIs.

Protection   For developing and implementing appropriate safeguards to ensure delivery of CI services. The high degree of redundancy that usually characterizes telecommunication networks is further emphasized, in order to implement high resilience solutions. Graceful degradation of performance, when under attack, takes advantage of Network Functions Virtualization (NFV) and Software Defined Networking (SDN) paradigms.

Detection   For early and timely discovering of physical and cyber security events. It includes continuous monitoring of the security status of the CI, operating in a highly dynamic environment with changing threats, vulnerabilities, technologies, business processes, and services. Key Performance Indicators (KPI) monitoring and interdependency models are further exploited to evaluate impacts, recurrent patterns, and the occurrence of complex events. RESISTO leverages on using sophisticated technologies, properly integrated with security solutions/components already available in the CI.

Reaction For orchestrating and implementing effective response to a detected event. RESISTO investigates the joint use of Security Function Virtualization (SFV) and Software Defined Security (SDS). The best response is achieved through tools for automatic impact assessment of the security risks and effectiveness of potential countermeasures.

Mitigation For developing and implementing appropriate actions to mitigate the threats' impacts and to restore as possible capabilities impaired due to security events.

19.2.2   The RESISTO Architecture and Key Elements

The platform integrates two control loops both running on top of the communication infrastructure and being interlinked with each other [18].

The Long-Term Control Loop (LTCL) is an offline procedure, following a well-defined methodology and supported by advanced tools, aiming to assess infrastructure vulnerabilities and cyber and physical threats and consequently to define assets configuration and interventions in order to improve CI resilience and robustness. For each loop cycle, a set of resilience indicators (RIs), relevant to critical threat event typologies, are estimated and stored in a knowledge base (KB). The LTCL is based on the risk and resilience assessment analysis and management process and tool, described in the previous section, which identifies and evaluates risks and suggests mitigation strategies on the CI configuration. A LTCL cycle is performed on a periodic basis or when particular events take place (new threats or discovery of previously undetected vulnerabilities).

The Short-Term Control Loop (STCL) is the runtime component of the RESISTO platform. It promptly reacts to detected cyber/physical attacks and events that may impact the operational life of the system. It enhances situation awareness and provides operators with a Decision Support System cockpit able to implement the best reactions to an identified adverse event with the aim of mitigating the event's effects and restoring standard operating conditions. The Short-Term Control Loop:

- Monitors the physical and cyber security status of the infrastructure in order to collect and/or detect anomalies, correlates the physical and cyber domain events, and provides early warnings on attacks or events adversely impacting security

- Evaluates the event impact to performance degradation of detected anomalies and attacks on the communication CI and interlinked CIs, based on cascading effects
- Supports decision-making providing a qualitative and quantitative What-If analysis tool in order to evaluate the best communication CI reconfiguration
- Drives reaction and mitigation through action workflows (as directives to intervention teams, physical protection devices activation) and, mainly, through orchestrated communication network reconfiguration and protection function activation

While the short-term loop provides tools for direct reaction against attacks in real time, the long-term loop leads to the identification of criticalities and definition of long-term strategies. The input data to the STCL and generally to the RESISTO platform include physical events (e.g., intrusions, damage) or potentially dangerous events (e.g., unauthorized UAVs); cyber-attacks; physical telecom CI monitoring data (e.g., power usage information and faults); and communication network monitoring data (e.g., traffic, alarms).

RESISTO acts complementary to the existing CI's security systems. Thus, in order to detect threat events and identify relevant hazardous data sources, it exploits and integrates various systems, both those already available and those introduced by RESISTO. The already available systems involve legacy Physical Security Information Management (PSIMs) system(s) Security Operating Centers (SOCs) or physical and cyber-attack detectors made available by the operator. Moreover, additional physical and cyber threat detectors are directly offered by RESISTO and are meant to detect more sophisticated kinds of threats as those emerging nowadays, i.e., using UAVs during malicious or terrorist attacks. The RESISTO threats detectors include airborne threats detection systems (namely, radars and acoustic sensors) for early detection against airborne threats; additional cyber threat detectors such as Open-Source Intelligence (OSINT)-based detectors; audio and visual analytics for identification and pattern recognition of physical intrusions; along with wireless devices for smart spectrum surveillance and/or blockchain functionalities acting as sensing networks in cases of intrusion through putting unauthorized devices into a telecom wireless network.

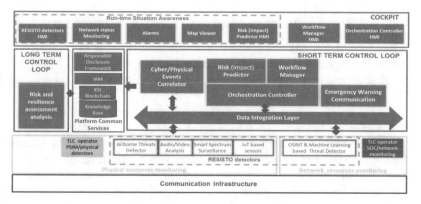

Fig. 19.3 The RESISTO high-level architecture and key elements

The RESISTO architecture is shown in Fig. 19.3. From a functional point of view, the input data are collected by the Cyber/Physical Events Correlator, a rule-based engine applying customized rules to correlate cyber-physical threat events and to generate and propagate alarms and externally detected and collected attack/anomaly events from apparently harmless events and monitoring data. The anomalies detected by the Correlator trigger the Risk Predictor which evaluates and highlights the impacts of the detected anomaly on the communication infrastructure and, mainly, on the services provided.

In parallel, the Correlator triggers the Workflow Management software engine to guide the operator during the reaction phase. Complex actions are performed by the Orchestration Controller built around the concept of SDS, taking advantage of NFV and SDN paradigms of the underlying communication network. Finally, the Emergency Warning Communication (EWC) function is activated when there is a need for sending instant messages, targeted alerts, and operating instructions to specific users present in areas where events like natural disasters and physical or cyber-attacks are occurring.

It is seen that the RESISTO solution aims to address all possible kinds of threats along with to derive to an innovative holistic solution for the CIs, availing all resilience cycle phases (prepare, prevent, detect, respond, mitigate) and covering both immediate and long-term responses and most importantly attempting to provide the needed correlation between physical and cyber threats as well as the impact on their propagation.

19.3   VALIDATION CASES: ANALYSIS AND DISCUSSION

The proof of concept of the above architecture is being held through certain cyber-physical threat scenarios, described in the following, being the most representative, sophisticated, and interesting cases when considering the modern types of attacks.

In the most devious attacks, rather than trying to gain full access into the system, an attacker may only want to open up a few strategic holes to the cyber domain of a network that will cause severe problems or failures to the offered services either immediately or at a later time. In the latter case, the attackers may perform reconnaissance and preparatory work on the digital front, before moving to actually perform the attack.

Thus, the attackers can exploit vulnerabilities in the physical domain of an infrastructure, to gain access to the cyber domain. These seemingly unimportant physical intrusions (unauthorized access to a building without obvious, direct or severe damage on the infrastructure) may be initially seen as physical assaults of a lesser importance in respect to their consequences on the cyber domain; especially when correlation between the physical and cyber intrusion events is hardly performed by the operator's existing security system. Thus, both events may not be given the proper attention.

The RESISTO Use Case for Cyber-Physical Threats

A cyber-physical attack takes place, targeting network equipment in a specific location that is physically protected by the telecom provider's security system. The physical attack (either by a hostile drone or by an intruder) is performed against the physical assets of the telecom provider. This physical threat is deliberately meant to enable a security threat in the cyber domain of the telecom provider's network. The telecom facilities are protected by the provider's existing security system, while the RESISTO platform, with its additional new sensors for detection, is also deployed. Two variation scenarios are envisioned, indicating the RESISTO added value to the provider's security systems:

First Scenario In this subcase, the attackers use a UAV to overcome the physical security (i.e., secure fence protected by the telecom operator's security system) and gain access to a network switch located inside a protected building and execute a cyber-attack. The UAV flies over the fence and approaches the building, ignoring its physical security. As it approaches, it is detected by the RESISTO airborne threat detector (radar and acoustic sensors); the detection system provides information about the path followed by the UAV and issues an airborne intrusion event to the RESISTO platform.

The drone connects wirelessly to the wireless network from the exterior of the building, gaining access to the network switch, initiating, i.e., a denial of service (DoS) attack, which targets the switch. Having detected the potential airborne intrusion, the RESISTO system identifies a potential security threat in the cyber domain, marking the cyber assets in the location as "compromised." Thus, it activates various cyber detectors of the provider's network to detect possible threats in the cyber domain. Subsequently, the DoS attack is detected, and a cyber-attack event is issued by RESISTO. Finally, RESISTO suggests a prevention/mitigation action, i.e., deactivation of the switch and redirection of normal traffic, neutralizing the attack.

Second Scenario In the second subcase, an attacker/unauthorized person breaches the secure perimeter, gains physical access to a protected building, and manages to enter the facility. The keycard access system is compromised, allowing the attacker to gain access to the building. Having entered the building, the unauthorized person gains access to an unattended computer and installs dormant malware that will be activated at some point in the future.

An audio/video analytics system (as a perimeter protection functionality complementary to the existing security system of the facility) is in operation for the detection and classification of this abnormal activity. The attacker is detected using data from the provider's sensors (i.e., cameras and microphones), which are processed by the sophisticated algorithms of the RESISTO audio/video analytics sub-system and a perimeter breach event is issued to the RESISTO platform. Thus, telecommunication assets in the vicinity are identified as "compromised." The RESISTO system activates various cyber detectors in the provider's network, which eventually detect the malware. A prevention/mitigation action is suggested, and the malware is removed. A potential threat in the cyber domain of the CI has been detected and eliminated by the prevention mechanisms activated by RESISTO.

In both scenarios, the intrusions initiate an attack in the cyber domain that would potentially cause a core network failure, either immediately or at a later time. Malware (active or dormant) can initiate a DoS to a server cluster, causing network traffic or partial shut downs. This attack will have an immediate effect to telecommunication assets, systems, and the offered services, along with impacts from the operational, economic, and societal point of view. Thus, a cyber-physical attack is executed by malicious artifacts or by an attacker targeting the provider's network, and it is being detected and neutralized by the unique capabilities of the RESISTO system. These are the integration of existing and new sensors along with the advanced functionalities and the decision-making mechanisms offered by the RESISTO system.

The use case concept builds on recent trends in airborne attacks, where airborne platforms, such as drones or small aircrafts, are used to not only perform physical attacks (i.e., bombing) and/or gather intelligence for physical security vulnerabilities but also gain access and compromise the cyber domain of the CI, directly attacking it, i.e., by connecting to the wireless network of a facility. This way, these scenarios represent realistic cyber-physical attacks that would perfectly fit an urban environment, where drones or UAVs are used for commercial purposes and can be concealed behind everyday activity that would not raise any kind of suspicion.

Both subcases cannot be detected and mitigated efficiently by conventional security systems. Although separate physical and cyber security mechanisms may be in place, the correlation between the events identified by RESISTO facilitates the efficient detection of the attack and enables its mitigation in its entirety. As it seems although both the physical location and the network were already protected by the physical and cyber detectors of the provider, without the RESISTO platform, the threats would not even be detected, let alone neutralized. Table 19.1 provides a summary of all response steps of the RESISTO solution.

19.4   Conclusions

The main objective of the RESISTO platform and the respective use cases is to enhance the resilience of the existing communication CIs toward both the domains of physical security and cyber protection. The focus is to advance the processes of detection and response and to result in new additional measures for mitigation and prevention, confronting threats that would have not been identified without the RESISTO system. It is

Table 19.1 The RESISTO response

Sequence	Action steps – analysis
Detection	Physical threats detected by the RESISTO sensing systems and legacy ones.
Reaction	*Correlation*: of the cyber-physical threat events, based on the RESISTO STCL correlation engines rules *Identification*: of the cyber assets in the location as "compromised" The STCL initiates various cyber detectors to detect the cyber malware. *Issue of event*: a cyber-attack event is issued by RESISTO *Countermeasures*: triggered by RESISTO, i.e., providing emergency signals *Notifications*: notifying the security operation center or the decision-making
Mitigation/ prevention	RESISTO STCL suggests deactivation of the switch and redirection of normal traffic (traffic rerouting). The malware is removed from the network *RESISTO LTCL iterations*: proposes disaster recovery plan (best practices and/or redundancy/resilience centers in respect to the assets affected)
End of cycle	RESISTO ensures communication continuity in the end

considered that a physical intrusion in a telecom operator's infrastructure can facilitate severe assaults in the cyber domain of a telecom network and vice versa. The main challenge is to perform this correlation in the short term and activate the respective response and mitigation actions; thus to prove that it is possible to detect the consequences of combined threats in short time and to use joint countermeasures.

It should be pointed out that these kinds of threats are not possible to be detected, correlated, and identified by the conventional security systems already used; instead they would be identified separately as only physical or only cyber ones, respectively. In such case, they would constantly cause security impacts requiring much more time and costs before they are finally identified and confronted, since the core of their creation would have remained undetected. RESISTO performs this correlation feature, enabled by additional sensors and algorithm framework, facilitating an effective detection of the attacks along with decision-making mechanisms for their response and mitigation.

The RESISTO proof of concept is being implemented to communication CIs. However, the fact that RESISTO can act complementary to conventional security systems and since other CIs greatly depend on telecom services (i.e., cyber domain or cloud data), it is evident that the above analysis can be adequately applied also in other kinds of critical infrastructures with similar implementation.

Acknowledgments This work is supported by the EU-H2020 RESISTO project, which has received funding from the European Union's Horizon 2020 Research and Innovation program under Grant Agreement No. 786409 (http://www.resistoproject.eu/).

REFERENCES

1. NFPA. (2010). NFPA 1600-Standard on disaster/emergency management and business continuity programs, MA, USA, 52 p. .http://www.nfpa.org/assets/files/pdf/nfpa16002010.pdf
2. ASIS, The Organizational Resilience Standard [ASIS SPC.1-2009]. (2009). Available at http://organizational-resilience.com/OrganizationalResilienceStandard.htm
3. ISO 22301:2012 – Societal Security – Business Continuity Management Systems – Requirements. (2012). Available at http://www.iso.org/iso/catalogue_detail?csnumber=50038
4. European Network and Information Security Agency (ENISA). (2011, February). Measurement frameworks and metrics for resilient networks and services – technical report.
5. Hayes, B., & Kotwica, K. (2012). Advances and stalemates in security. *Security Magazine, 34.*
6. Ohlhausen, P, et al. (2014). Effective, evaluated security metrics – persuading senior management with effective, evaluated security metrics. ASIS Foundation Report.
7. IES/NCES, National Center for Education Statistics, US Department of Education. https://nces.ed.gov/pubs98/safetech/chapter5.asp. Last accessed 2019/11/2.
8. Harris, S. (2013). Physical and environmental security. In *CISSP exam guide* (6th ed., pp. 427–502). New York: McGraw-Hill.
9. Hutter, D. (2016). *Physical security and why it is important.* GIAC (GSEC), SANS Institute. https://www.sans.org/reading-room/.../physical/physical-security-important-37120.
10. Niles, S. (2015). Physical security in mission critical facilities. White Paper 82, revision 2, APC White Papers, Schneider Electric's Science Center.
11. Omer, M., et al. (2009). Measuring the resilience of the trans-oceanic telecommunication cable system. *IEEE Systems Journal, 3*(3), 295–303.
12. Carlson, J. L., et al. (2012). *Resilience theory and applications.* Argonne National Laboratory, Decision and Information Sciences Division, ANL/DIS-12-1, Argonne, IL, USA.

13. Food Security Information Network. (2016). Measuring resilience. Crown Copyright. https://doi.org/10.12774/eod_tg.may2016.sturgess2.
14. Argonne National Laboratory. (2013, April). Resilience measurement index – an indicator of critical infrastructures resilience. ANL/DIS-13-01 Report, US Department of Energy.
15. Smith, P., et al. (2011, July). Network resilience: A systematic approach. Topics in network and service management. *IEEE Communications Magazine*, 88–97.
16. Häring, I., et al. (2017). Towards a generic resilience management, quantification and development process: General definitions, requirements, methods, techniques and measures, and case studies. In I. Linkov and J. M. Palma-Oliveira (Eds.), *Resilience and risk (NATO SfP and Security Series C: Environmental Security)* (pp 21–80). Dordrecht: Springer Netherlands.
17. Fehling-Kaschek, M., et al. (2019). A systematic tabular approach for risk and resilience assessment and improvement in the telecommunication industry. In *Proceedings of ESREL 2019 Hannover* (pp. 1312–1319). Singapore: Research Publishing.
18. The RESISTO Consortium: D2.6_RESISTO platform and tools reference architecture deliverable. http://www.resistoproject.eu/resources/. Last accessed 2018/10/11.

CHAPTER 20

Supporting Decision-Making Through Methodological Scenario Refinement: The PREVENT Project

Maria Kampa, George Kampas, Ilias Gkotsis,
Youssef Bouali, Anabel Peiró Baquedano,
and Rami Iguerwane

20.1 Introduction

Safety and security are of primary concerns for any transport system. This issue concerns both transportation nodes and terminals that can become a potential target for terrorism acts. Without a doubt, the establishment of a safe and secure transport environment is essential for citizens and transport operators across Europe.

M. Kampa (✉) · G. Kampas
Center for Security Studies, Athens, Greece
e-mail: m.kampa@kemea-research.gr; g.kampas@kemea-research.gr

I. Gkotsis
KEMEA – Center for Security Studies, Athens, Greece
e-mail: i.gkotsis@kemea-research.gr

335

Transport security can cover multiple dimensions of different threats and vulnerabilities from terrorist attacks to prevention of vandalism. Mass transportation systems hold a unique position as possible targets for attacks. They are built up as networks and feature a large concentration of people as well as a fundamental economic role. Moreover, increased security levels in air transport caused attackers to refocus on surface transport terrorism, including public transport. The threats of the entire transport supply chain and/or infrastructure also recognize many other forms like crimes committed on the terminals. It is a common understanding that emerging technologies can assist in creating a security transport ecosystem while reducing the duration and intensity of security checks and enhancing the capabilities of the transport operators in identifying and stopping potential attacks. In this regard the definition of future end users' requirements that allow an adaptation of the security system through a subsequent joint procurement is mandatory.

PREVENT project aims to map, through an iterative approach, the gaps and the needs of the transport operators around Europe in relation to security and propose the most promising one. These needs were identified in the format of an initial set of 12 scenarios, which were sequentially filtered to 6 by taking into account legal, procurement, and operational obstacles and constraints, as well as the economic component. Given the above, the end users group consisting of more than 30 public transport managers, operators, and security/police agencies from various EU Member States or others affiliated with the EU countries concluded in a shared challenge. Following this extensive analysis, the buyers involved in the project also decided as a next step that the purchasing activities of the desired solution shall be dealt with using a Pre-Commercial Procurement, meaning the purchase of R&D services from the industry.

———————————

Y. Bouali
Engineering Ingegneria Informatica Spa, Rome, Italy
e-mail: Youssef.Bouali@eng.it

A. P. Baquedano
Corvers Procurement Services BV, 's-Hertogenbosch, The Netherlands
e-mail: a.baquedano@corvers.com

R. Iguerwane
SNCF, Paris, France
e-mail: ext.rami.iguerwane@sncf.fr

The roadmap of actions, including the methodological approaches to verify the Common Challenge, is analyzed in the sections below. Section 20.2 will provide the outline of the methodology adopted. Section 20.3 will summarize the development of the security scenarios and the 12 to 8 scenarios' refinement; Sect. 20.4 highlights the refinement from 8 to the final 6 scenarios analyzing the different aspects that have been taken into account. Finally, Sect. 20.5 concludes with the main two results of the project, and Sect. 20.6 provides the conclusions of the aforementioned analysis.

20.2   PREVENT Methodological Framework

Scenario planning is a technique that can support decision-making by taking into account a number of uncertain and uncontrolled parameters that may have an impact on the implementation. The correlation between the scenario planning and decision-making has been established in several studies [6]. In this regard, this method was selected to help the practitioners select the most promising need after evaluating different aspects that could have an impact on their selection (Fig. 20.1).

As described in the image above, three progressive phases have been implemented:

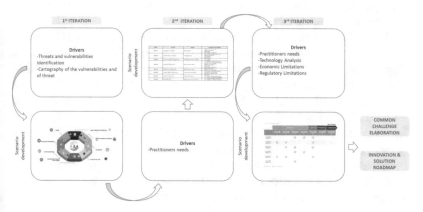

Fig. 20.1   PREVENT methodological framework

Iteration 1 12 security scenario definition. During this phase previously known and experienced, but also new and emerging threats are identified. The result of this phase is the definition of the initial set of the 12 scenarios.

Iteration 2 Refinement of security scenarios. Evaluation screening based on the practitioners' needs and second round of scenario screening (12 to 8). Final 8 scenarios were chosen through a voting-based refinement that involved all the project partners and external stakeholders.

Iteration 3 Final sorting of security scenarios. Evaluation screening through practitioners' needs, technological, economic, and regulatory criteria. Third round of scenario screening (8 to 6). Final 6 scenarios selection.

Project Outcome Common Challenge Elaboration. Based on the final set of scenarios, the end users via discussions session concluded to the most promising scenario, taking into account all parameters.

20.3   Security Scenarios Definition and First Refinement

The ultimate goal of building scenarios is to assess outcomes from alternative future trajectories, through model analysis and planning with stakeholders, to inform decision-making. In this regard, the elaboration of common security scenarios in the context of PREVENT was divided into the following logical steps.

As an initial step for the scenario development, PREVENT project focused on involving a substantial amount of stakeholders through the development of a group entitled "User Observatory Group" (UOG) which included practitioners from public transport operators (PTO) and law enforcement agencies (LEA), who have committed to take part in PREVENT's activities alongside with the Consortium partners. The active involvement of the UOG members and the Consortium partners to the project activities was ensured in order to create economies of scale and better analysis, to manage and spread the risks, and to foster the widest possible and collaborative uptake of the shared approach to security challenges in public transport in relation with terrorism.

The second step toward a concrete scenario development involves the identification of threats and vulnerabilities that the public transport operators face, through a security processes and practices analysis. In addition, PREVENT took into account other aspects like terrorists' attack patterns, the current European Transportation system – which facilitates the "free movement" between EU MS – and the security checks operations.

The aforementioned parameters allowed the preparation of an initial framework detailing a scenario attack through the elaboration of a certain storyline.

In this context, each PREVENT PTO and LEA has been invited to detail three attack events of high risk (Risk = Impact × Probability): one that occurred in the past, a realistic probable attack, and a complex high impact attack (Fig. 20.2).

Thus, based on the aforementioned attack events and the collaborative work, the first 12 scenarios have been developed including various aspects of an attack ranging from the threat, weapon type, the attack target, the location, and the rest factors as included in the image below. In this regard, the 12 scenarios defined were the following:

Scenario 1. An identified terrorist is crossing different European countries
Scenario 2. Stabbing attack in a PTO station
Scenario 3. CBRN attack in station with drones carrying the weapon
Scenario 4. Bomb attack in an underground station
Scenario 5. Hijacking of a train by a terrorist
Scenario 6. Several terrorists with weapons are using different kind of transportation
Scenario 7. Cyberattack on a PTO train dispatching, presumed isolated
Scenario 8. Bomb attack in a bus

Terrorist event that took place   Realistic terrorist event that could take place   Complex terrorist event

Fig. 20.2 Types of terrorist events considered in PREVENT

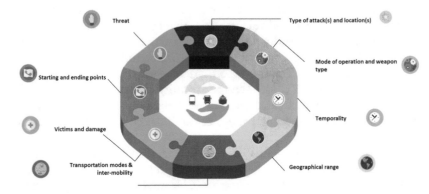

Fig. 20.3   Aspects of an attack analyzed within scenarios

Scenario 9. Left objects/baggage with explosive
Scenario 10. Vehicle crash in a crowd
Scenario 11. Sabotage: block fixed on rails to stop trains
Scenario 12. Massive shooting with rifles in a PTO station (Fig. 20.3)

Moreover, the scenarios included various gaps that need to be addressed in the European PTO environment. The total of 16 gaps have been identified and grouped into 4 distinct categories:

- Detection
- Tracking
- Protection
- Collaboration

allowing the elaboration of a scenarios and gaps matrix, where each scenario was associated with one or more primary and/or secondary gaps (Fig. 20.4).

For the refinement of this first set of scenarios, the methodology followed was quite simple, as the goal was the prioritization of the most crucial scenarios depending on the needs of PTOs and other practitioners. Thus, PREVENT initiated vote sessions and needs related discussions during project meetings where the PREVENT partners and the UOG members were invited to express their individual needs and capabilities into the scope of concluding eventually to the following eight security scenarios:

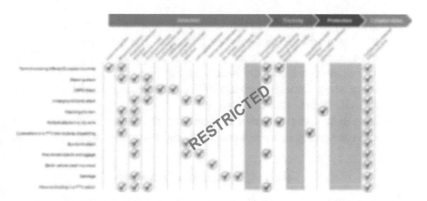

Fig. 20.4   Scenario gap matrix

Scenario 1. Mass shooting in a train station
Scenario 2. Unattended item(s) in a train station
Scenario 3. Terrorist crossing different European countries
Scenario 4. Reconnaissance before an attack
Scenario 5. Attack with a suicide vest in a subway
Scenario 6. Laying of an explosive material by a drone
Scenario 7. Bomb alert in a metro station
Scenario 8. Sabotage of the tracks

20.4   8 TO 6 SECURITY SCENARIOS

Besides the need's dimension, PREVENT partners decided to include for the second iteration of the scenario's refinement also three other dimensions that could potentially have an impact in the scenarios or even serve as a blocking point.

In this context, the next step included their progressive refinement leading to the selection of the six more promising ones by taking into account the four following essential components as analyzed in the proceeding:

- *Need:* Public transport operators' and security services' needs and expectations in the field of preempting terrorist attacks were evaluated.

- *Technology:* Technological readiness and innovation level of the identified available solutions (bearing in mind the definition of the state of the art of COTS technologies, related suppliers, patents and IPRs, as well as interoperability needs, benefits, and risks).
- *Regulatory:* Different aspects of the procurement legal background and GDPR aspects were analyzed.
- *Economic:* The procurement economic capabilities of the end users and the economic assessment of the eight scenarios were elaborated.

20.4.1   Technological Analysis

The initial step of the technological analysis was the definition of the available technologies for each gap identified in the scenario's definition phase, as included in the Scenarios and Gaps matrix presented in Sect. 20.3. In this regard, a list of technologies was provided, and the end users were called to evaluate and challenge the level of maturity of these technologies based on a benchmark methodology. For the benchmark, the partners adapted the Technology Readiness Level (TRL) scale [10] to the European PTO environment and to PREVENT's needs. Indeed, a technology such as facial recognition is technologically very advanced and even sometimes deployed in countries such as China,[1] so its TRL would be 9. But, principally due to regulatory obstacles, facial recognition is not so advanced in the European PTO environment, so its maturity would be probably 3 on a TRL scale adapted to it[2,3] Based on the results of the adjusted TRL maturity of the long list of technologies and the places of the PTO environment that each technology could be deployed, the ten most relevant technologies were finally selected according to the desire to implement them and the needs of the end users (Fig. 20.5).

Definition of the State-of-the-Art and IPR Search
Following the technologies identification, a state-of-the-art (SOTA) analysis was deemed necessary in order to plan accurately for a procurement. In particular, at that stage, such analysis is required to study the technologies that can best meet the Consortium's needs and their stage of

---

[1] https://www.businessinsider.com/chinese-company-claims-its-facial-recognition-95-accurate-masks-2020-3

[2] Article 10 Regulation (EU) 2018/1725.

[3] Article 9 GDPR, Article 10 Law Enforcement Directive.

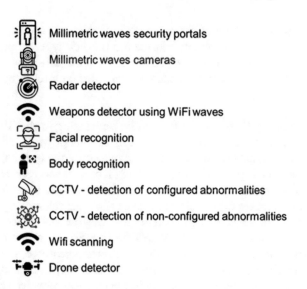

Millimetric waves security portals

Millimetric waves cameras

Radar detector

Weapons detector using WiFi waves

Facial recognition

Body recognition

CCTV - detection of configured abnormalities

CCTV - detection of non-configured abnormalities

Wifi scanning

Drone detector

Fig. 20.5 Ten selected technologies

development. It reveals whether a specific technology which could meet these needs is already available on the market or whether some degree of R&D is needed in order to further develop potential solutions. It informs the purchasing strategy toward either a procurement of R&D toward a desired solution (i.e., a Pre-Commercial Procurement, PCP), the modification or adaptation of existing solutions (i.e., a Public Procurement of Innovative Solutions, PPI), or the procurement of an identified Commercial Off-The-Shelf (COTS) solution.

In the PREVENT context, the state-of-the-art (SOTA) analysis included two activities:

(a) An evaluation of the available Commercial Off-The-Shelf (COTS) products which can satisfy the identified gaps and where relevant the integration effort required to reach the desired functionalities and interoperability. Consequently, the Consortium concluded to a list of products, equipment, and technical solutions available on the market that can be purchased, but do not fully cover the needs of the end users.

(b) A macro analysis of the total stock of relevant patents, standards, and literature to obtain information on their type, scope, breadth, content, radicalness, and technical relevance, as well as the associated institutions and related suppliers owning intellectual property rights (IPRs).

This second activity was performed by capitalizing in the iPlytics platform[4] and a keyword search that the applications provide. Therefore, a wide range of results of the world's state of knowledge in the field of the technologies analyzed under the PREVENT project was revealed. Out of this initial long list, the most relevant documents were identified on the basis of their degree of technical relevance and legal relevance.

The partners selected those patents that may be key to technological recommendations from the PREVENT project. So, an in-depth technical examination of the most relevant documents was performed by a technical expert partner, and the initial results were shortlisted. In addition, the most important intellectual property has been chosen, which in the context of the project may contribute to solving the problems defined in the security scenarios and meet needs articulated by PTOs and practitioners. Links between owners of key patents (or the most active patents in each topic) and manufacturers of solutions available on the market (described in the COTS analysis) were also compared.

The outcome of the analysis included the number of patents, the top 10 applicants, and the geographical location, among other important analytics deemed relevant for the scope of the project. Moreover, it should be emphasized that no close relations were identified between the manufacturers of technical solutions from the COTS analysis and global leaders in the field of their patents on individual topics. As a result, it was concluded that manufacturers rely on their knowledge or patent specific technical solutions that they can use in all their market products. A final recommendation was that it is worth following the global tycoons who are leading the world in CCTV technologies or radar technologies, because they can be potential contractors for future technologies.

Interoperability Needs
In order to define the interoperability needs and viable adoption models for PREVENT proposed technologies, the eight scenarios have been

[4] https://www.iplytics.com/

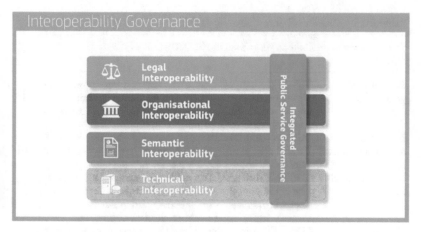

Fig. 20.6 Layers of interoperability of the EIF

analyzed and structured based on the European Interoperability Framework (EIF) of the ISA program. The EIF is "a commonly agreed approach to the delivery of European public services in an interoperable manner. It defines basic interoperability guidelines in the form of common principles, models and recommendations." This framework describes different layers of interoperability as shown in Fig. 20.6:

- Four layers of interoperability: legal, organizational, semantic, and technical
- A cross-cutting component of the four layers: integrated public service governance
- A background layer: interoperability governance

In the framework of PREVENT, these interoperability layers cover different aspects of interoperability and technology adoption requirements as described in Table 20.1.

20.4.1.1 EIF-Based Questionnaires
In PREVENT, practitioners played an important role in conducting key activities and providing domain know-how and expertise necessary to achieve planned goals. In this regard, both public transport operators (PTOs) partners of the project and members of the User Observatory

Table 20.1   PREVENT interoperability layers

Interoperability levels	Contextualization to PREVENT
Interoperability governance	PREVENT proposed technologies shall take into consideration existing regulations and policies on interoperability frameworks and propose recommendations in case there are gaps or constraints due to as-is situation
Integrated public service governance	PREVENT shall take into consideration the involvement of all potential users at national and European levels and describe different coordination and governance mechanisms among them
Legal and ethical interoperability	PREVENT shall take into consideration existing legal and ethical policies and strategies when framing interoperability and technology adoption requirements and propose recommendations of putting in place new legislation in case there are gaps or constraints
Organizational interoperability	PREVENT shall propose different interoperability configurations for potential users in different scenarios, both at MS and European levels
Semantic interoperability	All aspects of data and information exchange with regard to the adoption of proposed technologies will be addressed
Technical interoperability	PREVENT proposed technologies shall support service-oriented architecture (SOA) design paradigm using open and internationally accepted standards, a solution that makes PTOs independent of vendors, products, and technologies which offers great advantages and flexibility in shaping the adoption models' scenarios

Group have been involved in order to evaluate and define the interoperability and technology adoption requirements related to the proposed technologies. Two dedicated questionnaires have been prepared based on the EIF framework in order to collect the different aspects of interoperability: legal, organizational, semantic, and technological. The involved practitioners ensure a wide coverage of different public transport categories (train, metro, security forces in public transport), as well as different European countries (France, Portugal, Italy, Greece, Poland, and Switzerland). The abovementioned questionnaires have been submitted under two dimensions:

1. Coordination dimension: respondents were asked to provide their evaluations with regards interoperability requirements related to coordination/cooperation features of proposed technologies between different public transport organizations and and/or

authorities. Two levels of coordination were proposed: national level (within the same Member State borders and jurisdictions) and EU level (between organizations belonging to two or more different European countries).

2. Technology-based dimension: respondents were asked to provide their evaluations with regard interoperability requirements related to the proposed ten technologies as described in Sect. 20.4.1.

The results of the collected answers are briefly summarized in the below points:

- Interoperability requirements are very heterogenous among different organizations and through different Member States.
- In addition to EU regulations, each Member State has its own/specific laws and regulations which in many cases differ from other countries.
- For organizational interoperability, two major types of governance were considered depending on the level of autonomy the PTOs have or in the case they depend on national authority.
- Concerning the semantic interoperability, many data models exist throughout Europe, and there is a clear need for data model standardization in the domain of public transport.
- Data management is under the responsibility of the PTOs, who express major concerns on access rights when exchanging data with other organizations.
- With regard to technical interoperability, any new technology adoption will require consequent adjustment (in terms of hardware and software) of existing systems.
- Any new technology requires also coordination with other authorities and service providers.
- New technology requires setup of new security policy, specifying minimal security requirements that all users and entities must respect.

20.4.2 Regulatory Aspects

The regulatory environment was analyzed in order to identify the aspects that may have an impact on pursuing any of the scenarios as presented in Sect. 20.3. The major criteria to block and exclude one or more of the initial use case scenarios were based on the avoidance of conflicts among national regulatory frameworks. Particular attention was also paid to the impact of the GDPR on handling of information within and between practitioners, requesting support from the GDPR and Security Advisors working on the project.

The analysis was based on the input provided by the public buyers involved in the PREVENT project as partners or members of the User Observatory Group (UOG) regarding potential legislative obstacles to the implementation of any of the scenarios. The work focused on the regulatory aspects across the MS of the public buyers, to evolve from the different contexts and practices toward a single regulatory component that can be shared across public buyers for the deployment of the future procurement. The outcomes of the research were further analyzed in order to elaborate a conclusion on the most flexible and adequate legal framework.

The flexibility to deploy an R&D procurement procedure outside the scope of the defense and security procurement legislation (when costs and benefits are shared), the flexibility to conduct ad hoc joint cross-border procurement and to conduct market consultations, and the availability of fast court proceedings were the criteria selected to identify the most suitable public procurement legislation. Additional criteria, such as the willingness to act as lead procurers and the previous experience with the deployment of PCP, are also important to choose the applicable public procurement legislation.

Regarding the impact of the privacy regulation on the selected scenarios in the analyzed countries, it is important to underline that in the regulatory framework, the principle of accountability triggered a double obligation on the part of the data controller: to ensure the respect of the principles relating to the processing of personal data and, more in general, of the data protection law and to demonstrate and fully document such respect.

Therefore, with regard to the scenarios, there weren't absolute limitations or serious impediments on the privacy law side, but rather requirements that must be met in order to ensure compliance with the regulatory sources analyzed. However, particular attention must be paid to scenarios that involve the use of biometric systems, such as facial recognition where particular technical and organizational measures should be taken.

20.4.3   *Economic Aspects*

An economic assessment of the scenarios of the PREVENT project was performed. The goal was to identify the ones that are the most economically advantageous and to provide recommendations for the refinement of the scenarios. Following an extensive literature review, in order to identify the economic aspects of a scenario, the MCDA analysis [2–4] on the gaps addressed by the scenarios was considered the most suitable method (Fig. 20.7).

In this regard, seven criteria were selected in mutual agreement with the stakeholders, aiming at "measuring" the pros and cons of each gap:

1. The expected improvement of the services quality
2. The Gap Security Value, calculated based on the level of importance of each gap and its contribution to each scenario
3. The current disturbance in the activities of the PTOs deriving from this gap
4. The percentage of occurrence of each gap
5. The current cost of the currently used equipment
6. An approximate estimation of the current financial loses deriving from each gap
7. An estimation of the current price of the technologies available addressing the gaps

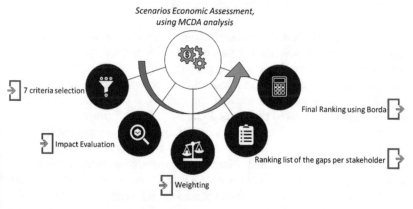

Fig. 20.7   Scenarios economic assessment using MCDA

Following the criteria selection, the proposed methodology involved two phases:

- Impact Evaluation. Impact evaluation is the phase where end users assigned a score s_i, to each gap and criteria C. All these values were normalized at the end, so that the best values became 1 and the worst values became 0.
- Weighting. Weighting indicates the importance of a criterion C in comparison with the other criteria. This was the outcome of the analysis made by requesting the end user to fill in a resistance to change grid.

After collecting the information requested above and making the necessary calculations, the evaluation matrix per end user was prepared. The overall preference score for each gap per end user was simply the weighted average of its scores on all the criteria. Letting the preference score for gap "i" on criterion C be represented by s_{ij} and the weight for each criterion by w_j, then n criteria the overall score for each gap, S_i, was given by:

$$S_i = w_1 s_{i1} + w_2 s_{i2} + w_n s_{in} = \sum_{j=1}^{n} w_j s_{ij}$$

(20.1)

Based on the weighted score above, the ranking list of the gaps per stakeholder was produced. The next step of this methodology is to merge the different lists ranking the gaps (one list for every participant) into a unique list.

The weighted scores, calculated earlier, reveal the most-preferred and the least-preferred solutions for a given participant. Of course, the most and the least-preferred gap vary greatly from one participant to another. In that context, it is of utmost importance to compute a unique list, which should somehow take into account the individual preferences of the participants, as expressed in their individual lists. For this purpose, the Borda method [9] was used: for a given solution, instead of computing the average of the scores from the individual lists, the respective "high-ranking score" (HRS) was calculated. Namely, the calculation made includes the number of times that this solution appeared in the top six of an individual's list, and this value was the HRS. Based on the HRS, the ranking of the gaps was calculated. This gap ranking was finally transformed in Scenario Value through the usage of the scenario gap matrix.

20.4.4   Security Scenarios Definition

To achieve a general overview of the aforementioned components, a table summarizing all the needs, technological and limitations analysis was used so as to facilitate the selection of the final set of scenarios (Table 20.2). On the basis of this exhaustive analytical work and with the continuous engagement of the practitioners, the initial scenarios were refined and adapted into six final scenarios:

Scenario 1: Unattended item(s) in a train station
Scenario 2: Reconnaissance before an attack
Scenario 3: Mass shooting in a train station
Scenario 4: Sabotage of the tracks
Scenario 5: Terrorist crossing different European countries
Scenario 6: Attack with a suicide vest in a subway

Each addressed shared technological anti-terrorist needs in public transport operators' environment, providing different target audiences with information about how the operational and technical challenges of the common security scenarios can be addressed from a technological point of view, taking into account the regulatory and economic capabilities.

20.5   Project Outcomes

20.5.1   Common Challenge Elaboration

As subsequent step and based on the six scenarios, the PREVENT project partners identified the Common Challenge as described below:
Enhancing security situational awareness through:

- *Timely automatic detection of unattended items in public transport infrastructure and in public areas in the vicinity*
- *Identification and tracking of perpetrators*
- *Advanced crisis management system*

This represented the most viable – in many terms – shared need among the different stakeholders. The exploitation of the vulnerabilities in relation to this need has a massive economic cost, which can be measured in

Table 20.2 Example table used for the final scenario's selection

Scenarios	Technology 1	Technology 2	Technology 3	Technology 4	Technology 5	Technology 6	Technology 7	Technology 8	Technology 9	Technology 10	Economical analysis (scenario scoring)
Scenario #1											
Scenario #1											
Scenario #1											
Scenario #1											
Scenario #1											
Scenario #1											
Scenario #1											
Scenario #1											
Technological analysis	■ *TRL =*										
	■ *X patents*										
Legal analysis	*GDPR compliant …*										

terms of indicators ranging from the cash value to financial losses, business interruption, and damage to property or in worst cases passenger losses. In this context, earlier detection of terrorists and potentially dangerous objects, tracking of detected individuals or situations and coordination of security forces' response, are critical actions that will mitigate the terrorism-related risks.

It is important to mention that this Common Challenge will serve as the basis for the subsequent procurement that will be undertaken on the form of a Pre-Commercial Procurement (PCP) since the need identified will be satisfied through the R&D services provided by the industry's side.

20.5.2 Innovations and Solutions Roadmap

One other important outcome of the aforementioned work was also the development of an innovations and solutions roadmap. This roadmap focuses on providing the PREVENT's consortium partners, as well as the largest European audience, with a consolidated overall picture of the main results obtained under the core activities of the project into a roadmap of solutions and innovations. Such a roadmap is meant to be a multidimensional and interactive map of innovations and solutions providing different target audiences with information about how the operational and technical challenges of the common security scenarios can be addressed from a technological point of view, taking into account the regulatory and economic capabilities.

In Fig. 20.8, the data structure of the aforementioned tool is presented. In particular, the five layers of the innovation roadmap and the rational and relationship between one layer another are analyzed:

Layer 1 – Security Threats Represents the list of identified threats and gaps related to security in public transport. Such threats have different security levels and may target people, infrastructure, transportation means, and/or related public spaces.

Layer 2 – Security Threats' Categorization:
- Detection: focuses on technology gaps that allow the detection of a potential threat in a PTO environment – abandoned items detection, weapons detection, explosive material detection, etc. They are in the scope of this benchmark.

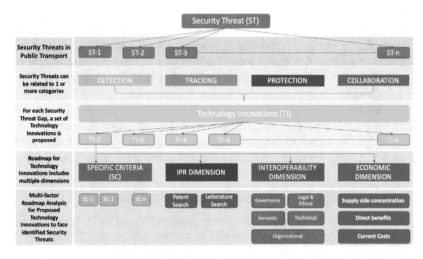

Fig. 20.8 Technology Innovations overall roadmap structure

- Tracking: focuses on technology gaps that allow the tracking of a person responsible for a threat or an attack that has occurred.
- Protection: focuses on technology gaps that allow a protection of strategic places in PTO areas.
- Collaboration: focuses on technology gaps that allow a better collaboration between PTOs and LEAs.

Layer 3 – Proposed Technology Innovations For each security threat/gap, a list of technologies and/or low TRL innovations is proposed.

Layer 4 – Roadmap Analysis Dimensions Each proposed technology innovation will be characterized under four dimensions – technology-specific criteria, IPR dimension, interoperability dimension, and economic dimension.

Layer 5 – Multi-Factor Roadmap Picture The four characterization dimensions are further detailed according to information collected from different sources under work packages 3, 4, and 5.

20.6 Conclusions

Given the increased complexity of the PTO security environment, this chapter contributes to a framework of decision-making based on scenario planning that stresses greater emphasis on the needs of the end users as well as other parameters that could potentially have an impact on the final selection. From the work undertaken by the Consortium, it is evident that technology, regulatory, and economic-related factors are vital data that end users should possess in order to bypass the uncertainties of any decision.

In this regard, PREVENT focused on satisfying the security needs and wants of a variety of end user by designing 12 scenarios and developing a rich methodology in order to refine them and to ensure that the final selection made will represent a viable in many terms need. The Consortium tackled several limitations on the decision-making mechanism ranging from the involvement of the end users and the efficient expression of their needs up to providing sufficient data for the definition of the Common Challenge of a future procurement, not only in security or transport field but also in every field that an innovative procurement is suitable to be conducted.

Acknowledgments

 PREVENT project has received funding from the European Union's Horizon 2020 research and innovation program under grant agreement no. 833444. The content of this chapter reflects only the author's view, and the European Union is not responsible for any use that might be made of such content.

Bibliography

1. PREVENT – PRocurEments of innoVativE, advaNced systems to support security in public Transport, Grant Agreem. No 833444 Horizon 2020 Eur. Union. (2019).
2. *Multi-criteria analysis: A manual.* (2009, January). London: Department for Communities and Local Government, ISBN: 978-1-4098-1023-0.
3. Noleppa, S. (2013, December). *Economic approaches for assessing climate change adaptation options under uncertainty: Excel tools for Cost-Benefit and Multi-Criteria Analysis.* Eschborn: Deutsche Gesellschaft für Internationale Zusammenarbeit (GIZ) GmbH.

4. Ribeiro, F., Ferreira, P., & Araújo, M. *Evaluating future scenarios for the power generation sector using a Multi-Criteria Decision Analysis (MCDA) tool: The Portuguese case.* https://repositorium.sdum.uminho.pt/bitstream/1822/26142/1/Ribeiro%202013.pdf. Last accessed 2020/07/20.
5. Hedel, R., et al. (2018). Assessment of the European Programme for Critical Infrastructure Protection in the surface transport sector. *International Journal of Critical Infrastructures, 14*(4), 311–335.
6. Chermack, T. J. (2004, April). Improving decision-making with scenario planning. *Futures, 36*(3), 295–309.
7. European Interoperability Framework (EIF) of the ISA programme. https://ec.europa.eu/isa2/sites/isa/files/eif_brochure_final.pdf. Last accessed 2020/07/20.
8. Organisation for Economic Co-operation and Development. Methodology for Assessment of National Procurement Systems (Version 4). 2006. Available at: http://www.oecd.org/dataoecd/1/36/37390076.pdf. The User's Guide (Section 1) of the document provides very helpful guidance on conducting procurement assessments. Last accessed 2020/07/20.
9. Saari, D. G. (1985, February). The optimal ranking method in the Borda count. IIASA Collaborative Paper. https://core.ac.uk/download/pdf/52944585.pdf. Last accessed 2020/07/20.
10. Technology Readiness Levels (TRL), EC. https://ec.europa.eu/research/participants/data/ref/h2020/wp/2014_2015/annexes/h2020-wp1415-annex-g-trl_en.pdf . Last accessed 2020/07/22.

CHAPTER 21

Securing the European Gas Network, the Greek Business Case

Ilias Gkotsis, Anna Gazi, Dimitrios Gritzalis,
George Stergiopoulos, Vangelis Limneos, Vassilios Vassiliou,
Eugenia Koutiva, Dimitrios Petrantonakis,
Eleftherios Lefkokilos, Evita Agrafioti,
Anastasia Chalkidou, Dimitris Drakoulis,
Anastasia Eleftheriou, Aspa Skalidi,
Panagiotis Demestichas, and Clemente Fuggini

21.1 Introduction

Modern and resilient European societies rely on the effective functioning of Critical Infrastructure (CI) networks to provide public services, enhance quality of life, sustain private profit and spur economic growth.

The first official effort for the protection of CIs on a European level took place in 2004 with the request by the European Council for the designation of an EU integrated strategy for the protection of CI. In 2006

I. Gkotsis (✉) · A. Gazi
KEMEA – Center for Security Studies, Athens, Greece
e-mail: i.gkotsis@kemea-research.gr; a.gazi@kemea-research.gr

© The Author(s), under exclusive license to Springer Nature   357
Switzerland AG 2021
B. Akhgar et al. (eds.), *Technology Development for Security*
Practitioners, Security Informatics and Law Enforcement,
https://doi.org/10.1007/978-3-030-69460-9_21

the EU set the parameters for the implementation of the European Program for the Protection of Critical Infrastructure (EPCIP) which adopted an "all-hazards approach" toward threats, including the protection against natural disasters, technological and man-made threats [1]. In 2008, the European Council Directive 2008/113/EC defined CIs in general as an asset, system or part of system which is located in a Member

D. Gritzalis · G. Stergiopoulos
DEPA – Public Gas Corporation S.A., Athens, Greece

Athens University of Economics & Business, Athens, Greece
e-mail: dgrit@aueb.gr; geostergiop@aueb.gr

V. Limneos
DEPA – Public Gas Corporation S.A., Athens, Greece
e-mail: v.limneos@depa.gr

V. Vassiliou · E. Koutiva
EDAA – Attiki Gas Distribution Company S.A., Athens, Greece
e-mail: v.vassiliou@edaattikis.gr; e.koutiva@edaattikis.gr

D. Petrantonakis · E. Lefkokilos
EXUS Software Single Member LLC, Athens, Greece
e-mail: dpetr@exus.co.uk; e.lefkokoilos@exus.co.uk

E. Agrafioti · A. Chalkidou
GAP Analysis S.A., Crete, Greece
e-mail: agrafioti@gapanalysis.gr; chalkidou@gapanalysis.gr

D. Drakoulis · A. Eleftheriou
Innovation Acts Ltd, Lefkosia, Cyprus
e-mail: ddrakoulis@innov-acts.com; aeleftheriou@innov-acts.com

A. Skalidi
WINGS ICT Solutions, Athens, Greece
e-mail: askalidi@wings-ict-solutions.eu

P. Demestichas
WINGS ICT Solutions, Athens, Greece

Department of Digital Systems, University of Piraeus, Piraeus, Greece
e-mail: pdemest@wings-ict-solutions.eu

C. Fuggini
RINA – RINA Consulting S.p.A, Genoa, Italy
e-mail: clemente.fuggini@rina.org

State and is essential for the vital functions of the society, and the health, safety, security, economic or social well-being of people, which (infrastructure) if disrupted or destroyed will have a major impact on the society and functions of a Member State [2]. Additionally, this directive focuses on the European Critical Infrastructures (ECI) of Energy and Transport. The Energy sector is further analysed to the sub-sectors of Electricity, Oil and Gas. Further to the above, several initiatives and regulations specifically on the Gas sector focus on security, pinpointing the need for enhancing protection of such CIs, such as the European Energy Security Strategy [3], the stress tests on the resilience of the EU gas system [4] and the EU Regulation 2017/1938 on Security of Gas Supply [5].

All the aforementioned effort at strategic level finds fertile ground for further development, given the real security incidents in Gas CIs. Due to their distributed nature and often completely publicly known routings, the gas grids are prone to physical attacks, cyber-attacks (e.g. SCADA manipulations) and cyber-physical attacks. Statistics on main physical hazards, natural (e.g. earthquake, ground displacement, landslides) and man-made (e.g. sabotage and attacks) have been provided by EGIG over the last 50 years. A total of 1366 incidents to gas network have been reported from 1970–2016 resulting to an annual primary failure frequency of 3.1×10^{-4} per km, for transmission pipelines [6]. Main causes of incidents have been identified in "external interference" (e.g. digging, piling or ground works by heavy machinery) and "ground movement" (e.g. dike break, mining), both characterized by potentially severe consequences. In 2016, energy was the industry second most prone to cyber-attacks, with nearly three-quarters of US O&G companies experiencing at least one cyber incident [7], while the cost of Cybersecurity breaches for companies in utilities & energy reached $14.80 million in 2016 being the industry sector with the second highest annualized costs [8].

Moreover, as interconnections of gas elements, interfaces with other grids, automated monitoring and regulation loops are increasing, besides cascading consequence effects, also emergence of novel types of threatening behavior are expected. Taking account of these challenges, there is an increasing need for joint physical, cyber and especially cyber-physical threat risk analysis and management comprising preparation, prevention, detection as well as optimized response, recovery and restoration to a better system state as before the event.

21.2   SECUREGAS PERSPECTIVE

As already indicated, the Gas network and infrastructure represents a significant example to be protected, made secure and resilient to both physical and cyber threats. The complexity of the gas network, its difference among transportation lines, the peculiarity of the areas crossed (remote or densely populated), the various production and storage facilities make it a strong, relevant and challenging environment to cope with.

SecureGas project focuses on the 140.000 km of the European Gas network covering the entire value chain from production to transmission up to distribution to the users, providing methodologies, tools, and guidelines to secure existing and incoming installations and make them resilient to cyber-physical threats. Three business cases, addressing relevant issues for the gas sector and beyond, have been identified so that to ensure the delivery of solutions and services in line with clear needs and requirements, focused on risk-based security asset management of gas transmission and distribution networks; impacts (economic, environmental, and social); and cascading effects of cyber-physical attacks on interdependent and interconnected European Gas grids; integrity, and security, through the operationalization of resilience guidelines, of strategic installations across the EU Gas network. SecureGas tackles these issues by implementing, updating, and incrementally improving extended components, integrated and federated according to a High-Level Reference Architecture built upon the SecureGas Conceptual Model (CM), a blueprint on how to design, build, operate, and maintain the EU gas network to make it secure and resilient against cyber-physical threats.

SecureGas adopts a comprehensive, yet installation-specific, approach to the security and resilience of gas CIs. The Conceptual Model and the High-Level Reference Architecture provide a blueprint and the rules for its implementation on how gas installations and systems have to be planned, designed, constructed, operated, and maintained to be secure and resilient against the combination of cyber and physical threats. The project addresses the resilience of gas CI and aims at integrating the resilience capabilities (plan/prepare, absorb, recover, and adapt) in the disaster risk management cycle (preparation, response, recovery, and mitigation) within an asset life-cycle perspective, thus securing to achieve resilience across the various phases of the life cycle of an infrastructure.

SecureGas extended technical components have been selected on the basis of complementary to address various type of threats and cascading events. As such a cyber-physical correlator will be exploited, countermeasures for attacks to the SCADA system will be investigated, and landslide and seismic events will be assessed as potential threats to gas pipelines. SecureGas will perform a characterization and ranking of the complete range of cyber, physical, and cyber-physical risks for gas grids. A dedicated approach looking at interactions among safety and security in communicating information to the public starting from existing means and methods used by gas operators in liaison with public authorities will also be investigated. The participation of end users ensures that SecureGas is validated in the most complex and realistic scenarios.

21.3   METHODOLOGY

Resilience is appropriate to be treated more as a concept [9], characterized by four main capabilities – Plan/Prepare, Absorb, Recover and Adapt – and four main socio-technical system dimensions, physical, information, cognitive, and social, to be defined and described for each of the aforementioned capabilities. In practical terms robust and resilient CI means that an existing and new Gas infrastructure will have to resist to hazards and absorb their impacts more efficiently and more effectively; accommodate and recover the effects of a hazard more efficiently, timely, and safely; and be designed/restored to coordinate more efficiently across the various phases of a disaster risk management cycle.

In order to effectively enhance the resilience of a gas CI, SecureGas will follow a business case (BC)-driven approach, with three main phases as depicted in Fig. 21.1 and further analyzed below:

21.3.1   Phase 1: Construct/Develop

In this phase, initial actions include a state-of-the-art analysis in order to identify the existing limitations and barriers in practice as well as the current challenges, constraints, and needs of the infrastructure managers and operators in terms of security against both physical and cyber threats.

Moreover, technology providers and BC owners will interact in order to define how the various components can be integrated into the High-Level Architecture and how they will interact to serve the purpose of achieving the required level of service (LoS). A Concept of Operations (CONOPS) will guide the implementation and integration process. As such the High-Level Architecture will implement the features of the

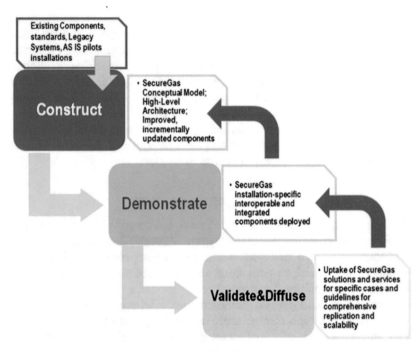

Fig. 21.1   SecureGas methodological approach

Platform as a Service (PaaS) concept so that to ensure flexibility, modularity, third-party interoperability and integration, customization, and contextualization to the BCs.

The SecureGas CM and CONOPS approach and application relate to the necessary inputs such as user requirements, technical requirements, potential threats to be covered, performance indicators of solutions, system specifications as well as the expected further use of the CM and CONOPS during development, implementation, testing, validation, and improvement. Main advantages are expected from the CM and CONOPS approach for the SecureGas High-Level Reference Architecture (HLRA) development as well as the specification of the technical security solutions in all three application cases.

21.3.2 Phase 2: Demonstrate

This phase consists of the BC development including adaptation and customization of the components to pilots, pilot execution, and performance evaluation.

In this direction, initially SecureGas components will be customized, integrated, and deployed, as previously developed and incremented, into the three BC. Indeed, according to the respective scenarios, the components built in phase 1 will be selected and be customized in line with the needs and constraints and then be integrated and unified according to CM and HLRA. This means that, up to the level required and defined, the extended components will be adapted and integrated to the BC existing infrastructures (physical and IT) and to the legacy systems in place. To allow this implementation is made consistently and effectively, the HLRA will be built on the concept of Platform as a Service (PaaS).

The deployment of the aforementioned extended and integrated components in the BC will happen through piloting, consisting of testing and demonstration. In parallel, supportive and collaborative activities will be implemented, such as learning and knowledge sharing, discussions, and assessment of best practices.

21.3.3 Phase 3: Validate and Diffuse

This last phase has twofold objectives: on one side to perform a cumulative and summative evaluation of SecureGas CM, HLRA, components, and their deployment into the BCs and on the other side to make sure that SecureGas outcomes and core principles are diffused beyond the consortium reaching a wider community of users.

Regarding the first objective, SecureGas will follow a four-step validation-evaluation approach:

1. Set the context: Identify the actors involved, identify requirements and process, define objectives, and identify criteria.
2. Plan the BC: Define scenarios, analyze criteria, and select evaluation tools.
3. BC implementation: Plan the BC, conduct validation, and assess data quality.
4. Assess and report results: Assess results, and prepare validation and evaluation report.

For what concerns the "diffusion" aspect, SecureGas will make sure that the project outcomes reach a wider community as possible of stakeholders by defining and implemented an effective and efficient dissemination and communication strategy. This will be supported by the establishment of a project Stakeholder Platform (SP), consisting of representatives of gas CI operators/owners, companies delivering services for gas network security, and policy makers. Moreover, SecureGas will establish liaison activities with CEN/CENELEC on the basis of the agreements made with the European Commission in supporting the implementation of standardization activities, processes, and outcomes in European research projects [10].

21.4   TECHNICAL COMPONENTS

The SecureGas technical partners will customize, integrate, and deploy the extended components, as previously developed and incremented, into the three BCs. According to the foreseen scenarios for each BC, the components built in phase 1 of the aforementioned methodology will be selected and be customized in line with the needs and constraints as defined by the scenarios and then will be integrated and unified according to CM and HLRA. In the following sections, the set of components that will be integrated and address the needs of the Greek BC will be presented.

21.4.1   Joint Cyber-Physical Risk and Resilience Management

The joint cyber-physical risk and resilience management component (RMG) aims at enhancing the security and resilience of gas CI networks, covering the main principles imposed by resilience and disaster risk management cycle. Indeed, the RMG component aims at providing support to the operators before, during, and after an incident occurrence.

Preincident The RMG component will operate as a standalone tool, supporting gas CI owners, operators, security managers, and decision-makers in identifying, assessing, and evaluating risks (physical and cyber) that may compromise gas network integrity. For the evaluation of security breach scenarios' risk level, the RMG component adopts the all-hazard approach and follows the requirements of a risk management framework, fostering thus the targeted and efficient allocation of funds and resources toward security.

The main security risk assessment steps that will be integrated into the RMG component are:

(a) *Identification of CI assets:* Identification and ranking of all corporate assets along the gas network, on the basis of their criticality. Criticality is typically defined as a measure of the consequences associated with the loss or degradation of a particular asset and is related to the benefits arising from its prevented nonavailability. The assessment of asset's criticality requires the consideration of its role not only in the context of company's overall service delivery and mission but also in relation to the wider network. The criteria applied for asset criticality evaluation are, among others, the workforce at the specific asset, service delivery, interdependencies, and potential domino effects of a security incident to adjacent CI assets.

(b) *Identification of threats:* Identification of possible sources of threats to each critical asset (identified in the previous step) and development of security breach scenarios. The RMG component will provide an extensive threat catalogue, covering physical and cyber aspects of security, to enable user have a broad view of potential threat sources that could target his network.

(c) *Risk analysis and evaluation:* Assessment of the likelihood and consequences of potential security breach scenarios. For those threats that are non-probabilistic (e.g., intentional/man-made threats), their likelihood of occurrence is estimated considering parameters such as:

- Feasibility, which is the overall likelihood of a certain type of attack occurring regardless of the target and relates to the ease of attack's execution from the attacker's point of view (e.g., knowledge about the asset, difficulty in obtaining the type of weapon).
- Target attractiveness, which is the likelihood that a certain asset would be targeted for a specific type of attack and is related to the features of a particular asset that make it more or less likely to be attacked (e.g., potential for casualties, symbolic value).
- Vulnerability, which is regarded as the probability of the successful completion of the attack and is closely related to the security systems deployed in the CI as well as to the existing security procedures.

The consequences of each scenario are assessed through the evaluation of impact-related criteria such as number of casualties, number of fatalities, cost of asset loss, environmental impact, out of service time, etc.

Analyzing input information and drawing on pre-established model parameters, the RMG component will inform the user on scenarios' risk level. Appropriate recommendation will be provided in terms of technical, organizational, and managerial solutions.

During an Incident Apart from serving as a standalone tool, the RMG component will also be integrated in the SecureGas solution, providing close to real-time risk assessment capabilities and offering advanced situational awareness in case of security breach incidents. To do so, the parameters applied for risk calculation of both probabilistic and non-probabilistic incidents will be modeled and integrated into the component, serving as a knowledge database and proving the cognitive ontology for addressing security breach scenarios. Thus, for every event detected by the SecureGas system (i.e., alert), the RMG component will provide an estimation on its risk level. That information will serve as input to the SecureGas UI, so as every alert appearing on the operational picture to be linked to a distinct risk level.

Post-incident The RMG component will provide targeted recommendation to the operator on the actions that need to be undertaken for the absorption of an event, the efficient response to it, and the fast recovery in case of disruption, incorporating the procedures and systems specified through operators' emergency plans and recovery actions.

21.4.2   *Cognitive Framework for Biometrics and Video Analytics*

This component's objective is to identify malicious physical presence near critical gas infrastructures. Data preprocessing and machine learning algorithms will be used to classify different kinds of objects detected from the cameras and input sensors within or near the CIs. The output of this task will be used to raise alerts when the detected object and its respective actions are putting the infrastructure at risk, for example, if someone attempts to climb the fence protecting a gas supply station.

A combination of cameras and other instruments, such as accelerometer, vibrometer, door latch, etc., will be used, in order to better identify, verify, and track the object/person by correlating information of their presence. The video analysis will be applied within the fenced-in area. Planned functionalities include:

- Detection of people that unauthorized enter the perimeter
- License plate recognition for vehicles entering from the outside area
- Face recognition for people entering the perimeter

The identification of authorized entry of persons will be done by matching with white lists where authorized persons are stored. The similar procedure is also used for vehicles by license plate recognition. In case of an unauthorized entry into the perimeter, an alert will be raised to the operator with information about which location is affected including a snapshot of the suspicious individual from the respective camera.

Additionally, Wi-Fi presence analytics will be exploited in this component. This solution has the benefit of tracking visitors through a unique identifier per device. Having a unique identifier is essential when not just tracking the number of people in a space. In order to recognize the suspicious connections, a white list system will be constructed on which we could enumerate the authorized connections and let the system alert when an unauthorized one gets into the network. An alert can be issued when an address does not belong to the white list or when a group of new addresses gather in a critical hub. However, it is important to note that white lists are not perfect/exhaustive and a determined attacker could bypass them. They can be used, as one tool among others, to complement the solutions suggested above. The result of this component will be the overview of the critical gas infrastructure's safety and security, via a combination of heterogeneous sensors and respective data processing.

21.4.3 Risk Aware Information to the Population

Mass notification requires a very good preparation and readiness level before any emergency occurs. In SecureGas the population is only indirectly notified by the organizations; the authority which has the competence to notify the population differs per country; however, in most cases it is either the civil protection authority or the Public Safety Access Point (PSAP) or the national "E-112" service.

The target of the component is to enable gas CI operators to (efficiently) notify authorities (civil protection, first responders, other CI operators) on an emergency situation. The informed decision of communicating this kind of information will be assisted by the alerts produced by the SecureGas platform.

When it comes to warn private, public, or civil authorities, one needs to make sure that panic is not provoked among the population.

Once those criteria considered and the need to warn gets compelling, we need to communicate as soon as possible in order to activate the reaction of the affected communities and thus minimize damage.

The message to send has to be specific, accurate, clear, and credible. By sending that message, we aim to maximize the response capability of the concerned authorities and enable them to notify the broader public as soon as possible by activating their own emergency plan communication system. The message must contain the following:

- The date, time, and location of the incident
- The type of the incident
- The current damage extent and the risks that might come
- Some suggested countermeasures
- The source of the information

21.4.4   Cyber-Physical Correlator

The cyber-physical event correlator component is a machine learning-based tool for advanced event processing which will be harnessed to monitor the resources of the SecureGas platform, as well as the results from the different components, aggregating the information in order to detect threats. The collected data will be processed, and machine learning-based event correlation will be applied to discover cyber, physical, or joint anomalies and threats for the operation of the gas infrastructure and network.

The input for this component will be aggregated data from multiple heterogeneous sources. This includes sensors that will be placed in the CI sites to monitor various parameters for physical security, data, and traces from cyber activity for the protection against cyber threats and legacy systems the operators already have, potentially gas grid SCADA, IT, and IoT Systems. Moreover, results generated from other components will also be incorporated. One of those components will be the algorithms for video analysis running as part of the intrusion detection tool. This will feed the correlator with information about potential detected persons or objects and face or license plate recognition performed on the video surveillance within the protected area.

The correlator will perform anomaly detection deriving from rules, labeled historical events, and automated artificial intelligence techniques. Additionally, parameter forecasting will be included to enable the prediction of future fluctuations of the monitored values and to facilitate preparation for the upcoming conditions as well as propose potential mitigation actions. WINGS will deploy its patented capability to detect anomalies projected to gas grids [11].

In order to maximize the accuracy of the component, labeled historical events will be given as input to facilitate the training of the model. Nevertheless, during the operation of the system, feedback will be acquired from the user to confirm the validity of the results and reveal mistakes. The feedback will also include classification of the detected events with regard to their nature (physical, cyber, mixed) and their significance. The vectors corresponding to anomalies and attacks, acquired either by threat databases or by previous events, will then be correlated to vectors produced by real-time monitoring of the infrastructure to improve the capability of the correlator to both detect and the classify new occurrences.

21.5   THE GREEK BUSINESS CASE

As already described in the methodology above, three BCs will be designed and implemented within SecureGas. In this section, the customization, deployment, and testing of the SecureGas HLRA and the extended components described in the previous section will be analyzed in the framework of the Greek BC, which involves two different gas CI operators (DEPA and EDAA). This will result in the deployment of a specific security solution (i.e., SecureGas service), integrated as far as possible into operations and evaluated by the BC owner (i.e., DEPA, EDAA) during pilots activities.

The steps to be followed include the definition and documentation of the as-is situation; the improvement potential regarding the implementation of SecureGas at DEPA and EDAA; the identification of CI assets and assessment of their criticality, identification of threats, target attractiveness assessment, and consequence assessment (Domino effects); as well as definition of requirements for information to the public.

This analysis will cover the transportation and distribution (midstream up to downstream) of gas at strategic (project planning), tactical (project risk assessment), and operational (distribution network) level.

21.5.1   End Users

DEPA (Public Gas Corporation S.A.) is the main NG and LNG importer in Greece and actively contributes in the strengthening of the Southern Gas Corridor, i.e., the realization of new alternative routing for the NG transportation from the Caspian Sea and Middle East to Europe. Furthermore, DEPA owns and operates two fast-fill compressed natural gas (CNG) and refueling stations for natural gas vehicles (NGV) – buses, trucks, and cars – in Attica, Greece. Both facilities serve as central filling stations of a fleet of approximately 600 buses and 100 garbage trucks in total. The performance of each station is described by the overall flow rate of 5,000 Nm^3/h, sufficient for 28–36 busses per hour. Stations are supplied by high-pressure pipelines. They consist mainly of an inlet section, a compression section, a storage section, and a dispenser filling section.

EDAA (Attiki Gas Distribution Company S.A.) is the distribution system operator (DSO) in Attiki, the most populated region in Greece, supplying NG to more than 350,000 households, 6,500 small commercial customers, and 200 large commercial and industrial customers, through a medium- and low-pressure network of more than 3,100 km. EDAA network is connected with the high-pressure network through 5 city gates peripheral in Attiki region, and along there are 150 distribution stations and ~180 valve pits, including connections with large customers also referring to CIs (e.g., hospitals) whose operational continuity would be comprised in case of a gas supply failure.

21.5.2   Greek Business Case Scenarios

In the Greek BC, two different types of scenarios will be used in order to validate and evaluate the SecureGas solution, in an effort to combine security and resilience aspects across both midstream and downstream gas infrastructures.

The first scenario type involves the simulation of a strategic risk assessment during life-cycle management of a hypothetical pipeline that is the only point for distributing gas nationwide. This strategic scenario engages multiple owners and operators to simulate key security and resilience issues and analyze potential threats and hazards affecting the delivery of natural gas related to spatial planning of gas networks, (gas) network unavailability risks, and diverse sources of threat. Output will focus on defining generic risks applicable to all modern in midstream architectures along with potential solutions and design security measures.

In the second type, the Greek Gas CIs participate in the analysis of attack scenarios involving downstream infrastructures. Potential attacks on downstream DSO infrastructures are reviewed along with their interconnections to other CIs. Validation scenarios include both cyber-physical and physical-to-cyber-attacks that target modern storage and distribution systems on industrial systems and networks, along with physical safety scenarios, such as combined terrorist attacks close to highly populated areas and sensitive receptors. This aims to provide a framework for detecting, assessing, and eventually responding to hybrid attacks to DSO infrastructures and interconnected CIs through the use of the SecureGas solution. Moreover, the proximity of strategic gas network nodes, distribution endpoints, and assets to populated areas and sensitive receptors as well as to other CIs is being taken into consideration for the Greek BC. It is deemed as one of the most important and integral parameters of the risk assessment and management procedure.

In the aforementioned scenarios, the following threats are some of those that are addressed:

- Geopolitical risks/threats
- Physical intrusion (person and vehicle)
- "Man-in-the-middle" attack to ICS/SCADA system
- Manual tampering for midterm corrosion combined with malware blocking control panel
- Physical intrusion to data/control rooms, combined with malware (e.g., through USB) to kick-start or block ESD
- Unavailability attack for a week, cascading failure to other CI sectors

21.6 Conclusions

In this chapter the authors have analyzed and presented the methodological framework and the details that will be followed in SecureGas project, in order to enhance the security and resilience of gas CIs, involved in different phases of the gas value chain, underlying also the necessity to focus on such CIs, given that the complexity of the respective networks and the peculiarities of such facilities makes them hard to cope with. More specifically the Greek Business Case was presented, where the implementation of the under development SecureGas solution will be tested and validated. The technical components that will be integrated and address the needs of the two Greek Gas CI operators were also reported.

At this stage of this research, the outcomes are preliminary, confirming though the necessity to have tailor-made solutions and approaches for gas CIs security. From the current state of analysis and developments, it appeared that actions such as the following should be implemented for tangible results:

- Demonstration of systemic security risk and resilience management approach including the combination of physical and cyber threats, their interconnections, and cascading effects
- Implementation of improved, integrated solutions to prevent, detect, respond, and mitigate threats and their customization to specific installations
- Assessment of the customized solutions through an extensive set of KPIs and establishment of a trustworthy mechanism for sharing information and best practices
- Identification and development of viable and effective security paradigms validated through the BCs

Acknowledgments The work presented in this chapter has been conducted in the framework of SecureGas project, which has received funding from the European Union's H2020 research and innovation program under Grant Agreement No. 833017.

REFERENCES

1. https://eur-lex.europa.eu/legal-content/EN/ALL/?uri=CELE X:52006DC0786
2. https://eur-lex.europa.eu/LexUriServ/Lerv.do?uri=OJ:L:2008:345:007 5:0082:EN:PDF
3. https://eur-lex.europa.eu/legal-content/EN/ALL/?uri=CELEX% 3A52014DC0330
4. https://ec.europa.eu/energy/sites/ener/files/documents/2014_stresst ests_com_en.pdf
5. https://eur-lex.europa.eu/eli/reg/2017/1938/oj

6. Gas pipeline incidents, 10-th report of the European Gas Pipeline Incident Data Group (EGIG). https://www.egig.eu/reports

7. https://www.us-cert.gov/sites/default/files/Annual_Reports/FY2016_Industrial_Control_Systems_Assessment_Summary_Report_S508C.pdf

8. Ponemon Institute. (2016). *2016 cost of cyber crime study & the risk of business innovation.*

9. Serre, D. (2011). *La ville résiliente aux inondations Méthodes et outils d'évaluation.* Université Paris-Est, 173 pp.

10. https://www.cencenelec.eu/research/tools/projects/ongoingENwork/Pages/default.aspx

11. Tsagkaris, K., Demestichas, P., Kotrotsos, S., Cardaris, D., & Margaris, A. *Early warning and recommendation system for the proactive management of wireless broadband networks.* United States, Incelligent P.C. (Athens, GR) 20180019910. http://www.freepatentsonline.com/y2018/0019910.html

PART V

Disaster and Crisis Management

Technological and Methodological Advances in the Protection of Soft Targets: The Experience of the STEPWISE Project

Stefano Armenia, Anita Schilling, Olivier Balet, Maureen Weller, and George Kokkinis

22.1 Introduction

The security situation in Europe has been deteriorating. While critical infrastructures are largely secured, hence making them hard targets for terrorist attacks, any easily accessible crowded place providing a high

S. Armenia (✉)
Link Campus University, Rome, Italy
e-mail: s.armenia@unilink.it

A. Schilling · O. Balet
CS GROUP, Toulouse, France
e-mail: anita.schilling@csgroup.eu; oliver.balet@csgroup.eu

M. Weller
Crisisplan BV, Leiden, The Netherlands
e-mail: weller@crisisplan.nl

G. Kokkinis
KEMEA – Center for Security Studies, Athens, Greece
e-mail: g.kokkinis@kemea-research.gr

© The Author(s), under exclusive license to Springer Nature   377
Switzerland AG 2021
B. Akhgar et al. (eds.), *Technology Development for Security Practitioners*, Security Informatics and Law Enforcement,
https://doi.org/10.1007/978-3-030-69460-9_22

probability of a successful attack has become a first-choice target. This includes public venues such as malls, transport infrastructures, market places, schools, hospitals, cultural or sport sites, places of worship, and other urban areas. The public character and the variety of venues makes the protection of soft targets a very hard task, since security and safety of people must be provided while at the same time preserving accessibility.

As pointed out in [1], there is now the need of protection of these soft targets, which traditionally would not have been entitled to the constant protection of law enforcement agencies (LEAs). Innovative security thinking therefore means putting the focus on both soft targets and hard targets. The key objective is however to protect individuals from serious assaults; the protection of private and corporate property is not relevant in the context of this chapter. As the protection of soft targets is a recent problem that emerged in the last decades, there are no established strategies on how to implement it yet. There is clearly no one-fits-all solution because each soft target has its own intricacies to consider. The approach to design a security plan is however sometimes pursued in an unstructured way [2] and is clearly also influenced by common practices of LEA as well as budget and availability considerations [2], which may negatively impact its quality. Yet, it is common knowledge that the activity of planning and training for preparedness leads to better decisions, especially in critical moments, and to a more effective use of resources. This is where the STEPWISE project aims to make a difference, by providing a methodology and the software tools to approach security planning in a structured way, thus helping with the creation of more solid security designs.

22.2   Assessing the Security of Public Spaces and the Protection of Soft Targets

Prior to designing any security measures for a public space, the potential weaknesses of the soft target have to be identified. The features and modalities of the venue and of the event that takes place should be assessed against any potentially exposed vulnerabilities. The physical security has to be taken into account along with all other aspects that embody the soft target for this risk assessment: first, it should be clarified what is to be protected [1]. This is a shared mission among many security stakeholders both from public and private sectors. Second, vulnerabilities need to be identified and assessed with respect to their potential severity. Third, potential threats need to be identified together with potential specific targets and their likelihood of occurrence. Fourth, the potential impact of the assault should be assessed regarding possible causalties. The broad security plans should in fact prioritize saving lives and minimizing harm [3]. Although a violent assault may entail damage to property, which may represent significant costs, the focus of the paper is exclusively on the safety of individuals, which is a priority for LEA. Last, the consequences of an incident should be identified and assessed in their severity and other relevant aspects.

Clearly, methodologies should assist security planners to pursue a systematic and systemic approach to security design. A structured approach, guided by a methodology, helps to keep the focus on the relevant aspects and to avoid neglecting others. An important point often overlooked is the human factor. As people are usually under time pressure, some tasks need to be prioritized more than others, and some may be entirely forgotten or erroneously deemed to be too resource-intensive to be evaluated.

In the scope of risk assessment, but also when eventually designing security measures, all available information and data should be exploited to have a comprehensive understanding of the soft target, i.e., its normal functioning and the deviations that may be caused by the event being planned. Elements that should be considered are, for example: (1) layout of the location and topology from a strategic and vulnerability point of view; (2) analysis of processes - e.g., what staff does, who does it in particular, and when; (3) supply chains, (4) interdependencies; and (5) available data from similar events and their lessons learned.

In conclusion, a large amount of data needs to be considered for a profound risk assessment and a comprehensive discussion on potential security measures. LEAs would strongly benefit from a decision support tool that bundles all this information in one place, and that aids in taking better informed decisions on this basis. At present, security planning seems to be largely a pen and paper task for LEAs, which bear inherent problems of communication difficulties because people literally do not see the same thing. Here, even just a mere visualization of the venue can clearly contribute immensely to reduce confusion and help users to "be on the same page."

There is no one-fits-all solution for the protection of public spaces. It is in fact unfeasible for LEAs to protect all public spaces due to technical and mostly economic reasons [2]. Consequently, there is a clear need for innovative and collaborative tools that make it easier for LEAs to work cross-sectorally and to devise solid security concepts. The economic aspect governs the choices which are made for the security measures. Providing sufficient security while respecting budget constraints, and while at the same time fulfilling the demand to plan flexible, foolproof, and creative measures [1] is clearly a hard task. Decision Support Systems are targeting this difficulty and are designed to provide the support that is needed in the respective knowledge domain, so to help users make better decisions and better use of their resources.

22.3   State of the Art on Decision Support in the Security Domain

The task of management aims to bring an effort to its ultimate good conclusion, by deciding the needed courses of action at certain turning points where a decision from someone having a systemic perspective is needed. This is true in all fields of the human knowledge including ensuring the security and safety of people.

All managerial tasks have the latent need to be supported in their effort to do a better job, and a relevant, effective, and useful Decision Support System (DSS) can play a crucial role in helping "self-confident professionals" to rely on the evidence provided by facts [4] and not on wrong mental models or beliefs that are backed only by expertise (which of course could be biased by one's experiences). As reported in [5], most of the early authors refer to the Gorry and Scott Morton [4] paper ("A Framework for Management Information Systems") in Sloan Management Review in 1971 as the true starting point for DSS technology. Scott Morton's doctoral thesis [6] at Harvard Business School in 1971 already outlined the first concept of what can be seen as a "management decision systems". Later on, a paper from Sprague [7] ("Framework for the Development of Decision Support Systems"), in MIS Quarterly in 1980, summarized all the essential elements for the design, development, implementation, and use of DSSs. As still reported in [5], early case studies analyzed by Keen [8] showed several benefits identified by DSS users, among which cost and time savings thanks to faster and more effective response.

These and other similar general benefits still appear today in the literature, even if the underlying DSS technology has changed several times and even if such technology gets labelled differently with respect to the core idea of DSS. In fact, the DSS architecture mainly builds on three components: (1) an interface between the user and the functional routines, (2) a data manager, and (3) functional routines.

In [5] the authors report that Sprague [9] collected the following "DSS characteristics" from several other authors:

1. DSS aim at the less well-structured, underspecified problems of upper level management.
2. DSS combine the use of models or analytic techniques with traditional data access and retrieval functions.
3. DSS focus on features which make them easy to use by noncomputer people in an interactive mode.
4. DSS emphasize flexibility and adaptability to accommodate changes in the environment and the decision-making approach of the user.

Thus, a distinctive feature of the early descriptions of DSS is that it should support all phases of decision-making (hence "decision support").

Additionally, [5] reports about another fundamental and distinguishing feature of a DSS, that is, its iterative design [9], a typical four-steps process of information systems development (1. analysis, 2. design, 3. development, and 4. testing), which repeats iteratively.

Under a wider perspective, we must notice that modern DSS are not only based on IT but also on methodology. The engine of the decisional process can be represented by the diagram in Fig. 22.1, in which we did not represent explicitly the technological stack, which by the way can be quite complex in data representation and fruition, because we wanted to draw the reader's attention on the fact that calculus methods are a key element in modern DSS.

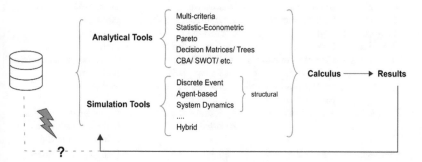

Fig. 22.1 A generic representation of the decisional process as presented in [10]

Moving from this perspective, it is also worth mentioning that such methodologies rely on a number of underlying decision-making theories/models, whose evolution is captured in Table 22.1, which correlates various decision-making theories, modes, and attributes. In [21] the authors describe the evolution of such decision-making tools/techniques characteristics that, over the last 40 years, have evolved from a mere descriptive/diagnostic power in operational/tactical environments to the prescriptive/predictive power with a focus on strategic decision-making, capable of an effective response also to potential future

Table 22.1   A general approach to decision-making

Decision-making models			
Analytic		Reflective Thinking [11]; Organizational Decision-Making [12, 13]; The Knowing Organization [14]	Strategic, Informed Decision Making for the Future [15] Intelligent Organizations [16]
Rules-based		Organizational Decision-Making [12, 13]; The Knowing Organization [14]	
Belief-driven		Behavioural Decision Theory [17]	
Cognitive	Naturalistic Decision-Making [18]; Rapid Processing Decision Theory [19]	Learning Organization [20] The Knowing Organization [14]	
Decision-making attributes			
Timing	Immediate	Short-term	Long term
Type	Critical, urgent	Operational, tactical	Strategic
Environment	Dynamic	Recurring	Uncertain
Objective	React	Explain, optimize	Predict, act
Technology	Mental simulation of options using leading practices and pattern matching	Applying logic or rules, plus computerized, probabilistic information processing	Simulation and decision tools with impact analysis
Knowledge	Tacit knowledge	Tacit and explicit knowledge	Convergence of explicit knowledge with tacit and cultural knowledge
Strategies	Heuristic	Algorithm	Convergent analytics

Adapted from Ref. [21]

environments (hence with the aim to build/increase resilience for the system under analysis).

As a growing number of companies/organizations are making use of DSS (which – as seen above – can be very different among themselves), it is interesting to note that some maturity models have been proposed, with a specific reference to the most currently diffused model, which is based on data analytics [22]. In this work, we will just cite the Gartner maturity model [23] (see Fig. 22.2) that shows four levels of maturity: descriptive, diagnostic, predictive, and prescriptive. Such levels can be differentiated based on how much automated the decisional process is or how much is left to an external intervention by the decision-maker.

It is now worth mentioning a few of the latest innovative approaches in building DSS. Again, in their work, in [5] the authors report that the D2I joint industry and university research program (cf. [5]) proposed a vision that is built around human and system joint intelligence for digitalization: in other words, DSS use fast, automatic algorithms for large, well-structured, datasets and combine this with knowledge from seasoned context experts. In order to make it work, human users need context relevant advice (in real time, with real data and information) that is adapted to their cognitive abilities and background knowledge (i.e., advices they can understand and use): they ultimately argue that this could be the mission statement for the DSS of the 2020s.

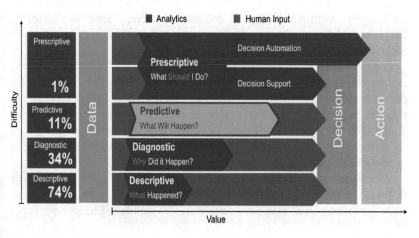

Fig. 22.2 Gartner's analytics maturity model [23]

The idea of digital coaching systems got started a few years ago (cf. [24]) as an answer to the demand on human operators to master advanced automated systems in complex and very large industrial process systems. Digital coaching works well with data that is collected from digital devices, instruments, tools, monitoring systems, sensor systems, software systems, data and knowledge bases, data warehouses, etc. and which gets then processed to be usable for the digital systems that will guide and support users.

A well-known attempt of this has been undertaken with the so-called Industry 4.0 framework [25], which essentially heavily relies on the digital twin concept. The digital twin is basically a digital replica of the organization under analysis and provides the possibility to act under the cyber-physical perspective that means allowing actions undertaken in the virtual environment to affect the physical world. The digital twin concept is strictly connected to the digital coaching model, as it requires to master the transition from data to information and onwards to knowledge, also known as "digital fusion": a) data fusion collects and harmonizes data from a variety of sources with different formats and labels, b) information fusion uses analytics to build syntheses of data to describe, explain, and predict key features for problem solving and decision-making, and c) decisions then get "actuated" in the physical replica/twin generating the expected/simulated impact (see Fig. 22.3).

Another decision-making framework that integrates the knowledge from the process-generated information and the knowledge from the individuals in an organization has recently been proposed in [10] and, as for the above mentioned digital coaching systems, aims at adapting IT-generated content to the cognitive levels of the users, in a sort of triple-loop learning process.

A DSS has to be distinguished from a decision-making system, where the system actively recommends a result or solution. The creation of a decision-making system is feasible if the domain knowledge is well defined, structured, and narrow. For example, medical expert systems can be counted to this group: a catalogue of clinical symptoms can be turned into an ontology to automatically diagnose an illness based on the provided symptom information [26].

The domain knowledge regarding the security of soft and hard targets is, however, not as well defined and certain aspects are in constant evolution. Therefore, only DSS are relevant in this domain. The development of DSS for the security domain poses a significant challenge: due, in fact, to the sensitive nature of information on terrorist attacks and soft target

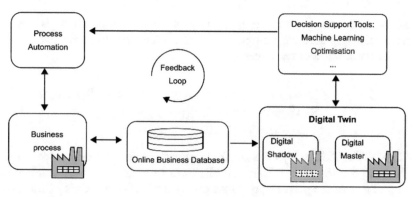

Fig. 22.3 The digital twin decision-making cycle [25]

security, data is not necessarily readily available and may be available to authorized individuals only. It is likely that a number of DSS in the security domain exist for the protection of public spaces; however, online information is sparse, such that the majority may be a confidential in-house development of private security companies. To our best knowledge, a dedicated DSS for the protection of soft targets that would be available for LEAs does not exist yet.

There is an ISO standard (ISO 31000) [27] on risk management, which aims to help organizations to perform well in an environment full of uncertainties. It has already been applied to the assessment of urban spaces in the scope of the EC project DESURBS [28]. ISO 31000 does not contain specific recommendations and is a general-purpose risk management tool. It can be argued that safety risks for people are a special case and should be excluded from general-purpose risk management, as any risk to people is unacceptable.

In recent literature [1, 2] regarding the protection of soft targets, the paradigm DDRM is predominant. The acronym stands for:

- Deter: Effective preventive measures should be put in place together. The significance of psychological obstacles must not be underestimated.
- Detect: Ensure that any suspicious action is rapidly detected.
- React: Response protocols should be in place to react quickly to a detection.
- Mitigate: Strategies to mitigate an incident if it happens.

Nevertheless, [2] reports that the protection of soft targets is often approached in an unstructured way, which often leads to ineffective measures and a waste of resources.

22.4   BEYOND THE STATE OF THE ART ON DECISION SUPPORT IN THE SECURITY DOMAIN: THE STEPWISE PROJECT

The STEPWISE project is an EU-granted project funded under the Internal Security Fund Police (ISFP) from DG-Home, which builds on top of one of the partners' previous experiences, CS GROUP's "Crimson by DIGINEXT" platform, using virtual reality and digital twin principles for the supervision and protection of sensitive sites, and developing new capabilities for a five key phase methodology:

1. **Assessing vulnerability of public spaces**. A city has countless public spaces, and public events can take place in parallel creating several potential soft targets that need at least minimal protection. Prioritization is essential. STEPWISE provides a visualization of the venue in 3D and traditional cartography in 2D. Advancing the current paper-based manner of assessing maps helps to minimize confusion about the locations and improves the overall quality of the discussion. Users are guided by an adaptable checklist that leads through the process of vulnerability assessment.
2. **Creating innovative security plans**. The risk assessment of the previous stage represents the input for this phase where practitioners are requested to jointly formulate security measures to eliminate the identified vulnerabilities and adjust security concepts and measures to minimize their exposure to potential threats. Again, this stage is guided by a checklist that can be customized by the user and which helps to identify the key aspects, measures, and operations that are to be taken.
3. **Devising effective response plans**. Risk assessments and security plans are focused on prevention, helping to minimize the possibility of an assault. Clearly, good prevention and deterrence measures cannot entirely eliminate the risk. It is therefore necessary to investigate potential incident scenarios, which is done in this stage. Security professionals emphasize that the process of planning for an incident itself is just as important – or even more so – than the response plan that is produced. As reported in [1], without any kind of plan or protocol, even minor

incidents may cause severe confusion and the inability to react adequately in a critical moment.

4. **Assessing plans and designs against various scenarios.** In the 3D environment that STEPWISE provides, the security and response plans can be tested. The user can play through different situations supported by 3D equipment on the 3D map and annotations as text or drawings. Security measures are notoriously difficult to test, and a test on the public space would bother the population apart from the investment of time and resources. An evaluation of the security measures in the digital mock-up of the venue represents therefore a great advantage.

5. **Enhancing preparedness of first responders by virtual reality training**. Training is undoubtedly a key to a successful incident management. STEPWISE therefore provides a training option with virtual reality (VR) headsets. Users can immerse into the VR environment to familiarize with the layout of the site and to study the topology of the buildings. In addition, users are also able to solve problems and take decisions in a serious game context.

As shown in Fig. 22.4, the 3D map is a central element in the interface where users can work collaboratively and add annotations as text, drawings, or 3D models that represent equipment, such as cars or barriers.

Fig. 22.4 The STEPWISE platform for decision support in security planning

This represents a digital playground where potential threat situations can be created and discussed with remote colleagues that have the same view on the 3D scene. Users can create checklists for each of the methodology steps. All the checklists are stored on the server, where they can also be archived. In this way, a checklist can be used as a template or simply inspiration for the protection planning of another public space.

The STEPWISE 3D mock-up module is another component of the toolkit, which enables the creation of 3D building models as IFC files [29] – a BIM [30] compliant file format. 3D building models are mostly available for newly constructed buildings, but for older houses, they remain a rare resource. The 3D mock-up module therefore provides an easy to use building editor where a user can rapidly model a 3D building based on available floor plans to be used with the STEPWISE toolkit for the risk assessment and the planning of the security measures.

The STEPWISE VR training module provides training capabilities to teach the topology of the site and to increase the preparedness of trainees. However, this module is not intended to be used while performing operations, which is currently beyond the scope of the STEPWISE project.

22.5   LIMITATIONS AND FUTURE DEVELOPMENTS

A key aspect of the structured approach to planning security measures for the protection of soft targets is the visualization of the venue, notably the realistic virtual representation in 3D. Building models with topologically correct indoor modelling is, however, still quite a challenge. In fact, as for more recently constructed buildings, 3D models could be available, as the architectural and construction process nowadays works already with 3D models and ideally follows the BIM concept, it is easy tounderstand that the larger part of a European city consists usually of older and even historic buildings, for which a digital replica often does not exist. Where it is not the aim of STEPWISE to develop a professional architectural software, the STEPWISE 3D mock-up module provides however a potential solution to rapidly sketch and edit 3D building models, but its applicability is a function of the complexity of the building that is to be modelled. Consequently, a trade-off has been made to ensure high user-friendliness such that buildings with simple and largely regular interior topology can be modelled. Further development, at the moment beyond the scope of the STEPWISE project, may enable the modelling of complex buildings and highly irregular structures. At the same time, there are open source tool available, like FreeCAD [31], which already allow for the modelling of complex buildings with a typical CAD approach [32].

In contrast to 3D buildings, a lot of datasets on city-relevant topics has been made available in the public domain recently by open data initiatives, e.g., [33]. It would be desirable to include this data in STEPWISE as another aid for the user. A lot of infrastructure information [34] is also present in the OpenStreetMap [35] data. However, the current interface does not allow the direct exploitation of it in the cartography view of STEPWISE. These are definitely valuable approaches to make the most use of the public data and support the decision-making process for LEA which could be included in future releases of the STEPWISE platform.

As a last consideration, it is worth noting that not only LEA may profit from STEPWISE. As pointed out in [2], most soft targets are privately owned venues such as entertainment parks, commercial centers, private academic institutions, stadiums, and museums. Some incidents may be too insignificant or out of scope for a LEA to intervene. However, they may still pose a risk to the security and safety of people. STEPWISE may also be of interest for soft target owners to devise their security and response plans. As it is quite commonplace knowledge nowadays, preparedness, which may be ensured on all levels, helps raising awareness and minimizing confusion, thus ultimately and generally leading to better reactions in critical situations.

22.6   Conclusions

The protection of soft targets is a tremendously hard task due to the variety of public spaces and economic considerations, while at the same time the open and welcoming character of the site needs to be maintained. There is currently no common methodology that helps LEAs to get to a reasonable protection plan in a structured way, although the pressure to provide effective security measures in the face of the terroristic attacks that Europe has experienced in the last decades is clearly high. All of this points to the fact that better decision support is needed to create security designs which make effective use of available resources. The STEPWISE project aims at filling this gap by providing a methodology for systematic security planning together with supporting software tools. STEPWISE provides 3D visualization of public spaces, which advances not only the current pen and paper-based planning of most LEAs but also minimizes confusion regarding locations, thus enhancing the overall quality of the security discussion. This collaborative tool ultimately allows to jointly formulate security designs without a physical presence at the real venue. In addition, 3D buildings can be created rapidly by means of an easy to use modelling application, and are then used in the toolkit to recreate the venue in the 3D environment. Moreover, the users can familiarize themselves with the site and gain related competence by improving their skills inside the STEPWISE VR training module. In brief, STEPWISE is a decision support toolkit that could significantly aid in the protection of soft targets for LEAs.

Acknowledgments This project has received funding from the European Union's Internal Security Fund – Police under grand agreement no. 815182 (STEPWISE). The STEPWISE Project is available at the following URL: https://www.stepwise-project.eu/

REFERENCES

1. Kalvach, Z. (2016). *Basics of soft targets protection – guidelines* (2nd ed.). Prague: Soft Targets Protection Institute.
2. Larcher, M., & Karlos, V. (2018). Protection of public spaces. In *BauProtect 2018*. Munich.
3. Australia-New Zealand Counter-Terrorism Committee. (2017). *Australia's strategy for protecting crowded places from terrorism.* Canberra: Commonwealth of Australia.
4. Gorry, G. A., & Morton, M. S. S. (1971). A framework for management information systems. *Sloan Management Review, 13*(1), 55–70.
5. Carlsson, C. (2018). Analytics mobilized with digital coaching. *Intelligent Systems in Accounting, Finance and Management, 25*(1), 3–17.
6. Morton, M. S. S. (1971). *Management decision systems: Computer based support for decision making* (Doctoral dissertation). Harvard University, Cambridge.
7. Sprague, R. H. (1980). Framework for the development of decision support systems. *MIS Quarterly, 4*(4), 1–26.
8. Keen, P. G. W. (1981). Information systems and organizational change. *Communications of the ACM, 24*(1), 24–33.
9. Sprague, R. H. (1981). Decision support systems: A tutorial. In D. Young & P. G. W. Keen (Eds.), *DSS-81 transactions* (pp. 193–203). Atlanta.
10. Armenia, S. (2019). Smart model-based governance: Taking decision making to the next level by integrating data analytics with systems thinking and system dynamics. In S. E. De Falco, F. Alvino, & A. Kostyuk (Eds.), *New challenges*

in corporate governance: Theory and practice (pp. 41–42). Naples: Virtus Interpress.

11. Dewey, J. (1910). *How we think*. Lexington: D.C. Heath & Co..

12. Simon, H. A. (1977). *The new science of management decision* (2nd ed.). Englewood Cliffs: Prentice Hall.

13. March, J. G. (1994). *A primer on decision making: How decisions happens.* New York: The Free Press.

14. Choo, C. W. (2005). *The knowing organization: How organizations use information to construct meaning, create knowledge, and make decisions* (2nd ed.). New York: Oxford University Press.

15. Podolak, I. (2015). *How health organizations can apply scenario based planning and analytic technologies to make strategic consequential decisions in a time of uncertainty* (Doctoral Dissertation). Brock University, Oakville.

16. Schwaninger, M. (2008). *Intelligent organizations: Powerful models for systemic management.* Heidelberg: Springer.

17. Edwards, W. (1961). Behavioral decision theory. *Annual Review of Psychology, 12*(1), 473–498.

18. Kahneman, D., & Tversky, A. (1979). Prospect theory: An analysis of decision under risk. *Econometrica, 47*(2), 265–291.

19. Klein, G. (2008). Naturalistic decision making. *Human Factors, 50*(2), 456–460.

20. Argyris, C., & Schön, D. A. (1978). *Organizational learning: A theory of action perspective.* Massachusetts: Addison-Wesley.

21. Podolak, I., Ayanso, A., Connolly, M., Law, M., & Cosby, J. (2017). Convergent analytics and informed decision-making: A retrospective multimethod case study project in Kenya. *Health Policy and Technology, 6*(2), 214–225.

22. Rajterič, H. (2010). Overview of business intelligence maturity models. *Journals of Contemporary Management Issues, 15*(1), 47–67.

23. Lindeman, T. (2017, December 5). *A human alternative to pattern recognition.* https://www.linkedin.com/pulse/human-alternative-pattern-recognition-tim-lindeman-%E6%9E%97%E5%BE%B7%E5%A8%81-/. Last accessed 2020/10/26.

24. Fern, A., Natarajan, S., Judah, K., & Tadepalli, P. (2014). A decision-theoretic model of assistance. *Journal of Artificial Intelligence Research, 50*, 71–104.

25. Rodič, B. (2017). Industry 4.0 and the new simulation modelling paradigm. *Organizacija, 50*(3), 193–207.

26. Vihinen, M., & Samarghitean, C. (2008). Medical expert systems. *Current Bioinformatics, 3*(1), 56–65.

27. ISO 31000 Risk Management. https://www.iso.org/iso-31000-risk-management.html. Last accessed 2019/12/11.

28. DESURBS – Designing Safer Urban Spaces. https://cordis.europa.eu/project/id/261652. Last accessed 13/12/2019.

29. ISO 16739-1:2018 Industry Foundation Classes (IFC) for data sharing in the construction and facility management industries – Part 1: Data schema. https://www.iso.org/standard/70303.html. Last accessed 2019/12/11.
30. Autodesk, Building Information Modeling, White Paper. http://images.autodesk.com/apac_grtrchina_main/files/aec_bim.pdf. Last accessed 2019/12/11.
31. FreeCAD Homepage. https://www.freecadweb.org/. Last accessed 2019/12/11.
32. Sculpteo Blog – CAD vs 3D modeling software: what is the difference? https://www.sculpteo.com/blog/2019/03/19/cad-vs-3d-modeling-software-what-is-the-difference/. Last accessed 2019/12/11.
33. EU Open Data Portal. https://data.europa.eu/euodp/en/home. Last accessed 2019/12/11.
34. Open Infrastructure Map. https://openinframap.org/about.html. Last accessed 2019/12/11.
35. OpenStreetMap. https://www.openstreetmap.org/. Last accessed 2019/12/11.

CHAPTER 23

Trials: New Method of Assessing Innovative Solution in Crisis Management

Joanna Meitz, Jakub Ryzenko, Marcin Smolarkiewicz, and Tomasz Zwęgliński

23.1 INTRODUCTION

Current and future challenges due to increasingly severe consequences of natural disasters and terrorist threats require the development and uptake of innovative solutions that are addressing the operational needs of practitioners dealing with Crisis Management. Project DRIVER+ (Driving Innovation in Crisis Management for European Resilience) [1] focusing on delivery full service as a completely new environment (test-bed), which supports practitioners in finding (Portfolio of Solutions), testing (Trial)

J. Meitz (✉) · J. Ryzenko
Space Research Centre of Polish Academy of Sciences, Warsaw, Poland
e-mail: jmeitz@cbk.waw.pl; jakub.ryzenko@cbk.waw.pl

M. Smolarkiewicz · T. Zwęgliński
The Main School of Fire Service, Warsaw, Poland
e-mail: msmolarkiewicz@sgsp.edu.pl; tzweglinski@sgsp.edu.pl

and assessing (Trial Guidance Methodology) innovative crisis management solutions.

DRIVER+ project has three main objectives:

1. Develop a pan-European test-bed for crisis management capability development:
   - Develop a common guidance methodology and tool (supporting Trials and the gathering of lessons learned.
   - Develop an infrastructure to create relevant environments, for enabling the Trialling of new solutions and to explore and share CM capabilities.
   - Run Trials in order to assess the value of solutions addressing specific needs using guidance and infrastructure.
   - Ensure the sustainability of the pan-European test-bed.
2. Develop a well-balanced comprehensive Portfolio of Crisis Management Solutions:
   - Facilitate the usage of the Portfolio of Solutions.
   - Ensure the sustainability of the portfolio of tools.
3. Facilitate a shared understanding of crisis management across Europe:
   - Establish a common background.
   - Cooperate with external partners in joint Trials.
   - Disseminate project results.

23.2   TRIAL GUIDANCE METHODOLOGY

The Trial Guidance Methodology (TGM) has been applied in four Trials and systematically evaluated [2, 3]. The TGM give step-by-step guidelines to carry out robust assessment of the solutions through recommendations from the preparation phase until evaluation results [4]. Trial Guidance Methodology offers several tools to support and to enable its implementation as:

Training Module – providing education, practice and assignments via e-learning and face-to-face workshops. Modules cover all aspects of organising a Trial and are delivered as a complete training package.

Trial Guidance Methodology Handbook – it's a practical guide providing detailed explanation of the preparatory six-step approach and the execution and evaluation phases.

Trial Guidance Tool – a web-based software tool developed to support Trial owners and high-level crisis managers in the implementation of the TGM through the Trial phases.
Trial Action Plan – a comprehensive co-working template and checklist to plan and prepare a Trial. Records efforts, circulates decisions and aids assessing progress.

The TGM describes all steps and needed roles to prepare a Trial. TGM consists of three distinct but connected phases:

Preparation phase: The objective of this phase is to design a Trial. The design follows an iterative and nonlinear six-step approach. It starts with the identification of the objectives and the formulation of research questions. In the Trial, there is need to address the questions through an appropriate data collection plan as well as through evaluation approaches and metrics to analyse the data collected during Trial. To do this, realistic scenarios must be developed, and solutions to be trialled must be selected to figure out if they can be innovative.
Execution phase: This phase is much more than just the actual Trial. Before execution, there is a need to check if everything what is needed to gather relevant data is prepared. After checking and testing, everything is ready to run your Trial.
Evaluation phase: This phase amounts to a systematic assessment of the potential added value of the solutions that were trialled. When the analysis is done, the next step is to sum up the results, providing evidence of the impact of the solutions, and to disseminate the results within and beyond of crisis management community (Fig. 23.1).

Trial Guidance Methodology also defines roles and responsibility between expert responsible to prepare a Trial. The whole team is called *Trial Committee* and splits to a several different roles to cover all aspects needed to create, execute and evaluate a Trial. Trial Owner is responsible for the overall management and success of the Trial. Trial Owner is heading and coordinating the Trial Committee and its coordination meetings. Ideally, he is a member of organisation that is the "gap owner". The Trial Host is the provider of the premises the Trial takes place and is responsible for the infrastructure and the support. Solution Coordinator is the organiser of the socio-technological solutions applied in the Trial and hence responsible for the integration and deployment of the solutions. The

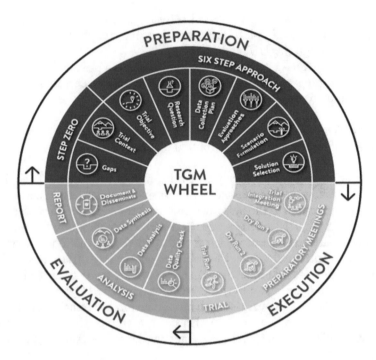

Fig. 23.1 Trial Guidance Methodology wheel. Phases and steps to prepare Trial [4]

Test-bed Guidance Coordinator ensures the correct utilisation of the Trial Guidance Methodology in the Trial. Trial evaluation coordinator is to ensure a high evaluation quality; the evaluation coordinator needs to carefully question and verify the overall test-bed application from the very beginning up to the end of a Trial. The test-bed infrastructure coordinator is heading the technical test-bed support for the Trial. He is responsible to deliver the complete technical test-bed infrastructure connecting the solutions with the Trial participants and other platform systems provided by the Trial Host.

23.3 VALIDATION OF TRIAL GUIDANCE METHODOLOGY

23.3.1 Evolution of the Methodology

During the project Trial Guidance Methodology was evaluated, and feedback was shared and implemented in the next versions. Methodology is still under development, and validation process is still ongoing, but first outcomes and results look promising and showing a good approach in this process.

The time between the first two Trials was intense and entailed an in-depth assessment of the lessons identified and learned. The main feedback in interest of TGM developers was a question whether methodological approach was helpful to assess potential innovative solutions for CM development.

While it turned out that the TGM is appropriate for this specific purpose, in the Trial Committees, specific needs emerged, and a significant amount of support was deemed necessary to evaluate the solutions in a systematic way.

The relevant stakeholders involved in Trials were not always familiar with unavoidable processes and steps that lead to Trials. To make sure that implementation of TGM is going well and learning process is processing on the both sides – practitioners and TGM developers – several activities was to gather feedback, and lessons learnt was taken. The main was a focus group discussion after each Trial to gather all thoughts from Trial Committee. Also during implementation on each phase, several internal meetings, informal discussions and working sessions with the Trial owners have been done.

To collect data required for Test-bed Guidance Methodology evaluation used for preparation, execution and evaluation phases, and to demonstrate the level of achievement and success reached in the Trials, validation of TGM was done in a frame of focus group discussion after Trial execution and a set of questioners after finalising each of the phase.

Collection of data for validation of the test-bed methodology approach was done from evaluation of:

Guidance Methodology Handbook – guide how to design and evaluate Trials.
Guidance Tool – support implementation of the TGM.

Trial (demonstration) infrastructure where stakeholders collaborate in testing and evaluating new CM solutions.

It was important that the data has to be acquired and analysed in the way to receive results and formulate conclusions regarding the test-bed parts mentioned above independent to:

V1. Type and level of quality of solutions which were evaluated during the Trial
V2. Crisis management procedures (or their variants designed only for the purpose of the Trial) supported by solutions
V3. Level of competence of practitioners and end users being involved in the Trial
V4. Level of competence of experts being involved in the preparation and execution of the Trial

In drawing conclusions, it was taken into account that the two groups (the practitioners involved in the project and the team designing and implementing the Trials) may have different levels of competence including not only the level of professional, but also different knowledge of the usability and implementation of the methodology (TGM) and technical or organisational skills.

23.3.2   General Approach to TGM Evaluation

From the perspective of an idealistic general approach to formulate conclusions independent on variables V1, V2, V3 and V4, data collected during evaluation of the test-bed methodology would be cross-analysed. Results of experts' estimations from qualitative survey would be transformed to quantitative parameters to use statistical methods to obtain reliable conclusions about the effectiveness and improvement of the test-bed methodology [5]. On Pic. 1 the idea of approach to the evaluation method for analysing an effectiveness and improvement of the Trial Guidance Methodology is presented. On Tables 1–3 the scheme of idealistic general survey for evaluation of the test-bed methodology approach consisting of the Guidance Methodology Handbook, the Guidance Tool and Trials 1–4 infrastructures (environments) is presented (Fig. 23.2).

The concept of idealistic survey for evaluating the Guidance Methodology Handbook component of the test-bed is based on two ways

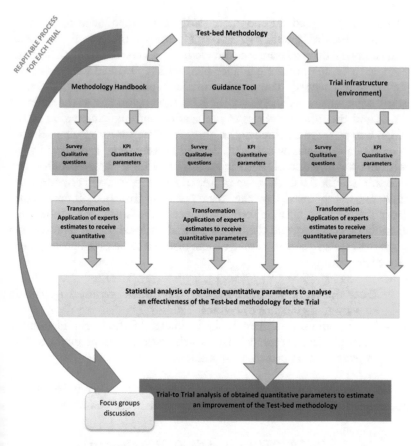

Fig. 23.2 Approach of the evaluation method for analysing an effectiveness and improvement of the test-bed methodology [5]

of measurement: subjective and objective observations as qualitative methods. For subjective observations making a separate analysis among groups (Trial owner's crew, end users and practitioners, solution providers, Guidance Tool designers, test-bed infrastructure users (coordinator's crew), Guidance Methodology Handbook creators (to allow comparison between group of creators of the Handbook and its users)) involved in preparation and execution of the Trial was planned. The research methods planned to be used were individual depth interview (IDI) with additional technique (extended questionnaire with open-ended, semi-open-ended and closed-ended questions). For this purpose, the Key Performance Indicators (KPIs) relevant to the design process (improvement from Sample to Sample) of the guidance methodology manual were observed, defined and analysed. The research methods planned to be used were a statistical comparative analysis of the KPIs (corresponding to the different Trials and the different stages of preparation and implementation of the Trial) and a statistical comparative analysis of the survey results (selected questionnaire interview answers). The KPIs were to be measured by (or with the participation of) specific groups involved in the preparation and execution of the Trial (the team of the Trial owner, solution providers and end users supported by the team of the Test-bed Infrastructure Coordinator, the Guidance Tool developer and the Guidance Methodology Handbook creators), and then collected and analysed. Based on these two approaches, also general questions were planned to be asked groups mentioned above:

- Does the Guidance Methodology Handbook provide a systematic step-by-step guidance to conduct Trials?
- What is an improvement of the Guidance Methodology Handbook from the last executed Trial (which stages/parts of methodology implementation process were influenced the most and how)?

What is the advantage of using the Guidance Methodology Handbook for evaluating CM solutions within an appropriate environment?

23.3.3   Elements of the TGM Evaluation

The general approach presented in section 23.3.2 Evaluation of the Testbed Methodology was idealistic in nature and after the Evaluation Team realised that the amount of data to be collected was too large due to time constraints and the fact that the elements of the Testbed Methodology were not complete and would be developed step by step in a learning by doing approach, with each subsequent Trial it would become more and more mature, it was decided to change the concept. Therefore, a more simplified procedure based on an idealistic general approach was planned and carried out [5]. The basis of this procedure and the proposal of realistic Trial conditions for the evaluation of the Test-bed Guidance Methodology are presented below, taking into account:

- What elements of the Test-bed Guidance Methodology can be evaluated
- Who may be questioned to collect data necessary for evaluation
- When evaluation data can be collected, together with conclusions regarding the connection of these aspects with each evaluation process described in Trial Guidance Methodology

Finding the set of initial KPIs measuring the performance of the Test-bed Guidance Methodology against existing methodologies was challenging regarding the fact that there is no standardised (baseline) methodology for crisis management capability development based on systematically conducted Trials and evaluation of solutions within an appropriate testing environment. In this area DRIVER+ is an innovative and pragmatic approach. Additionally differences existing among CM systems functioning in different EU countries together with dissimilarities among organisational systems of certain services (fire brigade, police, emergency service, etc.) make the comparison of KPIs' values very difficult and less reliable.

It is therefore proposed to focus on measuring subjective KPIs – parameters that describe (using a limited scale) the subjective assessment that users of the Test-bed Guidance Methodology perceived of the components of the actual performance indicators. The purpose of this measurement is to answer the question: "To what extent does the Test-bed Guidance Methodology improve performance in:

- Resource Management (both human and physical),
- Time management, Financial Management, compared to the methodology used in your country or organisation?".

Subjective KPIs can be measured using the individual depth interview (IDI) with use of additional technique.

Concluded conditions of the realistic survey for evaluating the Test-bed Guidance Methodology were as follows. Test-bed as a whole (Trial Guidance Methodology Handbook (TGM), Trial Guidance Tool (TGT), Trial (test-bed) infrastructure (TBI)) was evaluated throughout (individual depth interview (IDI) with use of additional technique (extended questionnaire), focus group and individual interview (InI). Each component of the test-bed methodology was evaluated by using certain techniques (templates) corresponding to the features of the component. In individual depth interview (IDI), Focus Group evaluation, participants, validators and observers took part and for individual interview (InI) Guidance Tool designers, Trial Guidance Methodology Handbook developers and test-bed infrastructure coordinators.

The assumption for data collection was that answers for certain questions collected during surveys (IDI) from validators together with impression measure (InI) among partners responsible for creation of each component of the Test-bed Guidance Methodology will give an input to the evaluation of the methodology. Answers for certain questions collected during surveys (IDI) from participants and observers will give an input to the Evaluation of the Trial. And answers for certain questions collected during surveys (IDI) from validators and observers will give an input to the subjective KPIs analysis for the purpose of the Evaluation of the Trial.

After each step of the survey results were discussed with focus groups and shared among DRIVER+ consortium members. Workshops after each Trial were executed to discuss outcomes of the Trial.

23.3.4   First Outcomes

Conducted discussions, meetings, individual depth interviews as a focus group discussions and collected data from the questioners at the each stage of preparation of each DRVER+ Trial shed light on a major need, having much more pragmatic guidelines to understand better who, when and how a task should be performed. While the TGM supporters responded timely and effectively to requests, it became clear that the initial version of the TGM [2] was not user-friendly and the delivery of a manual could not wait until October 2019, as originally planned. After feedback TGM developers decided to provide a handbook 9 months before the actual schedule.

Identifying one prominent aspect in this process was challenging as it was more a combination of different elements that led to the decision of designing a handbook early enough to be used in Trials 3 and 4. However, two considerations played a major role. First and foremost, what was missing in the first version of the methodology was a comprehensive reference to the pan-European test-bed as a whole. The rationale behind the TGM was explained along with the design, but the interrelation with the tools (intended in the broad sense of the term) was not captured in the first version. The reason is that the complexity of the relations and dependencies between DRIVER+ artefacts emerged mainly in the Trials. The necessity to offer the "full picture" through a less fragmented didactical approach was revealed both in the TGM team and when dealing with Trial owners. The second version of the methodology could not focus only on phases and steps; instead it should convey a vision of the conditions and the context in which a Trial can be carried out. The test-bed, in all its complexity, demanded more attention. While acknowledging that other deliverables are available to dig deep into, for instance, technical features, the TGM is the glue that keeps the pieces of the puzzle together. Hence, it was recognised as important to outline not only, for instance, the six-step approach in the preparation phase but also to indicate which tools can or should come into play in each step.

Second, having described the foundations of the methodology, the "how to put it into practice" needed to be rethought. To put something in practice, clear-cut information and answers were needed. With this in mind, it was decided to keep the explanations at a rather general level but to provide straightforward directions on who, what, how and when (also in terms of amount of time needed to carry out a task). Additionally, methods, tools, inputs and outputs should have been visibly indicated. To achieve this goal, a new layout was necessary as well as new ways of presenting content. After the decision of creating a handbook earlier each month after gathering more feedback, new versions appear. The newest version is available on public project DRIVER+ website. As the evaluation phase for last Trial is still ongoing, the last version of the handbook taking into account all lessons learnt and data analysed after all Trials and also validation of Trial Guidance Methodology results will be published in the reports and available at the end of April on the DRIVER+ webpage.

23.4 Trial Guidance Methodology in Practice: How to Create and Execute a Trial [5]

During the DRIVER+ lifetime, four Trials were organised and executed. Trial 1 in May 2018 in Poland, Trial 2 in October 2018 in France, Trial 4 (*The Netherlands Trial*) in May 2019 in the Netherlands and Trial 3 (*The Austrian Trial*) in September 2019 in Austria. Trials were planned in this order to enable also validation of the Trial Guidance Methodology as described in Sect. 23.3.2 [5].

As Trials have a scientific approach to measure improvements of trialled solutions into crisis management in objective way, below this approach and overall outcomes as a lesson learnt implemented till now into Trial Guidance Methodology after execution of four Trials are presented.

23.4.1   Step Zero

To create a Trial as regards Trial Guidance Methodology, there is a need to identify specific capability gaps and/or problems. Trial address specific crisis management gaps, for DRIVER+ Trials, these were gaps like the cross-border tasking and resource management, the high-level coordination, the volunteer management and the situation assessment and logistics. Gaps were identified and validated by practitioners during *gap assessment workshop* to make sure that planned Trials will correspond with practitioners needs across Europe.

After gap validation, discussion with practitioners was made to identify also interesting scenarios – a *Trial context* in which selected gaps would best reflect the needs of practitioners. For DRIVER+ Trials, gaps mentioned above were implemented in the following scenarios:

- Trial 1– cascading effects of a flood
- Trial 2 – multiple incidents caused by a scale forest fire
- Trial 3 – an earthquake
- Trial 4 – a flood and a power outage

After the identification of *Trial context* to create the DRIVER+ Trials, a *baseline* has been created for each of them. *Trial context* refers to socio-cultural and legal characteristics of the context in which the Trial is carried out (roles, responsibilities, legal constraints); also constraints should be taken into account. The *baseline* is a depiction of the as-is process that includes all roles, actions and information exchanges (including the means by who they are done) [6]. For DRIVER+ Trials, business process modelling notation (BPMN) was used.

When the *Trial context* and *baseline* were ready, *Step Zero* was finalised and the second part of the *preparation phase – Six-Step Approach –* started.

23.4.2   Six-Step Approach

Six-step approach is an iterative and nonlinear process including all activities to design a Trial. To define objective each of the Trial Owners use the SMART formulation. SMART stand for specific, measurable, achievable, reasonable and time-bound.

After first step, the next was to define *research questions*. The goal of defining research questions was to generate robust results regarding the added value of trialled solutions. This means that through defining research question and taking account Trial context, Trial owner with evaluation coordinator identifies research methods and data analysis techniques which will be used to measure impact of trialled solutions on crisis management.

To measure impact of trialled solutions on crisis management, collection of relevant data is important. To do this there are two things to need to be taken into account: which kind of data needs to be collected and how. The main part is to determine measures and metrics (KPI's) which can give the answer for the research questions. As different objectives and research questions were defined for each of the DRIVER+ trials, it is difficult to provide one best way. In each of DRIVER+ Trials, it was decided that observers will collect relevant data by observer support tool (OST) and paper questionnaires/checklist to support collected information from test-bed technical infrastructure. This approach is very practical because it is giving context for raw data collected by machines. Observers always can give a comment to what they observed, and this can give reasoning as *why practitioners reaction/decision was quicker or slower in specific situation.* Also, they are capturing practitioner-solution interaction.

There is a very important thing that needs to be highlighted here: observers and all measurements are not made to check practitioners and checking the correctness of their decisions during the Trial. Observers and all measurements are to collect the relevant data to check if the solution used by practitioners actually answers their needs.

DRIVER+ project creates a list of generics KPI's for three dimensions: Trial, solution and crisis management which are available in Trial Guidance Tool.

- Crisis Management Dimension – the most important dimension, as is the part where it is measured if new solutions have an impact on defined gaps.

- Trial Dimension – relates to the Trial organisation aspects for example the Wi-Fi connection, the number of participants or any technical issue.
- Solution Dimension – tackles all functionalities as well as the usability, novelty, simplicity, etc. of each sociotechnical innovation.

To make sense of the data collected during Trial, there is a need to identify different techniques and tools which can help in analysis. In most of the cases, the data analysis software is needed, but in noncomplex cases, basic software as Excel is enough. Everything depends on data collection plan and kind of data itself. In this step it is good to have a discussion with an expert to fit the best solution. On this stage it is important to make sure that the data which is planned to be collected during evaluation phase by using appropriate tools will be easy to draw conclusions. For DRIVER+ Trials it was decided that data analysis software will be used because of the big amount of different (logs from solutions, logs Inside8 test-bed, observers observations, feedback in questionnaires form from practitioners) data collected.

After creation of *Trial context*, all requirements are known – kind of Trial (table-top, field, etc.) and on which level (bronze, silver, gold). Scenario formulation is a process of creation specific situation which is needed to trigger gap. In this way the process of creation of innovation line is starting. Here is the first step of identification where impact of innovative solution is expected.

- Innovation line – the scenario line where innovative solutions (in first phase expected functionalities/futures) are connected with the story based on baseline.

For DRIVER+ Trials process of selection solutions was complex and detailed [6].

When potential set of solution is found, to make confirmation that solution is corresponding with a practitioner need is good to execute practitioner-centred review based on pre-assessment criteria developed by multidisciplinary CM practitioners. The second part is on the hands of Trial Committee as actual selection of the solutions, which includes future Trial-related consideration, like relation to gaps or the technical requirements. Below is presented set of different criteria in three-dimension approach used for DIVER+ Trials:

1. Can the solution be used to address the initial gap and to provide an answer to the main research question of the Trial?
2. Required skills of end users to use solution.
3. Functional adequacy with Trial scenario.
4. Setup.
5. Trial timing.
6. Adaption.
7. Information exchange.
8. Deployment.
9. Test-bed integration.
10. Assessment/evaluation/measurement.
11. Price.
12. Other aspects.

To make future selection easier, project DRIVER+ created Portfolio of Solutions (PoS) which is a webpage with still increasing number of various solutions responding on different crisis management gaps and CM functions. PoS includes all needed information at the beginning of the solution searching and preselection with information from evaluation from past Trials.

23.4.3 Execution Phase

The Trial Integration Meeting (TIM) aligns the perspectives of the practitioners, solution providers and Trial Committee. To draft the later Trial script, the participants discuss the integration of solutions into the practitioners' operations, the required information exchange as well as the data collection and evaluation criteria to address the Trial objectives.

Dry Run 1 is a first run of a Trial to check Trial design and test-bed technical infrastructure at the location where the actual Trial will take place. This is a step when all Trial organisers (Trial Committee, solution providers and other staff) checking if six-step approach were implemented correctly and everyone knows what is expected. The readiness of the solutions is checked by running a use cases focused on the functionalities which will be used during actual run and not on the scenario storyline.

This step contains the final tests and adaption of each Trial sub-system and should end with a complete Trial dry run.

From scenario perspective all injects need to be checked and tested if they triggered the gap. During checking of injects, first test for data collection should be made, to be sure if test-bed technical infrastructure, the solutions and/or observers will be able to "catch" data needed for further evaluation.

On this stage changes are acceptable if they do not change the main objectives of the Trial. It's recommended to do a to-do list with clear assignments and start the preparation of Dry Run 2.

The second run of a Trial (Dry Run 2) is a general test in preparation phase. As for the Dry Run 1 all test and Trial design need to be tested at Trial location to check also the *platform*. Everything that was changed after Dry Run 1 needs to be checked if implementation of those adjustments is feasible and correct. Also training on solution for the practitioners needs to be run. There are two ways to make a full rehearsal of a Trial, with and without practitioners, which will be players during actual Trial. It is not always possible to invite the same practitioner or observer to two meetings (Dry run 2 and Trial) as this requires their full commitment for at least 2 days per meeting and practitioners often have many other commitments. In this case, TGM allows to invite less experience practitioners as participants or to increase their (participants) number in case to ensure substitution. This solution implicated the need for a deeper solution and observer training before Trial run. Dry Run 2 as a full rehearsal of a Trial needs to be prepared as an actual event with all problems and bugs fixed. Scenario story, injects and data collection plan need to be ready and fully run.

The main goal of this Dry Run 2 is to ensure that all data can in fact be collected. Hence the main focus needs to be on the observer support tool, the data collection through solutions and test-bed technical infrastructure and that the participant questionnaires are ready and understandable. If something is not working, analyse if there is really a need to implement this and possibility to have the extra effort in getting it up and run.

After Dry Run 2, no changes can be made. Time until the Trial should be allocated for dissemination and communication activities, preparation of all necessary documents, instructions, presentations, etc. needed to promote the Trial as well to carry out the introduction, trainings (solution, OST, safety, etc.) and the Trial itself.

Trial run: This step describes execution of a Trial. The most important thing is to make sure during whole execution that all kind of data described in the data collection plan is collected. Also, solution training is important to give everyone enough time to familiarise the functionalities of the solutions. If some problem will appear as one of the solutions needs to be restarted, the Trial needs to be stopped; this is a Trial not an exercise.

23.4.4 Analysis

This task can be called a Trial evaluation, and it covers the evaluation of the level of achievement reached in each Trial regarding the planed goals. All measurements are made in three dimensions: solution dimension, Trial dimension and crisis management dimension. Each dimension can be analysed alone and also in relation to the others as they can interfere between themselves and in most of the cases can give the context of measurement. As the aim is to assess a solution in relation to a CM gap, it is very important to see how this was (maybe even negatively) influenced by the Trial or solution dimension. For example, it could be that a solution is very well capable of addressing the CM gap; however, during the Trial a breakdown of the system can occur due to a technical problem within the Trial location (Trial dimension). In this case the participants cannot see the whole potential of the solution. This is very important to consider during the analysis and evaluation and to ask how these disruptions influenced the overall setup and data collection.

After a Trial different kinds of data with various means (observer, test-bed technical infrastructure, questionnaire, etc.) were collected according to data collection plan. In this step it's needed to put all data in the same format to make it readable and to check deviations. During data quality check, deviations are the main points to identify. To be able to identify those deviations is good to answer for few questions: Is there data missing or broken? If so, is this data critical? If so, think of ways to regain it (repair or maybe there is possibility to ask a participant to fill in a dedicated questionnaire). Even if it is not critical, it is strongly advised to indicate where data is missing in evaluation.

After checking, the deviation data need to be structured. The structure can be extracted from data collection plan, for example, according to role, solution, research question or three dimensions: solution, Trial and CM.

Structured data need to be checked again to start to find patterns and second checking of missing data. To be able to go to the next step is good to look for things that don't fit those patterns. Check why they don't fit. Answering for questions also can help: Are there strong deviations? If so, is it recommended to find more data related to the aspect (maybe in the test-bed technical infrastructure)?. If there is no way to improve the data, indicate in the evaluation that the conclusions on this can only be limited. After answering questions the data set for analysis needs to be created, excluding irrelevant or poor-quality data but with indication why this was done.

After structuring data and patterns, identification data need to be visualised. Furthermore data is put in a first relation to defined KPIs:

Structure – start with the sessions of a Trial, the three dimensions and outcomes for the solutions.

Aggregate and visualise data – create relevant graphs or pie charts. First match to KPIs and metrics.

Patterns – what is standing out? Second match to KPIs and metrics. Here first conclusions may appear and give the trigger to dig deeper to see if assumptions turn into facts or into unexpected phenomena.

The data gathered and already analysed now needs to be put into the right context. A three-dimensional approach can be helpful here to see how gap has been addressed and what more needs to be done to reach that goal. It's good to have a discussion with the practitioners without showing them first conclusions. To make it easier, it is good to answer questions:

What stands out? What results are remarkable?

Are these results expected? Why or why not?

What are possible explanations for these results? Can putting them in relation to each of three dimensions be helpful?

What can be concluded based on these results? (Was the gap bridged? At least partly?)

Are the results generalisable to other teams/contexts?

What advice can be provided about the solution? Did it address to gap as expected?

23.4.5   Reporting

At the end of the Trial, it is recommended to put all structured data into document and create something sustainable. With this process Trial action plan as a structured document used trough whole process from the preparation till evaluation phase can help. It is good to write down lessons learnt with regard to Trials, etc. – for conducting Trials, for crisis management, for organisation, etc.

It is strongly advised to organise a meeting with all people involved in Trial to talk about results and discuss way forward. All results in relation to the Trial can be placed in the Trial Guidance Tool and be available for each solution, and all post-assessment information can be added for each solution in the DRIVER+ knowledge base and Portfolio of Solutions.

The main outcomes from four DRIVER+ Trials are available in the PoS and in more elaborated version will be available at the end of April 2020 on DRIVER+ webpage as the report D941.31–SP94 overall evaluation of the Trials and final demo [7].

23.5   Conclusions

To facilitate the uptake and implementation of results, also after the project duration, DRIVER+ is supporting the establishment of Centres of Expertise (CoE). A CoE is a practitioner-centred organisation that has close relations with (applied) research organisations, solution providers and policy-makers. A CoE acts as the primary contact point for practitioner organisations at the national or regional level, supports them in using Trial Guidance Methodology and other tools, maintains and updates these tools and exchanges lessons learned between other Centres of Expertise in the various European Member States. CoEs will provide a variety of services for practitioners interested in assessing innovative solutions, offering for instance access to testing locations (platforms) and support in TGM and test-bed implementation. It is expected that one of CoEs will be established in Poland, most probably as a cooperation between the Main School of Fire Service and the Space Research Centre of Polish Academy of Sciences.

Acknowledgments The DRIVER+ project has received funding from the European Union's Seventh Framework Programme for research, technological development and demonstration under Grant Agreement n° 607798.

References

1. The DRIVER+ project has received funding from the European Union's Seventh Framework Programme for research, technological development and demonstration under Grant Agreement n° 607798.
2. DRIVER+ report. D922.21 – Trial guidance methodology and guidance tool specifications (version 1).

3. DRIER+ report. D922.41 – Trial guidance methodology and guidance tool specifications (version 2) .

4. DRIVER+. (2019, Oct). *Trial guidance methodology handbook*, https://www.driver-project.eu/wp-content/uploads/2019/11/TGM-Handbook_final-min.pdf (15 December 2019).

5. Internal DRIVER+ document, prepared as the base for the DRIVER+ report: D941.31 *SP94 Overall evaluation of the Trials and final demo.*

6. D942.11 – Report on review and selection process, March 2019 https://www.driver-project.eu/wp-content/uploads/2019/02/DRIVER_D942.11_Report-on-review-and-selection-process_removed-solution-details.pdf   (15 December 2019).

7. D941.31 – SP94 Overall evaluation of the Trials and Final Demo, April 2020.

CHAPTER 24

EU Cities Vulnerability Assessment

Georgios Kioumourtzis, Patrick Padding,
Peter Van de Crommert, Pilar De La Torre,
Tatiana Morales, P. E. (Puck) van den Brink,
Jeroen van Rest, Graeme Voorthuijsen,
Ioannis Chasiotis, and Anna Gazi

24.1 Introduction

For decades, terrorism has been a reality in many European countries and a continuous threat to a great number of European cities. It seriously threatens the safety, the values of democratic states and the rights and liberties of citizens. Terrorism brought about long-term negative effects for cities and high social costs, not only from a financial but also from a

G. Kioumourtzis (✉) · P. Van de Crommert
DITSS - Dutch Institute for Technology, Safety & Security,
Eindhoven, The Netherlands
e-mail: georgios.kioumourtzis@ditss.nl; peter.vandecrommert@ditss.nl

P. Padding
National Police of Netherlands, Central Unit,
Driebergen-Rijsenburg, The Netherlands
e-mail: patrick.padding@politie.nl

© The Author(s), under exclusive license to Springer Nature
Switzerland AG 2021
B. Akhgar et al. (eds.), *Technology Development for Security
Practitioners*, Security Informatics and Law Enforcement,
https://doi.org/10.1007/978-3-030-69460-9_24

psychological point of view in the sense of an increased feeling of insecurity among locals and visitors [1].

Over the years, strategies to protect public space against terrorism have strengthened and evolved, mainly focussing on protecting critical infrastructures. However, terrorist attacks are evolving as well. By adapting to new contexts and opportunities, lately public space has turned into an attractive target for terrorist attacks. To illustrate, the latest terrorist attacks in European cities such as London (Westminster attack June 2017), Paris (Bars, Stade de France, Bataclan, November 2015), Manchester (Manchester arena bombing, May 2017), Stockholm (lorry attack April 2017), Berlin (Christmas market attack, 2016), Brussels (airport bombing, March 2016) and Barcelona (Van attack, August 2017) have occurred in public areas. These areas are considered as "soft targets". This means that crowded public places including the metro, shopping centres, sports stadiums, bars, restaurants, clubs and commercial sidewalks, easily accessible to the public, constitute an easy target for terrorists to do great harm. As per [2] a "soft target" is a person, thing or location that is easily accessible to the public and relatively unprotected, making it vulnerable to military or terrorist attack. Furthermore, threatening the safety of the public additionally affects the values of democratic states or the rights and liberties of citizens.

As stated by the European Commission in the Action Plan to support the protection of public spaces [3], "local and regional authorities are also important stakeholders in the protection of public space". EU Commission is thus committed to reinforce the involvement of these stakeholders by promoting the dialogue and exchange of perspectives between national, regional and local authorities and supporting the development of operational projects.

P. De La Torre · T. Morales
European Forum for Urban Security, Paris, France
e-mail: delatorre@efus.eu; morales@efus.eu

P. E. (Puck) van den Brink · J. van Rest · G. Voorthuijsen
Netherlands Organisation for Applied Scientific Research,
The Hague, The Netherlands
e-mail: puck.vandenbrink@tno.nl; jeroen.vanrest@tno.nl;
graeme.vanvoorthuijsen@tno.nl

I. Chasiotis • A. Gazi
Center for Security Studies, Athens, Greece
e-mail: i.chasiotis@kemea-research.gr; a.gazi@kemea-research.gr

In this context, the PRoTECT project aims "to strengthen local authorities' capabilities in public spaces protection by putting in place an overarching concept where tools, technology, training and field demonstrations will improve the awareness of local stakeholders towards improving direct responses to secure public places before, during and after a terrorist threat. This cross sectoral project is an initiative of the Core group of the European Network of Law Enforcement Technology Services (ENLETS) [4]".

In light of the above, local authorities responsible for the safety and security of their citizens must be aware of the vulnerabilities of their public spaces in order to be able to adopt appropriate measures to prevent and mitigate terrorist attacks and their consequences [5]. In this regard, DG HOME has developed the EU Vulnerability Assessment Tool (VAT) [6]. The EU VAT is part of the Commission's efforts to support local and regional authorities in the protection of urban spaces. The Commission continues to improve it by developing macros to have a complete and more useful tool.

As such, one of the project activities is to assess the EU Vulnerability Assessment Tool's (VA) quality by applying the EU VA Tool in five European cities (Malaga, Eindhoven, Larissa, Brasov and Vilnius), aiding these cities in assessing their vulnerabilities against terrorist attacks, and to give the resulting feedback about the use of the EU VA Tool to DG HOME.

24.2   EU VA Tool

In this section, the task of assessing vulnerabilities through the EU Vulnerability Assessment Tool (VAT) as well as the underlying conditions behind its use is thoroughly explained. Moreover, the chapter provides an indication on the stakeholder situations that dictate its use. In this regard, it is important to consider the vulnerability assessment in the context of the process of safety and security as a whole and the different steps in risk management, to understand how a vulnerability assessment fits in and what more to do for actual achieving protection for a specific public space. In this context, the use of tools such as the EU VAT enables the identification of vulnerabilities evident in specific sites against different kinds of terrorist attacks. As such specific sites that could be considered as soft targets may be highlighted while also indicating cases where existing measures against terrorism may be considered as efficient/effective.

24.2.1   Public Space of Interest

Municipalities are to some extent (as per case) responsible for the safety and security of people within the limits of their governmental jurisdiction, especially when considering public spaces. Public spaces, as the term implies, are generally open and accessible to members of the public, such as roads, parks and municipal buildings. There are cases where public spaces are being utilised for specific activities or events with specific social dynamics (e.g. concerts, festivals, transportation), resulting in congested areas with very high concentration of people. Consequently, a municipality can potentially consider these areas as areas of higher risk against terrorist attacks.

The activity, and the site where this activity takes place, could be (as per case) managed by separate organisations/owners or just one organisation/owner. This managing body is generally considered to be responsible for the security of the public taking part in the activity [5]. Given the above, a vulnerability assessment is a task that clearly falls within the responsibilities of entities managing the site of interest (or "operator"). In this context, the use of tools like the EU VAT segments the site in consideration into subsites as per the their proximity to the area where the activity takes place ("main site") and identifies vulnerabilities of these subsites on a more granular level. The following categories of main sites can benefit from the EU VAT (Table 24.1):

Table 24.1 Categories of main sites

Category	Examples
Transport hubs	Train station, bus hub, underground metro stations, etc.
Squares	Squares where many events take place, are next to important buildings, have regular big markets, festivals, etc.
Shopping areas	Malls, main shopping street in city centre, etc.
Nightlife areas	Area with a high density of bars, pubs and/or nightclubs, restaurants, coffee shops, small concert halls
Cultural venues	Concert hall, museum, monuments, sport events, stadiums, amusement parks, tourist sites, etc.
Business venues	Big hotels with meeting rooms, large offices, conference centres, etc.
Places of worship	Churches, mosques, etc.
Institutional venues	Public buildings, health buildings, education buildings, etc.

Aspects such as the selection of a specific public spaces of interest (PSOI) and the specificities that relate to s specific event that needs to be investigated are the responsibility of the site operators. However, EU VAT considers crowd density as being a highly relevant parameter and thus includes guidelines towards incorporating crowd density in the process of assessing vulnerabilities on a scale from two to five (person per square meter).

24.2.2 EU VAT Functionality

The EU VAT assists site operators/responsible entities in performing a vulnerability assessment for a specific PSOI. The tool considers pre-set vulnerability dimensions referring to the use of the PSOI. This chapter has been based on version 1.00 of the EU VAT produced by DG HOME.

In the context of the above, the EU VAT refers to a Microsoft Excel-based methodological tool containing six distinct sections (i.e. six worksheets). Each worksheet relates to the specific dimensions that a PSOI's security plan may be decomposed to and assesses vulnerabilities on the basis of applicable characteristics. Each worksheet denoted as "phase" refers to the following as per case:

- Phase 1: Access to the Venue
- Phase 2: Parking and Transport
- Phase 3: Approach to Venue
- Phase 4: Arrival at Venue
- Phase 5: Venue Security – No Access Control
- Phase 6: Venue Security – With Access Control

Note that Phases 1 to 4 correspond to surrounding sites. Phase 5 or Phase 6 refers to the main site. Figure 24.1 provides an example of the six phases (P1–P6) for a hypothetical concert venue in a park. Phase 5 and Phase 6 are alternatives of each other. As outlined above, the PSOI is distinguished in a main site and its surroundings (corresponding to the phases, outlined above). A graphical example of a hypothetical PSOI's analysis is provided below.

Looking at the tool's interface (Fig. 24.2), the four main sections refer to (i) threat types (Section 1), (ii) situations (Section 2), (iii) measure types (Section 3) and (iv) assessment (Section 4).

The general procedure for using the EU VAT is the following:

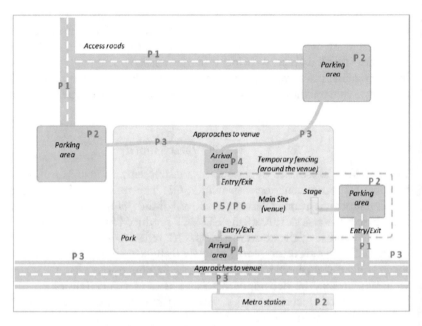

Fig. 24.1   Example of a PSOI and EU VAT phase designation

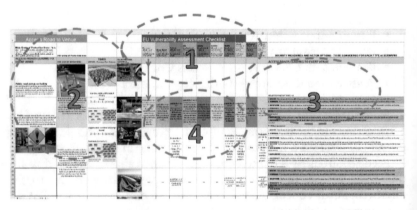

Fig. 24.2   EU VAT interface

- Decide which phase is relevant for the main site and each surrounding site (so the whole PSOI).
- Conceive viable attack scenarios from combinations of threat types (Section 1), situations (Section 2) and currently existing natural and emplaced security measures categorised as per the considered the measure types (Section 3).
- Estimate the consequence and probability of each attack scenario following the assessment suggestions (Section 4).

The tool is designed is such a way that users are stimulated to use their creativity and imagination in discovering possible attack scenarios, as opposed to a design whereby the user is simply asked a lot of detailed questions. The risk of the latter approach is that the right questions might not be asked and an important vulnerability is overlooked. Scenarios are thus presented in the tool as a mixture of possible threat types, images, situations, questions and examples – not necessarily complete in every detail and to be taken literally but to be used as inspiration in discussions within a team of experts. Once the attack scenarios and levels for the consequences and probabilities have been determined, the risk levels can be established by using any risk matrix.

24.3   Workshops and Results

This section provides an overview of the findings and the results of the vulnerability assessment workshops (VAW) that took place in the cities of Vilnius (Lithuania), Larissa (Greece), Eindhoven (the Netherlands), Malaga (Spain) and Brasov (Romania), between April and July 2019. Overall, the workshops in the five European municipalities were deemed successful. The purpose of the workshops in revealing and recording the vulnerabilities of the chosen PSOIs was achieved, and stakeholders gained awareness of the security status of the PSOI.

Feedback from participants regarding the tool and workshop methodology was positive, and further interest has been expressed regarding the improvement of cooperation between the stakeholders involved in the security of "soft targets". The findings revealed many applicable security and technological solutions. Recommendations for the improvement on the tool's methodology and the process of the assessment were recorded and reported to the European Commission.

24.3.1   Vulnerability Assessment Workshops

Vulnerability assessment workshops were conducted in each of the five cities that participate in PRoTECT project. Each workshop was attended by members from the municipality with a primary operational focus on civil protection, members from the local/national police, emergency/fire service and other experts whose fields of expertise covered among others the health energy and transport sector. All participants cooperated throughout the assessment of vulnerabilities exchanging views based on their operational context vis-à-vis a set of realistic threats. In total, 87 experts and stakeholders were involved in the five workshops "setting on the table" a wide spectrum of operating practices and gaps focussing on specific PSOIs that each selected for performing the VA.

24.3.2   Categories of Vulnerabilities Identified

Analysis that focussed on the PSOIs assumed for each of the VAW has considered that the probability and consequences of an attack may vary depending on (a) the phase being considered, (b) the existing security measures, (c) the concentration of people and (d) their location in relation to the main site. The attack risk level was ranked as low, medium or high according to the VAT and its associated risk matrix. Workshops concluded that for the current security and operational state of the PSOIs, the following attack types present the highest risk levels:

- Firearm attack (automatic firearms) – scenarios with individuals attacking crowd of visitors in areas of large crowd density with automatic weapons were analysed.
- Sharp object attacks (mainly knife) – this type of attack is considered more realistic in case of scenarios involving random criminal acts or theft rather than scenarios of organised terrorist attacks.
- Vehicle attacks – the scenario of a vehicle driven into the crowd of visitors at the main site.
- IED (improvised explosive devices) – the possibility of attacks with explosives against areas with high crowd density was examined.
- VBIED (vehicle-borne improvised explosive device) – the possibility of hidden explosives.

In some municipalities, the following types of attack where characterised as threats with high-medium consequences and low probability, thus of lower priority:

- PBIED (person-borne improvised explosive device)
- Chemical attack
- Biological attack
- Radiological attack

During the workshops, the managing bodies established the criteria influencing the consequences and probabilities, as follows:

- Crowd density – the fluctuation throughout the different hours of the day was taken into consideration.
- Proximity to the main site – the consequences and the probability of an attack vary depending on the location of the area examined in relation to the main site and the interconnection of the site.
- Access to weapons – how easy it is for a terrorist to acquire the specific weapons or materials needed for a specific type of an attack.
- Past events/experience – in some cases there was no experiences; in that case a comparison with other terrorist attacks occurred in other countries to estimate the impact and probability of the same attack if it was implemented on the PSOI analysed.
- Attractiveness of site – If a terrorist will choose to attack a site based on factors such as proximity to the main site, possibility of escape, surveillance of site, crowd density, etc.
- Existing security measures – how effectively can the currently implemented security measures mitigate the impact of an attack and if the existing security measures will prevent a terrorist from carrying out an attack.
- Ease of access to site – if adequate security measures are in place to prevent the terrorist for attacking a site.

Provided the above, the analysis conducted during the workshops based on the EU VAT concluded in a diversity of vulnerabilities faced by PSOIs even when considering the same set of threats. In this regard, the VAT, based on any assumptions made in the context of the points above will facilitate site operators to draw conclusions regarding the faced vulnerabilities for subsequently taking measures for their mitigation.

24.4   RECOMMENDATIONS

In tackling terrorism, local authorities and local stakeholders play a key role in terms of prevention of /immediate response to a terrorist act and crisis management. During the five VA workshops, it was acknowledged that local authorities have an important role in defining security plans and elaborating strategies to protect public spaces and specific events from any threat, which could jeopardise the security and safety of citizens. Municipalities in Europe have developed such strategies for managing large events in which they have defined the roles and services involved during a crisis. However, what was identified during these workshops is that the aforementioned strategies have not systemically integrated an assessment of the vulnerabilities against a terrorist attack, especially in cities, which have never suffered from such attacks.

Given the above, a summary of the recommendations addressed by municipalities for the security of "soft targets" based on the five VA processes and the experience of the participating stakeholders is outlined below.

These recommendations have also been validated by the knowledge sharing process during the world café sessions that was part of the first PRoTECT European Seminar that took place on 17th of July, in the city of Brasov, Romania. Eighty representatives from local police, national police, local governments, ministries of interior, civil protection and other sectors from 13 European countries participated. During the world café, challenges, measures and priorities regarding protection of public spaces were individually assessed by rotating groups of experts providing insights that were eventually aligned to findings of the VAWs.

24.4.1   Recommendations Stemming from VA Process

1. *Improving security strategies of public spaces of interest*
   - Elaboration/enhancement of the security strategy regarding a PSOI. In this regard, a vulnerability assessment should be integrated into the design of a PSOI security plan. It should take place recurrently (e.g. annually) or before a big event, led by the municipal staff.
   - Establishment of a map of existing measures protecting such spaces implemented by institutional actors as well as other measures and resources used by other relevant actors.

- Consideration of vulnerability mitigating actions in the context of long-term urban planning.
2. *Fostering coordination in protecting public spaces at a local level*
   - Information sharing between the municipal staff and the different security stakeholders regarding the security plan of a PSOI considering the existing security measures. Establishment of procedures for an effective cooperation.
   - Municipal staff and first responders, should enhance their communication culture and share technical and security information regarding "soft targets". Moreover, local stakeholders should inform each other about their critical assets and emergency response plans for better security planning and improved situational awareness.
   - Municipal staff, personnel from the local and national LEAs, transportation services, private security operators, first responders, national intelligence, counterterrorism agencies and any other relevant actors should participate in the vulnerability assessment and in the measures development process to mitigate a risk.
   - Information campaigns could be organised, in order to train the community on ways to support early warning of terrorist attacks.
3. *Promoting cooperation between the national and local levels*
   - Improving coordination capacity among local security actors also encompassing field training.
   - Standardisation of interoperability aspects including communications, information sharing and systems.
   - Field demonstrations and simulations of scenarios should be organised with the stakeholders, responsible for the security.

24.4.2   Experience from Other Disciplines

For the effective protection of public spaces, besides risk management, security professionals in charge of counterterrorism are making use of other crime prevention techniques that involve collaborative working and broad engagement of the community against terrorism and raising relevant awareness. Indicative approaches mentioned during the VA workshops and during the first seminar are as follows:

Community Policing is a strategy of policing that focuses on building ties and working closely with members of the community through interactions with local agencies and members of the public, creating partnerships and strategies for reducing crime and disorder. The concept of "community policing" is traditionally used by local law enforcement agencies focussing primarily in preventing and solving crimes that have a visible impact on everyday security of local community and affect citizen quality of life (e.g. burglary, theft and robbery). Community policing can be also used by local law enforcement agencies to tackle current terrorist threats. In this regard, this approach is used to minimise the spread of radical ideologies and as a form of gathering intelligence. Interaction between the police and the public can provide an important source of information for the intelligence process and thereby guide the actions of the police, both at local and national level [7].

Urban Planning, Design and Management is a multidisciplinary approach to prevent crime against the person and property and reduce feelings of insecurity, by incorporating evidence-based urban design, planning and management measures within proposals for urban development. Such measures generally seek to embed protective physical features and encourage prosocial behaviour through the design and management of a location.

European cities are exploring innovative solutions for security challenges in public spaces, ensuring the physical structures of crowded places and promoting safety of citizens. Security by design approach, as mentioned in the EU Action Plan to improve the protection of public spaces, is being used as a measure to increase security and promote public safety through the design of public spaces, lighting and public awareness campaigns as part of urban regeneration measures. Terrorist tactics have changed; the targets are less symbolic and more about inflicting maximum damage. This has led to deadly attacks in public spaces not previously optimised for security, such as the vehicle attacks in Berlin, Barcelona, Nice and Westminster Bridge. This creates a huge challenge not only for security professionals but also for designers, urbanists and planners.

24.4.3 Tackling Terrorism Without Fuelling Feeling of Insecurity

New modalities of terrorist attacks have led to an evaluation of counterterrorism protective security in cities. On the one hand, measures to protect public space against terrorism have strengthened and evolved mainly focussing on protective physical measures (e.g. barriers) in order to mitigate the impact of an incident. On the other hand, urban regeneration has mainly focused on inclusivity, liveability and accessibility. These conditions of an urban public space hardly conceal design measures against terrorism and lead to concerns about the exclusionary potential of counterterrorism features in certain locations. In this context, a challenge of blending protective counterterrorism security measures with urban design principles is generated. Protective security in this sense does not always enhance the feeling of safety and security and can have the opposite effect.

As explained in [8], during the project seminar keynote session (PRoTECT Seminar July 2019) and in [7], the question of whether the public and/or perpetrators should see the security measures (or not!) is an important dilemma. Showing all security measures for the public might result in more or less feelings of insecurity. This also depends on timing. For instance, roadblocks, set in place in squares and other crowded spaces in prevention of an attack, will probably be perceived as a necessary burden, and the same goes for security measures right after a terrorist attack or attempt. If security measures are applied for long periods, they could work as a constant reminder that terrorism is a realistic possibility and will thus increase fear and feelings of insecurity. This way, terrorist objectives to reach their goals spreading fear and terror would be facilitated.

UK [9] considers that when incorporating counterterrorism measures into the buildings, appropriate mitigation measures should be sought in terms of risk, cost, aesthetics and usability. Where the measures are appropriate, vulnerabilities of soft targets to terrorist attacks can be mitigated and can also reduce the impact of an attack. These measures must not go against necessary conditions of public spaces such inclusivity, liveability and accessibility and should consider the impact on feelings of insecurity.

24.5 Conclusions

This chapter presented the PRoTECT project outcomes on vulnerability assessments related to the protection of public spaces in five European cities (Eindhoven, Malaga, Larisa, Vilnius and Brasov) along with the findings of the 1st PRoTECT Seminar that took place on 17th of July in the city of Brasov in Romania.

PRoTECT so far has provided its beneficiaries as well as the community of users gathered at the first European Seminar, with tangible outcomes as well as strategic directions regarding the improvement of safety and security of public spaces.

The five municipalities are now more aware and better trained to assess vulnerabilities related to the protection of soft targets and will continue to receive the necessary support to identify the adequate solutions to mitigate such vulnerabilities. The dissemination and the use of the VAT and its manual is to be pursued throughout the project and extended beyond the project consortium so as to reach a truly European dimension mindful of the local dynamics that contribute to the protection of public spaces.

European Commission has been updated on the application process of the VAT at a local level, especially what has worked and what is to be improved in order to promote a comprehensive and standardised framework for the protection of public spaces among municipalities, LEAs, private operators and other relevant agencies.

The project consortium has gathered enough feedback from the VAWs participants to reinforce the process of VA and the proposed methodology in the manual. The consortium will ensure that the project results will nourish and strengthen the future activities of the project in order to deliver a positive operational and policy impact for municipalities willing to strengthen their capabilities in public protection.

Acknowledgements The work presented in this chapter has been conducted in the framework of PRoTECT project, which has received funding from the European Union's Internal Security Fund – Police under Grant Agreement No 815356.

BIBLIOGRAPHY

1. Cities against terrorism, https://issuu.com/efus/docs/cities_against_terrorism,last. Accessed on 2019/02//19.
2. Bennett, B. T. (2007). *Understanding, assessing, and responding to terrorism: Protecting critical infrastructure and personnel.* Hoboken: Wiley.
3. EU Action Plan to support the protection of public spaces (2017) - https://ec.europa.eu/home-affairs/sites/homeaffairs/files/what-we-   do/policies/european-agenda-security/20171018_action_plan_to_improve_the_protection_of_public_spaces_en.pdf
4. PRoTECT project web site, https://protect-cities.eu/
5. European Commission Staff Working Document "Good practices to support the protection of public spaces" (2019) – https://ec.europa.eu/home-affairs/

sites/homeaffairs/files/what-we-do/policies/european-agenda-security /20190320_swd-2019-140-security-union-update-18_en.pdf
6. DG HOME: Site assessment checklist master enlet1, (2019).
7. Van Soomeren, P., Davis, C., & Wootton, A. (2019). *DELIVERABLE 2.5 review of state of the art: CP-UDP. CUTTING CRIME IMPACT PROJECT.* Salford: DSP.
8. Van Soomeren, P., van Dijk, R., & Stienstra, H. (2019). *Design against terrorism: Soft targets and safe public places urban planning, design and management against ram raiders.* Amsterdam, . https://www.svob.nl/wp-content/ uploads/2019/04/Ram-Raiding-EN-20190614.pdf.
9. UK Cabinet Office. (2013). The Civil Contingencies Act, last accessed on 2019/02//19: https://www.gov.uk/guidance/preparation-and-planning-for-emergencies-responsibilities-of-responder-agencies-and-others.

CHAPTER 25

Risk-Based Methodological Approach for Planning for Emergency Sheltering due to Earthquake Disasters

Danai Kazantzidou-Firtinidou, Georgios Sakkas, Chrysoula Papathanasiou, and Georgios Eftychidis

25.1 Introduction

Natural hazards, such as earthquakes, floods, forest fires, volcanic eruptions, and landslides, have always been significantly affecting societies and even the evolution of civilizations, being a source of inspiration for mythic gods and stories (e.g., Enceladus in Greece or Ruaumoko in Maori, New Zealand for earthquakes, Mesopotamian mythic floods, etc.). Preventive and preparedness measures are taken by governments in order to minimize the risk and reduce the impacts of natural disasters. Nevertheless, extreme natural phenomena are able to greatly impact interconnected aspects of the societies, the livelihood, the habitat, critical services, the economy, as well the social structure itself.

D. Kazantzidou-Firtinidou (✉) · G. Sakkas · C. Papathanasiou · G. Eftychidis
Center for Security Studies, Athens, Greece
e-mail: d.kazantzidou@kemea-research.gr; g.sakkas@kemea-research.gr;
c.papathanasiou@kemea-research.gr; g.eftychidis@kemea-research.gr

429
B. Akhgar et al. (eds.), *Technology Development for Security
Practitioners*, Security Informatics and Law Enforcement,
https://doi.org/10.1007/978-3-030-69460-9_25

Since the measured impact of natural hazards on life and health, economy, society, and the environment is significant, there is a strong need to invest in all phases of disaster management cycle, in an attempt to mitigate consequences and ensure timely and effective response. Focusing on earthquakes, given their devastating potential, emergency planning should reflect the principle of protection of life and property, be in accordance with the current social situation, and be flexible and adaptable to the needs that rise during the course of the response. Following the definition of [1], "emergency planning is an exploratory process that provides generic procedures for managing unforeseen impacts and should use carefully constructed scenarios to anticipate the needs that will be generated by foreseeable hazards when they strike." It, thus, incorporates all procedures, distinct and interconnected courses of actions of different stakeholders, making use of available resources.

As a matter of fact, the capacity building of a state on civil protection requires the definition of roles and processes, as well as guarantee of availability of resources, as a prerequisite of the response mechanism at national and local level. Toward the estimation of the necessary resources and the structuring of the most appropriate sequence of actions, it is essential to perform *needs assessment analysis*, followed by *cost-benefit analysis*. This implies a close and dynamic collaboration between the scientific community and the civil protection authorities. Such a collaboration, on the one hand, enables policy and operational decision-makers to base their disaster risk management on scientific inputs and, on the other hand, provides realistic perspectives and data to research, thus supporting efficient estimation of actual societal needs [2].

Schematic distinction of sheltering policy after earthquake disasters follows the time frame of assistance: (a) emergency shelters, (b) temporary shelters, and (c) temporary housing [3]. *Emergency sheltering* refers to the first hours (usually 24 or 48 hours) after the occurrence of a catastrophic event and the provision of immediate assistance in public open spaces. *Temporary sheltering* refers to the time period of a few days up to few weeks, and depending on the extent of the affected area, seasonal and geographic conditions, the availability of facilities, and other circumstances, it is often provided in existing public or private facilities (e.g., schools, stadiums, ships, trains), in hotels far from the epicentral area, or in tent camps. Instead, *temporary housing* refers to the needs of a longer period, often up to a year or more, and it consists of temporary houses (containers or mobile constructions) in camp sites with facilities and services addressed for long-term needs, or state subsidies for renting. Most often, though, in disaster management planning with the term *emergency sheltering*, the comprehensive strategy and provision of sheltering assistance during the extended emergency response phase is implied. The latter incorporates the time slot since organized accommodation starts to be offered and until assisted population is channeled to temporary housing solutions or back to their homes (time period of a few weeks up to 2 months).

Table 25.1 enlists data of sheltering alternatives and quantitative information, when available, for recent important earthquake emergencies in Europe. It is evident that often a combination of solutions is opted for, considering weather conditions, the extent of the affected area and population, the availability of existing infrastructures, and the state's economic capacities. It should also be mentioned that most European countries (e.g., Greece [4]) consider the deployment of tent or container camps as the last sheltering option, when no other option is safe or efficient enough. Meanwhile, the recent example of the M6.4 Albanian earthquake should be mentioned. Among the 47,265 people directly affected, 10,225 were accommodated in 12 shelter camps, 3,613 in hotels throughout the country, and an unreported number privately arranged [5]. Immediate European assistance in tent camp material has arrived, with the Albanian government setting as priority their fast removal due to the heavy winter.

25.2   OPERATIONAL VALUE OF RISK-BASED NEEDS ASSESSMENT

During the immediate post-disaster phase, civil protection and governing authorities, at national or local level, often assisted by scientific experts, are in charge of deciding upon the most appropriate sheltering option and/or combination of them. Considering the required flexibility and the inherent urgency in sheltering provision, the authorities should be aware of the available facilities per administration unit, and most importantly, they should anticipate during the pre-disaster phase for provisions necessary for indoor or tent camps, at national or local level. Predesignation of appropriate sites for camps deployment is also an essential part of the local contingency planning and is ideally a combination of available sites meeting certain conditions (with regard to risk distribution, accessibility, and specific safety requirements).

Table 25.1 Sheltering data from past documented earthquake emergencies in Europe [6–10]

Emergencies	Homeless	People in tent camps	In other accommodation
Kalamata earthquake 1986	35,000	~100%	4 ferries: 400 families; 1 cruise: 1,000 people for up to 2 years
Aigio earthquake 1995	2,100	Assumed 20%	700 people in boat, others in 145 containers, solutions with tents, rent subsidies, containers
Kozani-Grevena earthquake 1995	12,000	1200 (10%)	Accommodation in churches, friends, clubs: later, subsidy for reconstruction or prefab houses
Athens earthquake 1999	70,000 households (~180,000 people)	~30% (inferred)	At later phase: 103 temporary housing camps with 6,854 containers in 33 municipalities, distributed to 5,500 beneficiaries; 30,000 households applied for rent subsidy
L'Aquila 2009	70,000	35,864 (50%)	6,956 in private accommodation (10%) 23,168 in hotels (30%) In total assisted: 65,988 (95%)
Emilia Romagna 2012	42,000 (~18–19,000 households)	Assisted in tents or covered structures: 16,000 (40%)	
Central Italy 2016		August: 4637 people in 43 camps and public facilities (e.g., sanitary buildings, sport centers) October: 31,763 assisted population, in tents, hotels, public facilities. Later: subsidies for self-arrangement, containers, mobile homes	

The percentage corresponds to the ratio of population that was estimated to be sheltered in tent camps with respect to the register of homeless people

Capacity building (resources for risk reduction) and capability planning (risk mitigation measures and concept of operations) for different aspects of emergency preparedness are essential steps of disaster risk management. The latter, according to international guidelines (e.g., [11]), should be supported by all necessary steps of risk assessment (i.e., risk identification, analysis, evaluation, and treatment [12]) and mapping, in order to address timely and effectively the specific identified risks. Scientific methods informed by operational and strategic reality are able to provide quantitative estimates, with a margin of error to be accounted for, due to inherent uncertainties at all steps of the methods. Both political and technical decision-makers are then accountable for defining the level of risk acceptance, allocating the corresponding budget and performing the necessary actions.

Understanding risk, with insights of the order of magnitude of the expected impact (at different terms), probability of occurrence, and spatial distribution of risk, allows authorities to make educated policies, create targeted plans with different levels of escalation, and eventually invest in prevention and preparedness with prioritized actions. It is by no means the unique and/or final step of disaster risk management and planning; it is an evidenced-based input allowing all relevant stakeholders to evaluate options, according to several other aspects, i.e., security, logistics, economy, and social tissue. In an attempt to develop a comprehensive process for the scientific support of the civil protection authorities at a strategic level, the implementation of the methodology is recommended for contingency planning within the urban tissue, as well as at national level, within a concept of national material procurement and optimization of distribution of resources throughout the country.

The methodological framework presented herein focuses on a generic risk-driven process that allows for the estimation of population that would need to evacuate their residences, due to earthquake-induced damage, and would finally seek state's assistance, as a function of their social characteristics and following past observations. Although needs estimates are often based on empirical approaches (e.g., past events), it is admitted by both the scientific community and authorities that risk-based scientific methods (e.g., risk assessment), when combined with past data, provide a more realistic estimation of requirements. The methodology described can be equally applied at local or national level, with different, yet consistent detail level of input data.

25.3 METHODOLOGY DESCRIPTION

Integrated Risk Assessment, proposed herein, incorporates the analysis of existing seismic hazard and, taking into account the different dimensions of vulnerability of the exposed assets (structures and people), estimates their consequences, in physical, human, monetary or other terms, as well as the associated occurrence probability. In Fig. 25.1, the flowchart

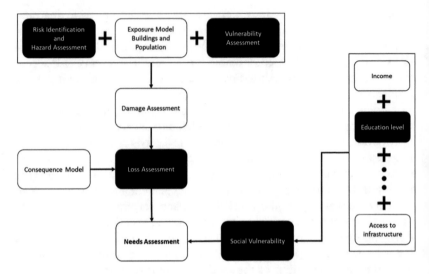

Fig. 25.1 Integrated risk assessment methodology followed for needs assessment

illustrates the steps of risk assessment, based on physical seismic risk modeling that yields loss outcomes in terms of displaced population and incorporation of social vulnerability weighting indices for estimating needs, based on population that will seek emergency accommodation.

25.3.1   Seismic Risk Assessment

The first step in risk assessment process is the risk identification and the estimation of its determining hazard. For the case of earthquakes, this analysis corresponds to *seismic hazard analysis* (*H*) which is usually performed as a probabilistic (probabilistic seismic hazard analysis (PSHA)) or deterministic (deterministic seismic hazard analysis (DSHA)) analysis. PSHA provides a probability estimation that selected seismic parameters (earthquake magnitude or macroseismic intensity) or ground motion parameters (peak ground acceleration, spectral acceleration, etc.) will be exceeded at a given site and in a predefined time interval [13]. PSHA considers all possible occurrences of earthquakes, surrounding a site, both in space and time [14], with the use of an earthquake catalogue. PSHA provides the rate of occurrence of earthquakes and particular levels of ground motions at the site [15]. DSHA on the other hand is based on developing seismic scenarios with certain seismic parameters (e.g., magnitude, fault geometry, focal depth) which could affect the site under consideration by studying the resulting ground motion at the site [15].

Next step is the *physical vulnerability assessment* (in structural terms) (*V*) of the building assets composing the exposure model. Seismic or structural vulnerability is the tendency of a building to suffer a certain degree of damage when exposed to seismic excitation of certain intensity. There are two main categories of vulnerability methods [16] with several available models in literature: (a) empirical or semiempirical, including expert elicitation, based on Damage Probability Matrices or vulnerability indices composed by statistical observations of recorded damage data on buildings of similar characteristics during strong past earthquakes (e.g., [17]), and (b) analytical/mechanical methods, employing parametric structural analyses of building models for the development of a set of fragility curves per building typology (e.g., global, [18]; national, [19]). *Fragility curves* are relationships expressing the conditional probability that different damage grades will be exceeded at specified ground motion levels. Having been derived accounting for all possible variations due to differences in geometric and material properties of buildings lying within a typology, typological fragility curves are important source of uncertainties. All methods have their weaknesses and strengths, and analytical methods are recommended, having been developed with reliable computational models, provided that they are validated with experimental and empirical data.

The *exposure model* (*E*) of an area refers to the building stock and respective population that is present in the study area and is exposed to seismic hazard (Fig. 25.2). Buildings belonging to the same building typology, being in the same geographic location (the level of precision of the location depends on the level of the analysis), compose an asset. Building typologies are defined according to common characteristics with regard to bearing material of the construction, lateral load bearing system, building height, construction period against seismic design code evolution in the country, and other structural information, if available, that are able to allocate to each typology (and thus building) a typological fragility curve. Most of the abovementioned information is given by the building census of statistical authorities, while further elaboration is required for aggregation of typologies. For analysis at local level, in situ inspection is also recommended for the creation of a more detailed exposure model and derivation of fragility curves representatives of the existing building stock

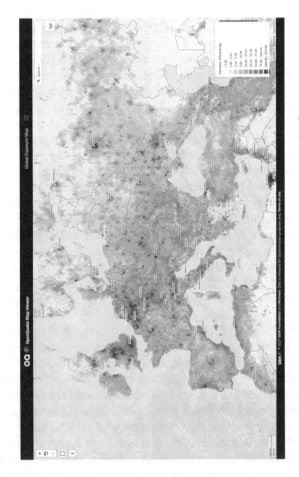

Fig. 25.2 Example of the exposure model of European countries prepared by [23]

(e.g., [20]). Day and night population per asset, with information from the population census, is also part of the exposure model, being crucial for the final estimation of risk.

Finally, the risk is assessed as the combination of seismic hazard and structural vulnerability. Risk, primarily, provides estimation of the physical impact on buildings, with the calculation of the expected damage under a specific earthquake scenario: $R = H \times E \times V$. *Damage assessment* is performed on a probabilistic basis; the outcome per asset consists of the probability of occurrence or exceedance of each damage grade. Depending on the vulnerability model selected, four or five damage grades are recognized, spanning between slight damage (for light nonstructural failures) and complete damage (for very heavy structural damage up to collapse).

In order to estimate the *loss* (L), i.e., the quantifiable consequences of seismic damage, herein in terms of displaced population, one or more *consequence models* are employed. These are expressed in terms of ratios per damage grade, reflecting the probable ratio of buildings that need to be evacuated per damage grade. Consequence models are mostly based on empirical observations and/or legislative actions. Two possible models are thereafter documented for the assessment of displaced population, in function to the damage grade and their probability of occurrence per physical risk assessment analysis.

According to SYNER-G Study (from L'Aquila data) [21] and engineering judgment, the 40% of moderately damage buildings (DG2) is expected to be characterized as "unsafe for use" ("yellow" or "partially usable") during post-earthquake inspection, while the 90% and the 100% of extensively (DG3) and completely (DG4) damaged, respectively, are characterized as "dangerous for use" ("red" or "unusable"). According to post-earthquake guidance (e.g., [22]), all buildings characterized "unsafe" or "dangerous" for use should be evacuated until containment measures are taken.

An alternative model adopts the assumption that all population (100%) from "uninhabitable" buildings is displaced, e.g., following the first-degree inspection guidance per Greek legislation. Often, and to the safe side, buildings being moderately damaged (DG2) are tagged as "uninhabitable" so that they are included into a second-degree, more detailed, inspection. Buildings extensively (DG3) and completely (DG4) damaged lie within the "uninhabitable" category too.

25.3.2   Social Vulnerability Modeling

The displaced population estimated from physical risk assessment is the first step for an integrated needs assessment. However, local and global experience has demonstrated that the actual number of people seeking state's assistance after large earthquakes is only a percentage of the total number of people potentially displaced by the same earthquake. Whether the assistance is provided by public shelters, e.g., in tent camps, or by other means of accommodation, the willingness of population to be benefitted varies accordingly. This behavior has been worldwide observed and is in general determined by the *social vulnerability* of the population, which is often omitted from capability building studies and planning. More precisely, the demand for sheltering varies in time, with a peak occurring around the fifth day from the event [24]. The dynamic needs of sheltering have been studied by [25], concluding on factors able to determine people's willingness, such as the degree of damaged residences and loss of access to safe key services.

As a matter of fact, data from moderate and large earthquakes of the twentieth and twenty-first centuries in Europe (Table 25.1) confirm that only a percentage of the homeless population sought shelter in public camps, being compatible to the general trend documented worldwide. The selection upon a ratio, based on historical observations, strongly depends on local situation and may be of an empirical approximation of the final population to be assisted and, thus, of needs estimate.

Various models and methodologies exist worldwide for the definition and quantification of social parameters, necessary for realistic planning of earthquake sheltering needs (e.g., [26–29]). Main parameters that affect the willingness and need of people to seek assistance are:

- Social and economic factors such as age, education level, household income and tenure, household size, perceived security of neighborhood, and ethnicity
- Access to critical services
- Geographical location of households
- Ratio of damaged buildings in the affected area
- Distance of household to the nearest shelter

Some of the abovementioned data are available from the population census, while others need to be retrieved from relevant socioeconomic studies. The correlation of the abovementioned social characteristics with the willingness and need of population to be sheltered in public settlements requires deep elaboration of social vulnerability metrics within the local social fabric. This is performed by clustering of the classes of the variables according to a common social behavior observed and attribution of an empirical vulnerability index per class. Equally, the weight factors per social variable, indicating the importance of the latter in the willingness and need of population to be assisted, should be defined. It is noted that different sheltering options also require different calibration of the social vulnerability model, as willingness and need of the population for sheltering are a function of the available options. Hence, the estimation of the integrated risk (#IR) is constructed in terms of displaced population (physical risk outcome #DP$_L$), reduced by a factor resulted from the convolution of weight factors per social variable (w_j) for k variables; vulnerability index per class within each social variable (v_i), for n classes per variable; and ratio of population per class of each variable (r_i) in the unit of study (25.1).

$$\# IR = \# DP_L \cdot \sum_{j=1}^{k} \sum_{i=1}^{n} w_j \cdot v_i \cdot r_i \qquad (25.1)$$

Considering the uncertainty and the dispersion each variable are related to the inaccessibility to certain information and significant influence of local conditions, simplifications of the model may be employed.

When several scenarios are analyzed, at local or national level, the consideration of more than one consequence and/or social vulnerability models further contributes to the variability of scenarios, yielding a range of results.

25.4 LOCAL CONTINGENCY PLANNING

As discussed in [30], the aforementioned risk assessment methodology may be followed at local level, within the urban tissue, so as to support the local civil protection authorities in their contingency planning, with spatial identification of the "most" and "less" affected zones. Safest zones could be then considered for predesignation of potential camp sites, in combination with camp specifications criteria (e.g., [31]) and multi-hazard evaluations. Concentration of resources for search and rescue activities and evacuation may be planned in probable most damaged zones. Damage and risk maps are appropriate tool for organization and execution of operational tabletop or field exercises based on scientific scenarios and realistically expected outcomes.

A more refined level of analysis is required, considering that input data of the exposure model can be provided at census tract or building block level. In order to minimize the associated uncertainties and assumptions, the most meaningful representation of risk maps is in terms of distribution of number of buildings per damage grade (Fig. 25.3, left), accounting for the probability each asset lying within each damage grade. For operational purposes, each study unit may be characterized by a unique damage grade being the mean value of the convolution of the probability per damage grade with the respective loss ratio of all buildings per unit. In correspondence to consequence models for population, loss ratio models define an estimate of the fraction of building that is expected to be lost per each damage grade. Integrated risk outcome in terms of displaced population may be estimated at building block level (Fig. 25.3, right), only to map areas with more or less affected population. Under no circumstances should absolute values be considered and employed.

25.5   Geographic Distribution of Capabilities

In most European countries, competences are distributed at different administrative levels of government, following the principle of subsidiarity, and capabilities have been so far built throughout the different levels. This means that local and regional governments are often equipped, even partially, with material needed for built-up sheltering options (e.g., tent camps), as part of their disaster management planning. The methodology presented above, as discussed, is strongly recommended to be employed for the needs assessment of the area under their jurisdiction. It is also evident that rarely can a local administration build capacity that render it independent in case of scenarios of significant severity, and it often counts for support from neighboring regions or from the central administration.

Should the needs assessment be performed at national level, in order to estimate requirements and build capabilities for strengthening the national civil protection mechanism, integrated seismic risk assessment is to be performed over a number of seismic scenarios (PSHA and/or DSHA). The final mandate of the decision upon final design figures for emergency sheltering belongs to the civil protection authorities and decision-makers,

Fig. 25.3 Distribution of damaged buildings per block (left) and displaced population per block (right) for a seismic scenario at Heraklion, Crete [30]

being part of a comprehensive strategy. Important is the estimated number of people for whom camp sheltering capacities should be anticipated, considering that material should be acquired a priori and stored in appropriate locations for guaranteeing timely and efficient response. With regard to the storage and maintenance of the material, there are often two main options: storage at one or two principle warehouses and storage in multiple locations with given areas of jurisdiction. When the second option is followed, the optimum allocation of quantities per storage center is recommended to follow a risk-driven approach described above, in combination with other logistic-related and cost-effectiveness issues. Seismic risk assessment at national level can be performed with the outcome of a PSHA analysis at national scale (e.g., for simplification the national hazard map of the country may be employed) and structural vulnerability and population exposed per administration unit. The risk outcome per unit may be expressed as a fraction of the total risk, and the total material can be distributed proportionally to the fractions per municipality. According to the jurisdiction areas per storage center, appropriate aggregations may be made. Although, interalliance among storage centers is a common practice, it is recommended to perform validation of the final quantities against probable seismic scenarios in the coverage area of each storage center and try to guarantee a minimum level of independence. In lack of risk assessment results, a simpler approach consists of the distribution according to population density, yet in this case, the crucial components of building vulnerability and seismic hazard are neglected.

25.6   Discussion and Conclusions

Historically in Europe, the highest needs in terms of shelters following natural disasters have been emerged due to earthquakes. Several alternatives are employed, according to the number of homeless population, the extent of the affected area, and other geographic particularities. Provision of emergency sheltering is one of the key pillars of civil protection mechanism, for which roles have to be well defined, competencies should be clearly distributed, and capabilities need to be built a priori, for an effective and timely response.

Toward the estimation of a realistic level of population in need, as well as distribution of earthquake impact, a risk-driven approach is proposed based on integrated seismic risk assessment. This is composed by two main modules: (a) the physical risk assessment, based on the seismic hazard of the area of study, the structural vulnerability of its building stock, and the exposed population, and (b) the social vulnerability model that employs socioeconomic factors for the approach of the willingness and need of population groups to seek state's assistance. The model is appropriately modified for different sheltering options that affect differently people's tendency to be sheltered. The latter component is often omitted from risk assessment studies, comprising large uncertainties, inherently related with social behavior, and requiring deep knowledge of the local societal fabric. Its importance is, though, highlighted, as it yields approximations in terms of population fractions that could significantly affect the final design figures.

Civil protection authorities and administrations may subsequently build capacities based on the latter figures, by acquiring equipment, training, and anticipating for human resources, based on the risk-driven needs estimates. Decision upon final figures is made at strategic level taking into account several other criteria, economic factors (for cost efficiency), logistics (for time effectiveness), and operational, political, and governance issues. Similarly, planning of material storage is a strategic decision, and criteria, such as risk distribution, interalliance among storage units and local administrations (principle of subsidiarity), logistics, geography, and cost, are to be co-evaluated.

The methodology described provides valuable information on the expected losses and is considered a valid tool for needs assessment at national or local level and contingency planning, enhancing preparedness and response. It is thus underlined that although past events are indispensable for calibrating models and results, designing and planning is recommended not to be based only on them. Future may evolve differently than the past, and a scientific risk-driven approach may only approximate it, with hazard and vulnerability estimations. The importance of close collaboration and exchange among scientists and decision-makers is, thus, highlighted, representing counterparts with distinct and complementary roles in disaster risk management process.

BIBLIOGRAPHY

1. Alexander, D. (2015). Disaster and emergency planning for preparedness, response, and recovery. In *Oxford Research Encyclopedia of Natural Hazard Science* (pp. 1–20). Oxford: Oxford University Press.
2. Dolce, M., & Di Bucci, D. (2015). Chapter 2: Civil protection achievements and critical issues in seismology and earthquake engineering research. In

A. Ansal (Ed.), *Perspectives on European earthquake engineering and seismology* (Vol. 2, 39, pp. 21–58). Cham: Springer.

3. Quarantelli, E. L. (1982). *Sheltering and housing after major community disasters: Case studies and general conclusions.* Columbus: Disaster Research Center, Ohio state University.

4. GSCP. (2018). *Civil protection action plan for addressing the risks of earthquakes (2018/09-03-2018).* General Secretariat for Civil Protection of Greece (in Greek).

5. WB, EU, UNDP, Government of Albania. (2020). *Albania, post disaster needs assessment.* World Bank, European Union, United Nations agencies, Government of Albania.

6. EPPO. (2007). *Earthquake – Knowledge is protection, earthquake protection at schools* (p. 105). Athens: Earthquake Planning and Protection Organization.

7. Pashalidou, M. (2011). *Experiences and lessons learned from catastrophic seismic events in Greece and the role of Local Government in their management* (Thesis). National Centre for Public Administration and Local Government, Athens (in Greek).

8. Regione Emilia Romagna. (2012). *I Danni del terremoto e le politiche messe in campo per affrontare l'emergenza e la ricostruzione.* Primo bilancio, Bologna (in Italian).

9. Imperiale, A., & Vanclay, F. (2019). Command-and-control, emergency powers and the failure to observed United Nations disaster management principles following the 2009 L'Aquila earthquake. *International Journal of Disaster Risk Reduction, 36,* 101099.

10. DPC. (2018). *I numeri del sisma in Centro Italia.* Dipartimento della Protezione Civile (in Italian).

11. SEC. (2010). *Risk assessment and mapping guidelines for disaster management.* European Commission Staff Working Paper 1626 final, Brussels.

12. IEC. (2009). *ISO31010: Risk management – Risk assessment techniques.* International Organization for Standardization, p. 176.

13. Cornell, C. A. (1968). Engineering seismic risk analysis. *Bulletin of the Seismological Society of America, 58*(5), 1583–1606.

14. Romeo, R., & Prestininzi, A. (2000). Probabilistic versus deterministic seismic hazard analysis: An integrated approach for siting problems. *Soil Dynamics and Earthquake Engineering, 20,* 75–84.

15. Bommer, J. J. (2002). Deterministic vs probabilistic seismic hazard assessment: An exaggerated and obstructive dichotomy. *Journal of Earthquake Engineering, 6*(S1), 43–73.

16. Calvi, G. M., Pinho, R., Magenes, G., Bommer, J., Restrepo-Vélez, L. F., & Crowley, H. (2006). Development of seismic vulnerability assessment methodologies over the past 30 years. *ISET Journal of Earthquake Technology, 43*(3), 75–104.

17. Giovinazzi, S., & Lagomarsino, S. (2004). *A macroseismic method for the vulnerability assessment of buildings.* 13th World Conference on Earthquake Engineering 2004, 896, Vancouver.
18. Martins, L., & Silva, V. (2018). *A global database of vulnerability models for seismic risk assessment.* 16th European Conference on Earthquake Engineering 2008, Thessaloniki.
19. Kyriakides, N., Chrysostomou, C., Tantele, E., & Votsis, R. (2015). Framework for the derivation of analytical fragility curves and life cycle cost analysis for non-seismically designed buildings. *Soil Dynamics and Earthquake Engineering, 78,* 116–126.
20. Lestuzzi, P., Podestà, S., Luchini, C., Garofano, A., Kazantzidou-Firtinidou, D., & Bozzano, C. (2016). Validation and improvement of Risk-EU LM2 capacity curves for URM buildings with stiff floors and RC shear walls buildings. *Bulletin of Earthquake Engineering, 15*(3), 1111–1134.
21. KIT. (2012). *Deliverable D4.7: Prototype framework for integration of physical and socio-economic models for estimating shelter needs and health impacts in earthquake disasters.* SYNER-G project.
22. DAEFK. (2016). *2nd Degree Inspection Form (Re-inspection Form): Guidelines for post-seismic inspection of buildings.* Dec DAEFK/6024/A42/21-12-2016.
23. Silva, V., Amo-Oduro, D., Calderon, A., Dabbeek, J., Despotaki, V., Martins, L., Rao, A., Simionato, M., Viganò, D., Yepes, C., Acevedo, A., Horspool, N., Crowley, H., Jaiswal, K., Journeay, M., & Pittore, M. (2018). *Global earthquake model (GEM) exposure map* (version 2018.1). https://doi.org/10.13117/GEM-GLOBAL-EXPOSURE-MAP-2018.1. Available at: https://maps.openquake.org/map/global-exposure-map/#3/32.00/-2.00.
24. Li, H., Zhao, L., Huang, R., & Hu, U. (2017). Hierarchical earthquake shelter planning in urban areas: A case for Shanghai in China. *International Journal of Disaster Risk Reduction, 22,* 431–446.
25. Wang, J., Zhao, L., & Gu, C. (2013). System dynamical study in needs of emergency shelters for earthquakes. *China Safety Science Journal, 23,* 121–128.
26. FEMA. (2003). *HAZUS-MH technical manual.* Washington, DC: United States Government.
27. SYNER-G. (2009). *Deliverable D4.1: Definition of a group of output indicators representing socio-economic impact from displacement in emergency/temporary shelter.* Systemic Seismic Vulnerability and Risk Analysis for Buildings, Lifeline Networks and Infrastructures Safety Gain, KIT-U.
28. Chang, S. E., Pasion, C., Tatebe, K., & Ahmad, R. (2008). *Linking lifeline infrastructure performance and community disaster resilience: Models and multi-stakeholder processes.* MCEER-08-0004.
29. Zhao, L., Li, H., Sun, Y., Huang, R., Hu, Q., Wang, J., & Gao, F. (2017). Planning emergency shelters for urban disaster resilience: An integrated

location-allocation modeling approach. *Sustainability, 9*(11), 2098. https://doi.org/10.3390/su9112098.

30. Kazantzidou-Firtinidou, D., Gountromichou, C., Kyriakides, N. C., Liassides, P., & Hadjigeorgiou, K. (2017). Seismic risk assessment as a basic tool for emergency planning: "Pages" EU project. *WIT Transactions on the Built Environment, 173*, 43–54. https://doi.org/10.2495/DMAN170051.

31. EPPO. (2013). *Shelter camps specifications.* Athens: Earthquake Planning and Protection Organization.

CHAPTER 26

FASTER: First Responder Advanced Technologies for Safe and Efficient Emergency Response

Anastasios Dimou, Dimitrios G. Kogias,
Panagiotis Trakadas, Fabio Perossini, Maureen Weller,
Olivier Balet, Charalampos Z. Patrikakis,
Theodore Zahariadis, and Petros Daras

26.1 Introduction

The European Environment Agency (EAA) reports that Europe is experiencing an increasing number of disasters, derived either from natural phenomena, technological accidents or human actions [1]. These disasters affect EU citizens, the EU economy and environment every year [2]. Over the period 1980–2016, the total reported losses caused by weather- and climate-related extremes in the EEA member countries amounted to

A. Dimou (✉) · P. Daras
Centre for Research and Technology Hellas, Thessaloniki, Greece
e-mail: dimou@iti.gr; daras@iti.gr

D. G. Kogias · C. Z. Patrikakis
University of Western Attica, Athens, Greece
e-mail: dimikog@uniwa.gr; bpatr@uniwa.gr

© The Author(s), under exclusive license to Springer Nature
Switzerland AG 2021
B. Akhgar et al. (eds.), *Technology Development for Security
Practitioners*, Security Informatics and Law Enforcement,
https://doi.org/10.1007/978-3-030-69460-9_26

447

436 billion EUR. The economic and societal impact will continue to escalate, as weather-related disasters alone could affect about two-thirds of the EU population annually by the year 2100, according to a recent data-driven forecast study [3]. First responders (FRs) are the people who are amongst the first to arrive and provide assistance at the disaster scene. First responders are typically professionals with specialised training, including LEAs, firefighters, emergency medical personnel, rescuers, K9 units, civil protection authorities and other related organisations.

Due to the nature of their work, first responders are often operating in risky and hazardous conditions disaster sites, like demolished, burnt or flooded districts, being exposed to non-visible threats such as very high temperatures and dangerous gases. Furthermore, first responders may experience incidents (e.g. sudden illness, dizziness or exhaustion strokes) during operations, which can prevent them from completing their mission but, more importantly, put their own health at risk. Overzealous first responders may often not notice early signs or choose to ignore them in favour of accomplishing their mission, which can lead to become additional casualties of the disaster [4].

Despite their willingness and proper training, first responders' capabilities may be limited by chaotic environments, making it extremely difficult for them to estimate the exact position of the victims, dangerous areas, other first responder teams or valuable resources. The overwhelming amount of information available to them may reduce rather than increase their situational awareness. Multiple displays and gadgets are adding clutter to their equipment. Autonomous vehicles are useful in disaster scenes, according to a member of the Reykjavik Search and Rescue (SAR) team,

P. Trakadas • T. Zahariadis
Synelixis S.A., Chalcis, Greece
e-mail: ptrak@synelixis.com; zahariad@synelixis.com

F. Perossini
KPeople Research Foundation, Mosta, Malta

M. Weller
Crisisplan BV, Leiden, The Netherlands
e-mail: weller@crisisplan.nl

O. Balet
CS Group, Toulouse, France
e-mail: olivier.balet@csgroup.eu

who claims that using drones has enabled them to respond faster and more accurately [5], but they lack in operational autonomy. Communication between first responders and the command and control centre is often obstructed by broken, overloaded or non-existent network infrastructure [6]. In addition to communication, cooperation and interoperability amongst first responders with LEAs and community members needing help or willing to help are often ad hoc and lack coordination. Often the problem lies not so much in the lack of resources and willingness to provide help but in the logistics to efficiently direct and deliver assistance to the right places where and when it is most needed. These problems raise the need to exploit rapidly evolving technological advances towards protecting first responders from multiple and unexpected dangers and provide solutions enabling them to operate in a seamless and efficient way in any environment and in cooperation with the community.

FASTER aims to address the challenges associated with the protection of first responders in hazardous environments, while at the same time enhancing their capabilities in terms of situational awareness and communication. FASTER will provide innovative, accepted and efficient tools covering (1) data collection providing a secure IoT platform for distributed, real-time gathering and processing of heterogeneous physiological and critical environmental data from smart textiles, wearables, sensors and social media; (2) operational capabilities providing flexible, multifunctional autonomous vehicles, including swarms of them, for extended inspection capabilities and physical mitigation; (3) risk assessment providing tools for individual health assessment and disaster scene analysis for early warning and risk mitigation; (4) improved ergonomics providing augmented reality tools for enhanced information streaming, as well as body and gesture-based interfaces for vehicle navigation and communication; (5) resilient communication at the field level providing haptic communication capabilities, emergency communication devices, communication with K9s and at the infrastructure level through 5G technologies and UAVs; (6) tactical situational awareness providing innovative visualisation services for a portable Common Operational Picture for both indoor and outdoor scenarios representation; and (7) efficient cooperation and interoperability amongst first responders, LEA, community members and other resource providers to request and deliver assistance where and when it is most needed using blockchain technology to give everyone involved the ability to write and read data (including sensor data) on an open-source platform to speed up disaster relief to a whole new level.

26.2   Overall Concept

FASTER aims to establish a new approach for disaster response in order to improve disaster resilience. This will be accomplished by the targeted employment and synergetic deployment of a set of appropriate and complementary technologies. Immediate response is a crucial part of the disaster management cycle for dealing effectively with disasters. Consequently, FASTER aims to improve the disaster response and monitoring capabilities by providing first responders with a suite of tools to augment their situational awareness and, as a result, enhance their safety and their operational capacity (Fig. 26.1). The focus of disaster response is mitigating the impact of the disaster and ensuring the safety of those in immediate risk. However, as this takes place during the emergency, it also includes the safety of first responders who provide the means and resources for effective disaster mitigation and protection of life. Their in-field effectiveness is critical to mitigation and ensuring a short and smooth recovery phase.

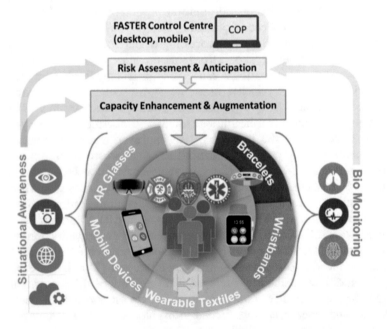

Fig. 26.1 FASTER core first responder capacity enhancement and augmentation tools

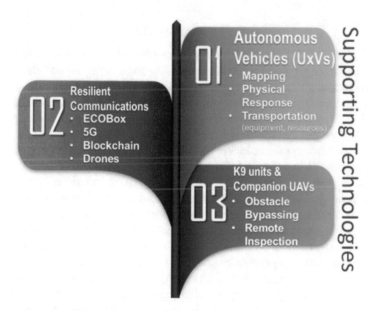

Fig. 26.2   Illustration of FASTER supporting technologies

FASTER's overall concept is illustrated in Fig. 26.2, where it shows that at the heart of FASTER's concept lies on the first responders that will be supported by a set of ergonomic and nonintrusive wearable devices that comprise sensors, actuators and displays, as well as artificial intelligence capacity. These will be responsible for assessing the situation, be it either individualised bio-monitoring of the first responders or local environmental sensing. Their purpose will be to deliver information either in a peer-to-peer manner amongst first responders or centralised points of presence.

The distinction made between these two schemes is necessary as disasters can manifest in various – typically uncontrollable – ways, necessitating the employment of centralised, decentralised and distributed (P2P) management schemes. To that end, FASTER will consider both edge-based and cloud-based processing and analysis technologies to realise a risk assessment and anticipation system that will reach decisions and analyse the overall situation to provide targeted information and instructions to first responders. These will be delivered by the same wearable devices to augment and enhance their operational capacities.

26.3   Technical Approach

FASTER will develop a set of tools towards enhancing the operational capacity of first responders while increasing their safety in the field. It will introduce augmented reality technologies for improved situational awareness and early risk identification, and mobile and wearable technologies for better mission management and information delivery to first responders. Body- and gesture-based user interfaces will be employed to enable new capabilities while reducing equipment clutter, offering unprecedented ergonomics. Moreover, FASTER will provide a platform of autonomous vehicles, namely, drones and robots, aiming to collect valuable information from the disaster scene prior to operations, extend situational awareness and offer physical response capabilities to first responders. FASTER will gather multimodal data from the field, utilising an IoT network and social media content to extract, either locally or in the cloud, meaningful information and to provide an enhanced Common Operational Picture to the responder teams in a decentralised way using Portable Control Centres. It will, additionally, use ledger technology to enable trusted communication. The whole system will be facilitated by tools for Resilient Communications Support featuring opportunistic relay services, emergency communication devices and 5G-enabled communication capabilities.

26.3.1   *Augmented Reality for Operational Awareness*

Mobile augmented reality (AR) can offer more efficient situational awareness and decision making to practitioners in critical conditions that require full attention and focus from involved first responders. FASTER aims to provide augmented reality (AR) technology delivering in real-time information gathered from the other FASTER components (e.g. alerts, team status and location, sensor values), filtering the information and providing targeted content to the AR user. AR will be supplied both through mobile phones and AR glasses (e.g. HoloLens) by superimposing the data to the real world. Many different factors may prevent first responders from reaching and visually inspecting unreachable and/or dangerous areas in disaster sites, such as ruins, obstacles and harmful or unknown environmental conditions. FASTER aims to extend first responders' visual perception by deploying lightweight and camera-equipped UAVs to explore otherwise inaccessible or potentially dangerous areas. These small-factor UAVs will comprise part of the first responders' gear and will be deployed on demand when and where necessary. FASTER will offer first responders an exocentric X-ray-like visualisation of occluded areas from their physical viewpoint, rendering the UAVs' video stream on AR devices. This will widen their field of view and offer the ability to make obstacles between the UAV and the responder partially transparent.

26.3.2 Mobile and Wearable Technologies

During emergencies there is a strong need of effective coordination between the control centre and in-field units. FASTER will design and implement a novel mobile application for first responders able to support inter-agency communication, manage mission tasking and progress monitoring and allow real-time reporting of incidents and of geolocated multimedia content to improve situational awareness. The mobile application will rely on a cloud-based back end and front end to provide data services and the user interface for decision-makers at control room, respectively. The mobile application will interact with other components to provide responders the latest available data, including the location of K9 units.

FASTER will design and develop a prototype regarding the use of sensors in wearable textiles that will be able to collect biometric data. Other sensors will be deployed on the first responder's uniform. All this data will be analysed locally using edge computing capabilities to enhance the information gained at almost real time. On top of this, the design of the solution for FASTER will have to follow the existing security standards for the first responder's uniforms, keeping in mind the protection of electronic parts of wearables, under extreme conditions that first responders may face.

26.3.3 Body and Gesture-Based User Interfaces

In order to improve the ergonomics of the tools, wearable devices will capture and identify arm/body movements exploiting artificial intelligence. FASTER will provide non-visual/non-audible communication capabilities, translating movements or critical readings from paired wearable devices to coded messages, able to be communicated to the team members on the field through vibrations on wearable devices. Given that often during operations communication infrastructure has collapsed, messages will be transmitted using IoT communication protocols (e.g.

Bluetooth Low Energy; BLE). FASTER will also enable UAV navigation through gestures.

In the context of FASTER, a novel wearable device for K9s will be developed, featuring sensors such as three-axis accelerometer and gyroscope, to extract valuable information about K9 behaviour and translate it to specific messages that can be transmitted wirelessly through IoT communication protocols to first responders. At the same time, the definition of a communication protocol will be studied that will translate the K9's behaviour (e.g. movement or bark) into a message addressed at the person in need in order to inform him/her about the K9's role and provide some useful tips that should be followed to facilitate the first responder's work.

26.3.4 *Autonomous Vehicles*

FASTER will also present a robotic platform, integrating different sensors (optical and thermal cameras, environmental, nuclear, biological, chemical, radiological and explosives) and, if required by the use cases, a robotic arm with several end-effector options (grippers or tools). Wireless communication capabilities will support data exchange of large amounts of data (including video) and enabling multi-robot cooperation. It will also feature advanced features such as operation control, with enhanced user interface and visualisation capabilities, localisation services and 3D map generation.

FASTER will employ an array of heavyweight drones of different sizes and payload capabilities that will be able to provide different services to first responders, such as mapping of the disaster area and physical operations like carrying heavy equipment and acting as communication nodes in an ad hoc network to provide resilient communication. The capability to operating in swarms in a coordinated manner following simple operational rules will also be provided.

26.3.5 *Resilient Communications Support*

FASTER will offer a novel, low-cost device, capable of delivering through broadcasting, critical information to first responders or instructions to civilians. The device will be able to send encrypted and signed messages in a massive way, able to cover large number of recipients. By deploying multiple devices, a mesh communication network could be provided, increasing the coverage area. Triggering message broadcasting will be feasible even from great distances, using the appropriate technology (LoRa or short wave to devices).

FASTER will also work on 5G network infrastructure to offer the means to manage and orchestrate resources of an edge cloud in the proximity to the geographic area under investigation and in accordance to the requirements of the rescue team. This will be achieved by leveraging the advantages provided by 5G technologies, including (1) high-speed, zero-latency network, (2) capability to extend allocated resources in real time and (3) steering traffic efficiently to cover the changing operational needs of the responders' teams in real time.

It has been shown that UAVs can be integrated into a cellular network to compensate cell overload or site outage, to enhance public safety in the failure of the base stations and to boost the capacity of the network. FASTER will provide a resilient communication service based on devices from the FASTER ecosystem and an augmented communication support through opportunistic relay services, e.g. swarm of drones' usage, ensuring the minimal acceptable network performance to provide the basic services in a crisis scenario.

FASTER will also develop distributed ledger technology that allows central systems of first responders and other relief mission participants, including social networks and IoT control systems, to connect via a distributed network. For the involved parties, this means (1) distributed power, (2) trusted interoperability, (3) ad hoc capabilities and (4) privacy yet can respond to needs they are capable of fulfilling.

26.3.6   Common Operational Picture

FASTER will develop a decentralised solution for Common Operational Picture that will be supported by services and tools to deal with an adaptive environment, considering contextual information and according to a shared situational picture. FASTER advanced visualisation tools will provide different types of information regarding the position of first responders, possible victims, evacuation and rescue routes, managing countermeasures, resources required and, at the same time, highlight dangerous areas and dangerous environmental conditions through an advanced geolocation information system visualisation layer for both 3D indoor and outdoor scenarios. FASTER Portable Control Centre will allow teams of FRs on the field to make efficient and effective decisions organising a proper response.

FASTER will harness information available in social media content, implementing smart filtering techniques that will exclude erroneous or misleading data, retaining only informative content related to emergency situations and enhance real-time situational assessment. Text mining and deep learning techniques will be used to classify social media posts according to the event type and map it into relevant categories. Deep learning models will be used to recognise scenes depicting emergency situations, such as floods, fires, and extreme weather events and also damaged infrastructures, from social media multimedia content.

26.4   Target Scenarios of the FASTER System

FASTER will be validated in three carefully selected scenarios that cover diverse disaster types and involve the tools deployed, namely, a collapsed building, urban flooding and an indoor disaster. The pilots will take place in Spain, Italy and Finland, respectively. FASTER will be demonstrated in a multi-storey building collapse case Fig. 26.3.

The scenario revolves around a structural failure that triggers the complete collapse of a building in an urban environment. Possible escalation factors in terms of the risks such a situation poses to first responders include building materials, building contents (e.g. storage of chemicals), presence of fire and weather conditions. These conditions may bring about physical hazards (e.g. unstable rubble pile, electrical equipment or sharp objects) or chemical hazards.

The second use case of FASTER consists of a major flood in a city with a high building density, which poses challenging hazards to first responders. Escalation factors may include disruption of services, dangerous debris carried by water, live victims trapped under water, looting and people stranded on evacuation routes.

The third application of the FASTER solution will be demonstrated in an indoor disaster scenario. An explosion in a populated building can include many hazards ranging from fires and their implications (e.g. heat or smoke), dangerous debris and hazardous materials, to the possibility of secondary attacks. Escalation factors may include secondary explosion, hostages, shooting, toxic chemical release, etc.

Fig. 26.3 Practice area for first responders in Madrid area, where the technologies of FASTER will be evaluated during the Spanish pilot

26.5   Exploitation Planning

FASTER is building a community that will consist of the aggregation of the relevant first responders and stakeholders in a systematic interactive approach. The project will use experience from the activities of the formation which operate in Member States of the European Union and carry out tasks for first responders, including in the field of fire protection as well as in the protection of persons and property. The firefighters, policemen, members of rescue teams and security engineers will share their experiences, and the final effect of their cooperation will be new technical, technological and organisational solutions in the field of response to crisis situations, transport (communication), construction, industrial and also natural disasters.

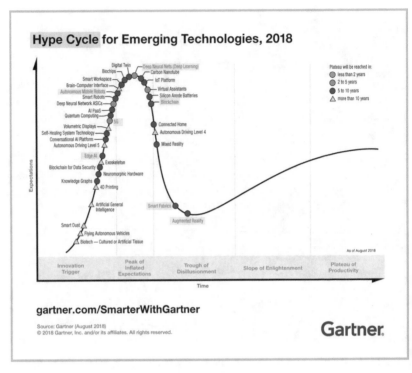

Fig. 26.4 Hype cycle for emerging technologies, Gartner

FASTER is expected to release technologies at an average TRL level 7. It is, however, important to have a forecast of future market need for FASTER outcomes. According to Gartner (Fig. 26.4), it is evident that most of relevant technologies proposed by FASTER to first responders are part of their hype cycle in 2018. As innovation proceeds, the array of mission-critical tools that can aid emergency services providers will continue to proliferate. By harnessing technology, FASTER will help provide first responders with crucial tools and information they need to operate. The project itself is planning to monitor existing players to better focus the research throughout the project duration.

26.6   CONCLUSION

FASTER is an ambitious project that aims to provide state of the art tools to first responders to improve their capabilities and safety. The envisioned toolset is covering a wide range of technological aspects, including communication, augmented situational awareness, remote operations, team monitoring and improved operational planning capabilities. All developed technologies will be evaluated by first responders within the project, as well as external ones, in three pilots addressing diverse scenarios in three different countries.

FASTER encompasses all the elements required to research and develop new technologies for first responders, actively involving them in the design and validation of the tools. The FASTER consortium is an interdisciplinary, while focused team consisting of eight experienced academic/research partners, three industrial partners, four SMEs and eight first responder organisations including law enforcement agencies, firefighters, medical emergency services, K9 units, disaster response teams and civil protection organisations. Moreover, in order to ensure the exploitation of the project results beyond the lifetime of the project, a community of stakeholders is built to support their uptake to the market.

Acknowledgement

 This research has received funding from the European Union's H2020 research and innovation programme as part of the faster (H2020-786731) project.

BIBLIOGRAPHY

1. Wehrli, A., Herkendell, J., & Jol, A. (2010). *Mapping the impacts of natural hazards and technological accidents in Europe.* Luxembourg: European Environment Agency (EEA).
2. EU Publications. *Overview of natural and man-made disaster risks the European Union may face.* https://bit.ly/2nNvGuN. Last accessed 17 Aug 2018.
3. Forzieri, G., et al. (2017). Increasing risk over time of weather-related hazards to the European population: A data-driven prognostic study. *The Lancet Planetary Health, 1*(5), e200–e208.
4. Slate. https://slate.com/technology/2013/05/rescuers-turning-into-victims-lessons-from-first-responders-on-saving-people.html. Last accessed 17 Aug 2018.

5. Tele Med Magazine. http://www.telemedmag.com/drone-use-soon-soar-search-rescue-operations/. Last accessed 17 Aug 2018.
6. Disaster Recovery Journal. https://www.drj.com/articles/online-exclusive/when-communications-infrastructure-fails-during-a-disaster.html. Last accessed 17 Aug 2018.

CHAPTER 27

The Architecture of EVAGUIDE: A Security Management Platform for Enhanced Situation Awareness and Real-Time Adaptive Evacuation Strategies for Large Venues

Dimitris Drakoulis, Dimitris Dres, Anastasia Eleftheriou, George Gkotsis, Dimitris Petrantonakis, Dimitris Kanakidis, and Paul Townsend

27.1 Introduction

An evacuation in response to a risk or threat is the movement of people away from a designated area that is under threat to a safer area. The need for evacuation can arise from naturally occurring events, human-induced events (both intentional and unintentional) and events caused by

D. Drakoulis (✉) · D. Dres · A. Eleftheriou · G. Gkotsis
Telesto Technologies, Athens, Greece
e-mail: dimitris@telesto.gr; jdres@telesto.gr; aeleftheriou@telesto.gr; ggkotsis@telesto.gr

D. Petrantonakis · D. Kanakidis
EXUS Software Single Member Limited Liability Company, London, UK
e-mail: dpetr@exus.co.uk; dkan@exus.co.uk

© The Author(s), under exclusive license to Springer Nature Switzerland AG 2021
B. Akhgar et al. (eds.), *Technology Development for Security Practitioners*, Security Informatics and Law Enforcement,
https://doi.org/10.1007/978-3-030-69460-9_27

461

technological failures. The safe evacuation is of paramount importance for the safety management of large facilities. This need is the most pressing in the case of sports stadia, which routinely host events that gather tens of thousands of spectators and have recently become targets of extremism and terrorism.

An evacuation plan consists of a footprint of the facility and the main safety features (exits, corridors, fire doors, extinguishers), indicating the routes for evacuation to safety zones for every part of the infrastructure and also including the emergency activation methods. The current evacuation plans are static, failing to effectively manage evacuation situations that evolve and change over time. Real-time, dynamic management of an evacuation process is of paramount importance and paper-based evacuation plans are of low value in actual stressful conditions, where human behaviour is unpredictable.

The technology used to assist evacuation incidents, in most cases, is limited to the CCTV monitoring of the areas of the stadium and communication with safety personnel located near the area where an incident occurs, using voice communication over UHF radio.

Current safety procedures are plagued by paper-based, outdated evacuation plans, insufficiently trained personnel and lack of sufficient situational awareness. There is an apparent need for a solution that will support decision-making, increasing the potential for an effective response and strengthening preparedness of the venue operators.

27.1.1   EVAGUIDE Platform

EVAGUIDE [1] is a security management platform for enhanced situation awareness and real-time adaptive evacuation strategies for large venues used for sports and entertainment events. The system aims to address the needs of the safety of large facility visitors during complex evacuation processes, following normal and abnormal events (crises) towards the creation of an easily deployable system that will be able to timely identify new threats, designate and sustain a location-based dynamic evacuation route (LDER) that improves all corresponding response times under any

P. Townsend
Crowd Dynamics International, Manchester, UK
e-mail: paul.townsend@crowddynamics.com

circumstances. Moreover, it will support the complete life cycle of evacuation planning, simulating complex scenarios, training of safety personnel and assessment of the performed actions. It is made up of a number of components, one of which is a mobile application that aids the evacuation of spectators and staff in different ways.

The mobile application for stewards is a mobile-based communication and dispatch mechanism for safety personnel inside the venue as well as in the general vicinity around it. It offers rich functionality based on a two-way communication mechanism, which is used during an emergency situation by the stewards and the security officer. Through the mobile application for spectators, the users are notified in case of an emergency about the situation. If an evacuation is required, they are informed about the optimum exit route and are guided through it.

An important prerequisite for the successful implementation of the mobile application is information about the position of stewards and spectators. Various indoor localisation technologies have been evaluated, concluding that the most appropriate one is BLE Beacons [2] as it is readily available for spectators and stewards alike.

27.1.2   Mobile Application for Spectators

General Description
In case of an emergency, the spectators are notified through the application about the situation. The messaging mechanism works through push notifications, meaning that the users are notified about the emergency, even if the application is not running.

If the situation escalates to the point of an evacuation being needed, the application supports the evacuation process by determining the actual position of an evacuee and indicating the route to the nearest safe exit.

It also helps identifying spectators who do not seem to progress with the evacuation, who may thus require assistance (this is also registered at the security operations centre to be managed by the safety command chain). It is expected that the security officer will gain increased awareness about the progress of the evacuation process, thanks to the connection of the mobile apps with the Common Operational Picture (COP). Even in the case that not all the spectators have downloaded or activated the mobile application, the trends obtained on statistical indices (percentages), based on the situation and whereabouts of thousands of users in an area, will still produce significant value for the situational awareness of the safety authorities.

Description of Features

The users of the app have access to the venue's map as well as the various points of interest (POIs) within it, such as the nearest cafe. By selecting a POI on the map of the stadium, they can navigate following the on-screen directions provided by the applications indoor localisation.

Users can also see their location inside the venue, based on the utilisation of the BLE Beacons infrastructure and the localisation algorithms within the app. This location is not stored at the platform in a normal situation; only during an actual evacuation will the locations of the spectators as calculated from their smartphones will be stored in the platform.

Spectators who wish to report an incident that could potentially represent a threat can do so using the application. The reported incident emerges on the incident reports' area of the COP screen as an icon, which the COP operator can select to get more details. COP offers increased awareness to the security officer compared to current CCTV-only systems, by taking advantage of the mobile-based communication and dispatch mechanism for safety personnel and first responders.

Communication

The application relies on the Internet connection for the following functionalities:

- Delivering of push notifications in case of an emergency.
- Downloading and displaying the venue maps from the map provider (only the first time the application is used).

Following the download of maps, the application may operate while off-line.

Additional Features

To increase the users' engagement with the application, we plan to offer tight integration with the mobile content the club offers its fans, e.g. game-specific content, for example, a video clip from a camera in the opposite side of the stadium from where the spectator sits. Special offers and discounts could also apply for the users of the application while shopping in the venue shops.

27.1.3 Mobile Application for Stewards

General Description

The mobile application for stewards offers rich functionality based on a two-way communication mechanism, which is used during an emergency situation by the steward to warn the security officer about existing or evolving threats to safety and used by the security officer to dispatch a steward to attend a threat.

The mobile-based communication and dispatch is used by stewards/field safety personnel in real-time evacuation; furthermore, it can also be used for personnel training purposes and stadium inspections from security authorities. A mobile (smartphone/tablet) application prototype has been developed within EVAGUIDE life span, in order to be used by the security personnel to report on potential threats and receive notifications to attend to incidents. The EVAGUIDE mobile application is complemented by components adapted from commercial off-the-shelf solutions (COTS), most significant being the COP.

The application allows stewards to have an overview of the venue map. It also displays the tasks that they need to attend to. The stewards can report incidents that are happening in the area of their responsibility via the application. This report creates a new pending task on the security officer's application.

27.2 PLATFORM DESIGN AND ARCHITECTURE

The EVAGUIDE system is composed of four main components (see Fig. 27.1 and further analysis below), namely:

- The Communication Middleware, to enable the different subsystems and sensing elements of EVAGUIDE to communicate between them
- The EVAGUIDE Core System, to handle the interconnection with sensing elements, the alarm raising mechanisms and the control of active exit signs and actuators, the algorithms for real-time crowd modelling and the calculation of the evacuation route in real time
- The mobile platform, for the safety of spectators as well as for the dispatch of safety personnel
- The Common Operational Picture, an off-the-shelf component that connects to the EVAGUIDE core to provide an intuitive picture of

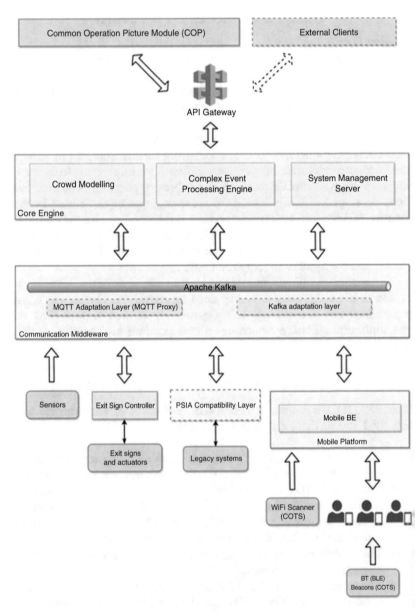

Fig. 27.1 System architecture, highlighting the core system components

the situation to support supervisors' decision-making at the stadium operation centre.

The system is complemented by an enabling infrastructure comprised of components and sensing elements distributed throughout the stadium:

- A resilient private Wi-Fi communication network used for guaranteeing priority access of security personnel to network resources in case of mobile network collapse
- Bluetooth beacons infrastructure deployed to support client positioning both outdoors and indoors
- A Wi-Fi scanner or other people-counting technologies, used to statistically estimate the number of subscribers
- Active exit signs and actuators

The system will finally attach to and collect generated data from existing sensing elements (temperature, smoke, fire, etc.) and legacy systems available on premise at the large venues where it is intended to be deployed.

27.2.1   Communication Middleware

The Communication Middleware enables the different subsystems and sensing elements of EVAGUIDE to communicate in an efficient and orchestrated manner. It is based on Apache Kafka, which enables the required connectivity between components for the flow of incoming information messages from sensors.

From publishers' side MQTT message protocol [3] is used to send messages from sensors to the MQTT broker. This message protocol is TCP-based and lightweight with minimal packet overhead, and it is appropriate for constrained devices with limited resources. The messages are forwarded from MQTT broker to Apache Kafka using connections established between them, each for every plane (status, data, control) specifically configured.

After analysing the functionality of the middleware from publishers' side, the consuming perspective should also be mentioned. At first MongoDB is used as a sink database to store messages that arrive at Kafka brokers for status and data plane. Kafka communicates with MongoDB by using Kafka adapters precisely configured for each plane. Additionally, a JAVA application which implements exit sign client is created in order to consume and represent status published messages of each connected device. Control messages are handled by the Systems Management Server which provides remote monitoring to the administrator.

27.2.2   Location-Based Dynamic Evacuation Route Component (LDER)

Normal stadium evacuation routes are static. They are fixed routes from each area to stairs, ramps and exit points of the stadium. The EVAGUIDE system makes the stadium evacuation strategy dynamic by taking the prevailing situation (numbers of people in different areas + incidents or congestion) and forecasting the congestion from the present into the future before an evacuation actually takes place. It would also dynamically change those routes given the situation, to optimise the evacuation time or reduce congestion.

A location-based dynamic evacuation route is the passage that spectators will take from a certain location to evacuate the stadium, which is dynamically calculated during the evacuation and can dynamically change during the evacuation as it progresses, if necessary. The LDER component takes data from the system measured from the current situation and uses crowd models to simulate the evacuation, forecasting congestion and to optimise the LDER for the stadium.

Crowd Modelling

The crowd model implemented in EVAGUIDE is based on the network model developed as part of the eVACUATE project. A network is created that comprises nodes (circular or rectangular) and edges (a line of certain width connecting two nodes). These are spaces where spectators can move around the stadium, and this network represents all possible routes that spectators use when evacuating the stadium.

The model is a mesoscopic agent-based model. Each agent represents one spectator in the stadium that will move through the network. The mesoscopic nature of the model implies that physical interactions between agents are not modelled, but a localised heuristic measures the density in the vicinity of the agent and adjusts speed of movement accordingly. Demographics are represented by the "speed vs density" profile of the agents, which can be calibrated for different audiences.

The model requires the following inputs:

- Location and number of agents for each edge and node in the network
- Which edges are blocked or reduced in capacity

The location and number of people are calculated by the complex event processing engine and are an interpolated count from the actual sensor data around the stadium.

Route Optimisation

The initial routing for all areas of the network is based on Dijkstra's algorithm [4] calculated using the distance to travel along each edge/through each node, allowing for each edge to have a cost associated with it that would make certain edges more or less attractive to travel along. This provides the most direct routes for agents without considering the congestion or capacity of routes.

The aim of this is to simulate the normal evacuation strategy of the stadium, which more often than not follows the normal egress patterns of spectators attending the stadium. To ensure that this strategy is followed in the initial simulation, large edge costs can be used to deter agents from taking a particular route unless no other is available.

When the initial simulation run is complete, the simulation can be rerun by increasing edge costs on routes with congestion that might be slowing the evacuation down. The simulation is rerun and compared against the original to see if the evacuation time has improved. This process continues iteratively for N runs, after which the most optimised route can be chosen.

A number of situations exist during evacuations whereby having a single route from any one point to another would not result in an optimal situation. When developing optimisation strategies for the stadium evacuation, it is important to bear in mind that perfectly optimum routing is not practical to enact by staff at the stadium. For instance, if the optimal routing strategy was to send 20% of spectators down path 1, 30% down path 2 and 50% down path 3 organising this on the ground is impossible. Therefore, if multiple routes from one area is the optimum, crowds would be split equally (50%/50%, 33%, 33%, 33%, etc.).

27.2.3 Systems Management Server

The Systems Management server's high-level functionality can be summarised as follows:

- Monitors the status of the EVAGUIDE subsystems in terms of connectivity and functionality
- Monitors the health of all the other systems attached to EVAGUIDE, the building management systems, CCTVs, smoke detectors, access control, ticket readers and fire detection systems
- Handles the discovery and registration management of sensing elements to the system
- Performs geolocation-based association of the exit signs and actuators with the LDER and controls them in terms of activation and management.

27.2.4   Complex Event Processing

The Complex Event Processor (CEP) constitutes the component of the core engine that is responsible for the real-time detection of hazardous events, data storage and provision of warnings and alerts. It is one of the core elements, which are closely interdependent with the crowd modelling server and the COP module of the EVAGUIDE platform. The CEP subsystem will not only aggregate and combine the different information sources of EVAGUIDE but will also monitor and generate meaningful insights per the operational status of a life threatening incident.

The CEP subsystem, based on the WSO2 Stream Processor, collects events with multiple messaging formats via multiple transports. It uses streaming SQL to process streams and detect complex events patterns, and it can also generate and notify the processed results as alerts instantaneously and visualise them via real-time interactive and user-friendly dashboards.

WSO2 Stream Processor is built as a lightweight, open-source, high-performance, stream processing platform which understands streaming SQL queries in order to capture, analyse, process and act in real time. This will facilitate the EVAGUIDE system with real-time, intelligent and actionable insights, while at the same time, its deployment ease allows for a multitude of different deployment schemes, aimed at adapting to different installation scenarios.

The state-of-the-art Siddhi stream processing and complex event processing engine [5] which lies at the core of WSO2 Stream Processor will allow the EVAGUIDE system to be enriched with build-reliable and high-performing streaming applications that will detect abnormalities in real time.

This also includes:

- An easy to use streaming SQL language specific to WSO2 (Siddhi)
- A variety of stream processing operators via Siddhi such as filtering, window operations, aggregations and summarisations, pattern machine and event correlations
- Many additional extensions that help developers support more complex use cases.
- Out of the box support for consumption of events and publishing alerts through connectors for well-known protocols such as HTTP, Kafka, MQTT and payload wise supporting XML, JSON, text, binary and key-value messages
- Out of the box integration with popular data storage systems, both SQL- and NoSQL-based ones

27.3 EVAGUIDE PILOT AT PAOK TOUMBA STADIUM

27.3.1 Configuration of the Pilot

The consortium had the opportunity to pilot test the operation of the system and gather initial results between 9 and 11 October 2019 at PAOK's Stadium ("Toumba") in Thessaloniki, Greece. The third floor of the VIP area of the stadium was selected as the demonstration area; it includes a VIP seated area, private VIP rooms and public space.

The system equipment was installed and tested during the first 2 days, while on the third day, the evacuation scenarios took place. Six cameras were placed at strategic locations along the exit route monitoring incoming and outgoing crowd flows at the VIP to and from the seats and the VIP restaurant areas. Sensors were installed to monitor temperature, pressure, light level and position of doors (open/closed). Twenty BLE Beacons have also been installed in the stadium infrastructure (e.g. seats, corridors) near to turning points.

Thirty four volunteers from PAOK's private security subcontractor actively participated in the system tests, providing valuable input to the EVAGUIDE partners. The volunteers were informed about the scope and objectives of the project.

27.3.2   Demonstration Scenarios

During the first phase, volunteers start in seated areas, and then at a specific point in time, they are asked to exit the stadium using normal routes. EVAGUIDE consortium team members timed the exit under normal conditions.

In the second phase of the scenario, the volunteers are asked to exit the stadium under crisis, without the use of the EVAGUIDE system; it involves a simulated incident from a torch lit during the evacuation that caused a fire. Before the scenario starts, the volunteers are asked to return to their seats and download the EVAGUIDE application for spectators. The EVAGUIDE counting module connected to the camera provides crowd counts during both the ingress and the egress stage. A steward is positioned next to the simulated incident with a mobile app. Because of the torch, a sensor detects a rise in temperature and the presence of smoke, and the respective rule displays the results on COP. The volunteers are then asked to exit the stadium. EVAGUIDE will only monitor the situation, record the times during each exit segment and not be involved. In this scenario, as the volunteers are unaware of the incident until the first one reaches the blocked point, they will need to retreat and select a different route.

In the third phase, similar to the second phase, the sensor detects high temperature and smoke (a spectator has lit a torch to celebrate a team's win and the thick smoke rendered the staircase inaccessible). This event was displayed on the COP and the safety manager decides to prohibit access to the staircase. Immediately the system will calculate an updated exit route; this is displayed on the COP. The safety manager decides to make this the new evacuation route. The EVAGUIDE facilitates the whole process by (1) indicating the exit route on the COP for the safety manager, (2) updating the dynamic signs to change their direction to indicate the new evacuation route and (3) sending push notifications to the spectator's app, informing about the evacuation and displaying the active exit route on a simplified diagram of the facility, respective to the individual's position (calculated by the application based on BLE Beacons received signals).

27.3.3 Results

The evacuation time under the optimum conditions was measured at 86 s required from the first person to exit to the last.

In the second scenario, where there is a blocked staircase because of the fire, confusion is created as the head of the group has to communicate the problem to the volunteers that follow, and the latter ones need to become the leaders of the group. EVAGUIDE is not offering the active route. As a result, the evacuation time was significantly increased 140 s. This is a rather underestimated figure given that the volunteers are familiar with navigating the stadium and finding alternative routes (being the stadium security personnel).

In the third scenario, the volunteers are asked to evacuate following the signs and/or the spectator's app, while the EVAGUIDE platform supports the evacuation process using the information from sensors and cameras and displaying up-to-date evacuation routes. In this case the evacuation time was measured as 97 s with the use of EVAGUIDE. Thus in the case of an obstructed exit, the reduced time with EVAGUIDE was measured as 30% or better.

27.4 Conclusions

The safe evacuation of large crowds from complex facilities is a common challenge for facilities across Europe and globally. EVAGUIDE is a security management platform for the safe evacuation from stadia and large facilities.

EVAGUIDE addresses the safety needs of visitors to large facilities during complex evacuation processes, following normal and abnormal events (crises), and creates an easily deployable system that in real time can identify threats, designate and sustain a location-based dynamic evacuation route and increase situation awareness and improve response times under any circumstances. Moreover, it supports the complete life cycle of evacuation planning, simulating complex scenarios, training of safety personnel and assessment of the performed actions.

In EVAGUIDE, the visual representation of the situation in real time is offered by an advanced user interface that uses a 3D model of the facility, which the operator can control to better understand the situation; this is the EVAGUIDE Common Operational Picture. All the information available from the CCTV cameras, the sensors, the legacy systems and the stewards, as well as critical information from the spectators, is optimally displayed. An intelligent engine raises alarms based on rules that combine data stemming from the aforementioned sources, while crowd simulations are realised in real time taking into account the number and location of spectators as well as other parameters (like blocked doors and blocked routes because of fire, congestion or collapsed structures) to calculate a location-based dynamic evacuation route that changes as the aforementioned parameters are altered. The optimal route to safety is communicated to evacuees via the proper activation of state-of-the-art active signs (using dynamic exit signs, media screens, PA system) that depict the calculated optimal routes within the stadium. The EVAGUIDE mobile application that interacts with the EVAGUIDE core platform offers spectators the opportunity to get accurate, dynamic, location-specific personalised directions to follow the fastest available route.

The system has been tested with a limited number of volunteers in PAOK's home stadium of Toumba, in Thessaloniki Greece. Results have been promising with improvements of 30% in evacuation times expected. Those improvements will be even more significant if compared to more adverse conditions expected in a case of low visibility and the panic that may ensue a real incident. Further to these already very promising results, the availability of an exit route specific to one's position which is continuously calculated in real time to reflect the potential obstructions represents a clear breakthrough over existing solutions.

Acknowledgements This paper is supported from the project EVAGUIDE ("A Safety and Security Management System whose main aim is to support the large facility owners and operators with Planning, Implementing, Simulating and Assessing complex evacuation scenarios"), funded by European Union's Horizon 2020 Research and Innovation Programme, under grant agreement no 831154. The authors would like to acknowledge the support of PAOK FC who granted access to the stadium and facilitated the installation of the components necessary for conducting the pilot and the overall organisation of the pilot demonstration.

REFERENCES

1. www.EVAGUIDE.eu
2. Kriz, P., Maly, F., & Kozel, T. (2016). Improving indoor localization using bluetooth low energy beacons. *Mobile Information Systems, 2016*, 1–11.

3. Hunkeler, U., Truong, H. L., & Stanford-Clark, A. (2008). MQTT-S—A publish/subscribe protocol for Wireless Sensor Networks. In *2008 3rd international conference on communication systems software and middleware and workshops (COMSWARE'08)* (pp. 791–798). Piscataway: IEEE.
4. Dijkstra, E. W. (1959). A note on two problems in connexion with graphs. *Numerische Mathematik, 1*(1), 269–271.
5. Suhothayan, S., Gajasinghe, K., Loku Narangoda, I., Chaturanga, S., Perera, S., & Nanayakkara, V. (2011). Siddhi: A second look at complex event processing architectures. In *Proceedings of the 2011 ACM workshop on Gateway computing environments* (pp. 43–50). New York: ACM.

CHAPTER 28

Stakeholders Involved in Hospitals' Crisis Management Processes

*Ilias Gkotsis, Vasiliki Mantzana, Anastasios Galanis,
Ioannis Galatas, Anna Tsekoura, Olivier Theveaneau,
Elodie Reuge, and James Philpot*

28.1 INTRODUCTION

According to the World Health Organization (WHO) definition "Hospitals complement and amplify the effectiveness of many parts of the health system, providing continuous availability of services for acute and complex

I. Gkotsis (✉) · V. Mantzana · I. Galatas · A. Tsekoura
Center for Security Studies (KEMEA), Athens, Greece
e-mail: i.gkotsis@kemea-research.gr; v.mantzana@kemea-research.gr;
i.galatas@kemea-research.gr; a.tsekoura@kemeresearch.gr

A. Galanis
General Hospital Asklepieio Voulas, Voula, Greece
e-mail: galanisanastasios@gmail.com

O. Theveaneau
Assistance Publique – Hôpitaux de Marseille, Marseille, France
e-mail: olivier.theveneau@ap-hm.fr

E. Reuge · J. Philpot
European Organisation for Security (EOS), Brussels, Belgium
e-mail: elodie.reuge@eos-eu.com; james.philpot@eos-eu.com

© The Author(s), under exclusive license to Springer Nature
Switzerland AG 2021
B. Akhgar et al. (eds.), *Technology Development for Security
Practitioners*, Security Informatics and Law Enforcement,
https://doi.org/10.1007/978-3-030-69460-9_28

conditions" [1]. Health sector is responsible for delivering services that improve, maintain, or restore the health of individuals and their communities [1]. These services are large and complex and affect and get affected by multiple interacting actors, such as doctors, nurses, patients, citizens, medical suppliers, health insurance providers, etc., with different backgrounds, knowledge, organizational beliefs, interests, and culture.

The two most critical hospital's assets are the patients' health and their electronic health records (EHRs) [2]. The first one can be affected in many ways, for example, turning off a critical medical device can cause a serious injury to a patient. EHRs contain a host of sensitive information about patients' medical histories, making hospital network security a primary IT concern. It has been reported that healthcare data is substantially more valuable than any other data, as the value for a full set of medical credentials can be over $1000 [3].

The healthcare sector is particularly vulnerable due to heavy involvement in patient personal and health information, time constraints, and complex day-to-day operations. It is one of the most targeted sectors; 81% of 223 organizations surveyed, and >110 million patients in the USA had their data compromised in 2015 alone [4], with only 50% of providers thinking that they could protect themselves from cyberattacks [4]. It has been reported that between 2009 and 2018, there have been 2.546 healthcare data breaches involving more than 500 records and resulting in theft/exposure of 189,945,874 records [5]. It faces unprecedented risks and compounding regulatory compliance requirements.

In addition to cyber threats, physical threats are increasingly growing, and even healthcare facilities are not immune to them. In an example, in 2018, at Mercy Hospital in Chicago, four people were killed in a shooting [6]. The man was able to make his way from the parking lot where the shooting started and proceeded inside the facility. Not only does an inadequate physical security leave employees vulnerable, but patients are also at risk. In fact, a study shows that hospitals are twice more likely to experience a physical attack incident than a cyberattack or breach [7]. In this direction, several low-cost and initial security measures may be applied, ensuring access controls such as requiring patients to be buzzed in past reception, proper security at entrances (especially to very sensitive departments, such as Department of Nuclear Medicine; Blood Banks using irradiators; or BioSafety Levels 3 (BSL-3), labs handling hazardous pathogens etc.), special access to certain floors through the elevator or from the stairway, etc. Badge tap, proximity badges, and biometrics are optimal but expensive solutions. Intrusion detection systems, security lighting, and video surveillance are generic categories of security measures that all contribute to a safer environment.

Any physical or cyber incident that causes loss of infrastructure or massive patient surge, such as natural disasters, terrorist acts, or chemical, biological, radiological, nuclear, or explosive hazards could affect the healthcare services provision and could cause overwhelming pressure to the affected health systems. In fact, the importance of physical and cybersecurity in healthcare has never been more pronounced. Now more than ever, medical organizations must be vigilant in establishing safeguards against physical and cyber threats, which is why it's imperative to have a solid understanding of the risks and protections available.

When physical or cyberattacks occur, a crisis management plan should be immediately executed in order to counter the problem and minimize the consequences. In crisis management, several internal and external stakeholders are involved, having different needs and requirements, trying to cooperate, respond, and recover from the crisis. The aim of this chapter is to identify and define those stakeholders and the respective processes usually followed during crisis management, through a case study conducted in a Greek hospital. Finally, the aforementioned findings and the conclusions that are drawn are linked with SAFECARE project (H2020-GA787005), during the framework of which this research was conducted, which project aims to provide solutions that will improve physical and cybersecurity in a seamless and cost-effective way and enhance threat prevention, threat detection, incident response, and mitigation of impacts, in healthcare infrastructures.

28.2   Crisis Management Process

In the following paragraphs, the normative literature related to the identification and definition of the stakeholders involved in crisis management process in the healthcare sector and the relative physical and cybersecurity rules and policies are described. Crisis management has been defined as "the developed capability of an organization to prepare for, anticipate, respond to and recover from crisis" [8]. The full cycle of crisis management can be described in four phases, as described below:

Preparedness: The aim is to prepare organizations and develop general capabilities that will enable them to deliver an appropriate response in any crisis. Preparedness refers to activities, programs, and systems developed before crisis that will enhance capabilities of individuals, businesses, communities, and governments to support the response to and recovery from future disasters.

Response: Response begins as soon as an event occurs and refers to the provision of search and rescue services, medical services, as well as repairing and to the restoration of communication and data systems during a crisis. A response plan can support the reduction of casualties, damage, and recovery time.

Recovery: When crisis occurs, organizations must be able to carry on with their tasks in the midst of the crisis while simultaneously planning for how they will recover from the damage the crisis caused. Steps to return to normal operations and limit damage to organization and stakeholders continue after the incident or crisis [9].

Mitigation: Mitigation refers to the process of reducing or eliminating future loss of life and property and injuries resulting from hazards through short- and long-term activities. Mitigation strategies may range in scope and size.

There exists an imperative need to reinforce our ability to manage crisis to a broad variety of threats – both physical and cyber – rising from the international environment. This led to the establishment of a comprehensive framework at national and European level concerning the security and protection of services and infrastructures that are essential for preserving vital societal and economic functions, health, safety, security, and the well-being of people against new emerging threats. The Council Directive 2008/114 [10] on the identification and designation of European critical infrastructures along with the 2016/1148 EU Directive on security of network and information systems (NIS Directive) [11] and the later 1082/2013 EU Decision on serious cross-border threats to health [12], in combination with the "Cybersecurity Act" [13], the General Data Protection Regulation (GDPR) [14] and the existing regulations on the "in vitro" and "in vivo" medical devices safety [15, 16], constitutes a set of integrity and continuity rules. They also set protective risk mitigation measures and incident response plans, based on an all hazards approach while supporting strategic exchange of information, facilitating the reinforcement of financial and human recourses and promoting the effective operational cooperation between all involved stakeholders at European level.

The efficient and timely identification of risks, threats, and vulnerabilities of the healthcare sector infrastructures and services the disruption or destruction, of which would have significant socioeconomic and environmental impacts, unquestionably requires communication, coordination, and cooperation at national level as well as, in order to deter, mitigate and neutralize any posed hazard and ensure the functionality, continuity, and integrity of all affected assets and systems.

Both generic and case-specific emergency preparedness and response plans exist at a national level outlining basic incident response procedures and the establishment of necessary security measures. These operational plans in combination with specific legal provisions lying under the national laws or after the transposition of the relevant European legislative framework into the national legal system designate specific individuals or in-hospital agencies, bodies, or committees that have the mandate to fulfill all the tasks and responsibilities related to the hospital emergency planning and response strategy. These tasks may include general or partial evacuation processes, security procedures that support the protection of the venue (e.g., emergency department, the triage area, other healthcare facilities, the morgue, etc.), and other sensitive, critical, or valuable assets and areas (e.g., computer room, central servers or blood bank, pharmacy, etc.) from unauthorized access. They also might refer to the implementation of operational procedures for securing premises perimeter by any unauthorized entry; deployment of entry and exit control measures; implementation of ambulance trafficking plan; enforcement of measures for the preservation of food, water and medical supplies and procedures for the protection of IT infrastructure, pharmacy, and blood bank stockpile; etc.

28.3   RESEARCH METHODOLOGY

The *research design* proposed is the first independent part of the empirical research methodology, as illustrated below. The starting point is to review the literature, thus developing an understanding of the research area under investigation. This led to a specific research area and identified a research need. Based on the need of the empirical study, it was decided that the research design would utilize a case study strategy through the employment of qualitative research methods.

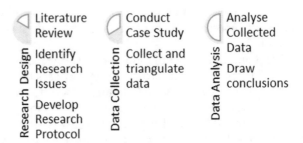

Fig. 28.1   Empirical research methodology

Therefore, various *data collection* methods such as interviews, documentation, and observation were used. In the context of this research, interviews constituted the main data source in the case. Multiple stakeholders were interviewed through structured interviews, followed by open discussion in some cases. Using an interview agenda that was designed for this case, the interviewees replied in specific questions regarding the identification and evaluation of the crisis management process followed by different stakeholders in their healthcare organizations and external interconnected ones. Open discussion was used in some cases, following the structured interview, in order either to clarify issues that derived from structured interviews but also to gather general comments and the perspective on health sector security. In addition to the interviews, data were collected through several sources like archival documents, minutes for meetings, consultancy reports, and the website of the organization; *analysis* and conclusions were drawn (Fig. 28.1).

28.4   The Case Study of a Greek Hospital

Based on literature review and case study conducted, in the following paragraphs, we identify and define stakeholders involved in crisis management in the healthcare setting and describe their respective processes usually followed. The specific case study was conducted on the basis of the methodology described in the previous chapter. The case of a Greek hospital was studied and is analyzed in the following paragraphs. In the context of the hospital infrastructure management factors considered in this chapter and in alignment with ENISA's "Good Practice Guide on Vulnerability Disclosure" [17], it was chosen not to identify this hospital.

The so-called hospital (for confidentiality reasons) provides health services and covers the needs of a large area (consisting of several municipalities but also external visiting patients) with approximately 2 million residents. It has several departments, covers most medical specialties, and is staffed by well-trained medical, nursing, paramedical, and administrative personnel. It aims to provide patients high-quality services and improved building infrastructure and operate in an efficient, effective, and modern way through the use of information systems and medical equipment and improve staff's terms and conditions of employment.

The hospital has an integrated information system with several subsystems, such as medical and laboratory, nursing, administrative, financial, and technical. The hospital has developed its own website and links to national web-based applications such as electronic prescription, insurance capacity, clinical examination, cloud applications, telemedicine systems, digital surgical and robotics equipment, etc. Moreover, the hospital has several infrastructures that include among others central air conditioning, fire detection and extinguishing systems, access control systems, and patient prioritization systems. These infrastructures include power supply, power generators, uninterruptible power supply systems (UPS), servers, PCs, laptops, storage media (hard-disk drive, redundant array of independent disks (RAID), optical drive, Universal Serial Bus (USB) flash drive, etc. and all kinds of peripherals, printers, storage, backup devices, active and passive network equipment, and all kinds of network and telephone devices.

The users of the aforementioned systems are data registrars, subsupporters, developers, trainers, analysts, network specialists, and project managers. Critical roles in the organizational structure include the commander, clinic and nursing directors, heads of departments and departments, and security personnel. Consequently, administrators, doctors, information scientists, nurses, support maintenance companies, citizens, patients and indirect government agencies, and insurance agencies are involved.

The hospital has implemented several security measures and technology solutions to deter, detect and react to physical attacks, such as (a) fences/walls, (b) guards, (c) building control, (d) intrusion detection and access control, (e) video and audio surveillance systems, and (f) physical security information management (PSIM) systems. As the hospital may also be subject to cyberattacks, additional cybersecurity measures have been adopted, such as (a) data protection, (b) network monitoring, (c) intrusion response systems, (d) endpoint monitoring, (e) authentication and access control systems, (f) software development based on privacy by design techniques, (g) IoT sensors for health, and (h) artificial intelligence (AI) techniques.

28.4.1   Hospital Physical and Cybersecurity Crisis Management Process

Healthcare organizations' security stakeholders are individuals or organizations that may contribute to, be affected by, or get involved in issues related to security planning, response, or recovery in any given emergency situation or posed threat. Through the interviews conducted, in the following paragraphs, we identified and defined 18 stakeholders involved in crisis management process in the hospital and whether they are involved in cyber and/or physical attacks (as displayed in Table 28.1).

Interviewees stated that security stakeholders can be categorized according to their involvement and perceived proximity to the healthcare organization into internal and external, as further analyzed below.

Internal stakeholders are these entities designated with duties and responsibilities within the organization's environment, play a role to its performance, and can affect or can be affected by all the decisions made.

The Data Protection Officer's (DPO) primary role is to ensure that processing personal data of its staff, customers, providers, or any other individuals follows the applicable data protection rules.

Physical security manager/security personnel main role is to develop and implement security policies, protocols, and procedures, manage training of security officers and guards (internal and external), and plan and coordinate security operations and staff when responding to alarms and emergencies, all related to the physical part of security.

IT security manager/security personnel is responsible for leading and managing all the relevant activities of the Information Security Risk Assessment and Security Operations team (implementation, installation, monitoring and service/support of healthcare IT infrastructure such as networks, platforms, applications, devices, etc.; develop, assess, update, and enforce security plans and policies in accordance with IT policies, standards, and compliance requirements; respond to cyberattacks and mitigate cyber risks; provide reports on security issues/threats; and train the IT personnel).

Technical manager/technical staff is also stipulated as internal stakeholder. The technical staff not only can identify the sensible technical

Table 28.1 Hospital's stakeholders involved in physical and cybersecurity crisis management process

Hospital safety and security stakeholders		Cyberattacks	Physical attacks
Internal	1. Data Protection Officer	V	
	2. Physical security manager/personnel		V
	3. IT security manager/personnel	V	
	4. Technical manager/technical staff	V	V
	5. Security and safety teams		V
	6. Crisis Management Team	V	V
External	7. Interconnected Critical Infrastructures and related Organizations	V	V
	8. Law enforcement agencies	V	V
	9. Fire brigade		V
	10. Emergency medical services (ambulance)		V
	11. Other healthcare control centers		V
	Centers for Disease Control and Prevention		
	National Health Operations Center		
	Greek Atomic Energy Commission		
	12. General Secretariat for Civil Protection and Administrative Regions of Greece	V	V
	13. Ministry of Health	V	V
	14. HNDGS/Directorate of Cyber Defense	V	
	15 National Intelligence Service	V	V
	16. EETT	V	
	17. ADAE	V	
	18. Hellenic Data Protection Authority	V	V

components for a health structure, such as energy, elevators, technical gas/fluid, temperature, air control systems, or building management but also is responsible to manage physical access rights, hospital Supervisory Control and Data Acquisition (SCADA) systems, natural hazards, and safety events to healthcare organizations infrastructures and processes.

Security and safety teams are responsible for safeguarding the Hospital against physical attacks: (a) technical assets (e.g., gas, electricity, water), (b) hazardous materials (e.g., radioactive, diagnostic or therapeutic materials), (c) personnel and patients, and (d) against natural disasters and firefighting. These teams are continuously trained and participate in tabletop and field exercises and simulations with patients, staff, fire brigade, volunteers, etc.

Crisis Management Team (CMT) "focuses on detecting the early signs of a crisis; identifying the problem; preparation of a crisis management plan; encouraging the employees to face problems; and solving the crisis" [18].

On the other hand, the external stakeholders' category includes individuals or groups outside the organization that can affect or can be affected by it, as they are conjoint into an interdependent relationship. These are described in the following paragraphs:

Interconnected/interdependent critical infrastructures and related organizations include all types of CIs (as described in the EU Directive 114/2008, the NIS Directive, and national policies that are further analyzed in Sect. 28.2 of this chapter), Member States, national authorities and EU officials related to CI resilience or healthcare programs and regulatory work. These entities also support incident management for physical and cyber threats and respond against respective security events.

Law enforcement agencies' (LEAs) mission is to ensure peace and order as well as citizens' unhindered social development which also includes general policing duties and to prevent and interdict crime.

Fire brigade provides fire, rescue, and assistant services and deploys operational procedures during natural or man-made disasters (e.g., structured fires; technological disasters; earthquakes; floods; chemical, biological, radiological, nuclear (CBRN) threats; etc.) which also fall into this category.

Emergency medical services (EMSs) refer to rescue and emergency services that provide medical response to injured or ill people at the scene of the accident.

Other *healthcare control centers* identified through the interviews conducted are the following: (a) Centers for Disease Control and Prevention; (b) National Health Operations Center (NHOC), and (c) Greek Atomic Energy Commission.

General Secretariat for Civil Protection and Administrative Regions of Greece is the body responsible for promoting the country's civil protection relations with relevant international organizations and relevant civil protection agencies in other countries.

Ministry of Health's role in crisis management process is to support, coordinate, and formulate *crisis management process in healthcare organizations*.

Hellenic National Defense General Staff (HNDGS)/Directorate of Cyber Defense is responsible for defending against acts of cyberwarfare and for the coordination of cyber defense exercises [19].

National Intelligence Service (EYP) is designated as National Authority Against Electronic Attacks (national CERT), competent for preventing and statically and actively dealing with electronic attacks against communication networks, information storage facilities, and computer systems [20].

Telecommunications and Post Commission (EETT) is the national regulator for electronic communications [21].

ADAE is the Hellenic Authority for Communication Security and Privacy [21].

Hellenic Data Protection Authority is a Greek authority responsible for the protection of personal data and privacy of individuals constitutes a fundamental human right. Data protection law grants the data subjects, i.e., individuals, certain rights and imposes certain responsibilities on data controllers, i.e., anyone who keeps personal data in a file and processes it [22].

28.4.2 Crisis Management Process and Stakeholders Involved in Healthcare Organizations

With regard to the crisis management process, from the interviews conducted, it appeared that all external and internal stakeholders get involved in the process either by executing sub-tasks, coordinating, managing, or by regulating and just getting informed. In the following paragraphs, the crisis management process and the stakeholders involved are presented as explained by the interviewees. More specifically, they stated that four phases, namely, preparedness, response, recover, and mitigate, were also followed by the hospital in managing crisis, which are analyzed below:

Preparedness is a continuous cycle of planning, organizing, training, equipping, exercising, evaluating, and taking corrective actions that internal and external stakeholders should cooperate closely to ensure organization readiness. Interviewees mentioned that that there exist different national crisis management security plans, namely, Sostratos, Xenocrates, and Perseus, that describe, for example, the hospital's spatial planning; describe emergency actions to be followed through coherent flow chart; set hospital evacuation committee; describe the evacuation process or patient transfer to other hospitals; set epidemic prevention measures; level of readiness, and activation; etc. [23]. These plans and procedures are part of the regulatory framework that the hospital follows and the environment

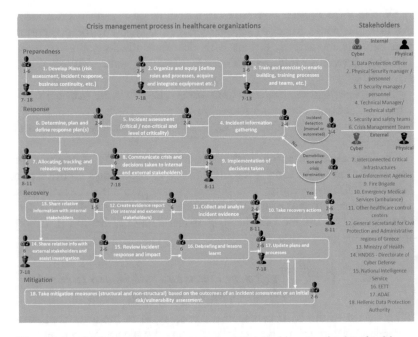

Fig. 28.2 Crisis management process and stakeholders involved in healthcare organization

that it operates, independent to any internal plans and procedures, as part of the national health sector and network; and depending on its focus, it is formulated by different external stakeholders. In addition, training and exercising plans are organized and prepared by external stakeholders, as displayed in the following Fig. 28.2; and all internal stakeholders should get involved. Training and exercising are the cornerstone of preparedness which focuses on readiness to respond to all-hazard incidents and emergencies.

Response initiates when an incident is detected by an internal stakeholder with a manual or automated way. Internal stakeholders should start gathering information that will be used for the initial assessment of the incident. Information gathering and assessment are crucial and continuous steps of this phase, as they highly depend not only on the source, quality, relevance of it but also on the capacity of stakeholders involved in analyzing, interpreting, understanding, and adding value to raw information. Based on the criticality of the incident, Crisis Management Team should be informed and triggered; and CMT should determine, plan, and define which response plan(s) should be activated (e.g., ambulance trafficking plan, evacuation, business continuity, etc.); resources should be allocated and released and actions should be assigned and tracked. In addition, relative information (that can be used for management, informative purposes) should be communicated on time, accurately, and precisely to internal and external stakeholders, in order to manage crisis management process and protect the brand and reputation of the organization. The aforementioned steps could repeat, till resources return to their original use and status (demobilization) and crisis terminates.

Recovery consists of those activities that continue beyond the emergency period to restore critical community functions and begin to manage stabilization efforts. This phase starts after the response phase termination and is directly affected by decisions made as part of the response. The recovery team should decide the recovery actions to be taken, by coordinating closely with the Crisis Management Team. Moreover, the evidence from the incidence should be collected (in cooperation with relative stakeholders, e.g., law enforcement agencies, fire brigade, etc., depending on the nature of the incident) and analyzed; and an evidence report should be created. Relative information should be shared with internal and external stakeholders and investigations should be assisted. Moreover, as crisis serves as a major learning opportunity for both individuals and organizations as a whole, the overall process should be reviewed, and plans, procedures, tools, facilities, etc. should be evaluated, to identify areas for improvement. Following the evaluation lessons learnt should be identified, and recommendations/changes should be made.

Mitigation is the process to reduce loss of life and property by reducing the impact of crisis. It involves structural (such as change the characteristics of buildings; flood control projects, raising building elevations, etc.) and nonstructural measures (adopting or changing physical and cyber access control codes, training, insurance, discussion, planning, etc.). Nearly all internal stakeholders should get involved in the mitigation phase.

Finally, interviewees mentioned that the following functionalities could support them in cooperating timely, effectively, and efficiently in crisis management process:

- Should be able to exchange on-time, accurate, and precise information on hospital suspicious behavior, threats, and events with internal and external stakeholders.
- Should be able to exchange on-time, accurate, and precise information on security events with internal and external stakeholders, in order to support incident response (e.g., provide mobile power generators, triage tents, etc.).
- Access control system should be integrated with other security systems, to enhance information provided (e.g., CCTV with access control and personnel shift timetables).
- Cyber and physical security monitoring systems should be managed in a holistic and integrated way, to enhance crisis management process.

28.5   Conclusions

In this chapter the authors have analyzed and presented the security and crisis management stakeholders and processes in the healthcare setting. It is important to focus on such an approach since the stakeholders involved in this significant area are many and with many different interests. The proposed approach is novel in terms that is combining a crisis management process with stakeholders approach, and it is applied in an area, which lacks of research and can support healthcare organizations in crisis management. Hence, the authors suggest that a combination of the crisis management process with stakeholders approach can be used in other sectors as well. However, to apply this combination in other areas, the stakeholders (of the sector under investigation) should be first understood and identified.

The outcomes of this research presented herein are based on a real-life case study. This is one of the limitations of this work as the data and the observations derived from this case cannot be generalized. Nonetheless, it is not the intention of this chapter to offer prescriptive guidelines to crisis management in healthcare but rather describe a case study perspective that allows others to relate their experiences to those reported. Thus, this chapter offers a broader understanding of the phenomenon of crisis management in the area of healthcare. Last but not least, from the analysis it appeared that it would be interesting to analyze the process in more detail and examine how the different security operational centers run and affect the process. Moreover, a more detailed categorization of the stakeholders could be useful.

Acknowledgments The work presented in this chapter has been conducted in the framework of SAFECARE project, which has received funding from the European Union's H2020 research and innovation program under grant agreement no. 787002.

References

1. World Health Organization. (2019). *Health systems.* [Online]. [Cited: 09 01, 2019]. http://www.euro.who.int/en/health-topics/Health-systems/pages/health-systems.
2. ENISA. (2016). *Securing hospitals: A research study and blueprint.* Independent Security Evaluators. [Online]. https://www.securityevaluators.com/wp-content/uploads/2017/07/securing_hospitals.pdf.
3. Sulleyman, A. (2017). NHS cyber attack: Why stolen medical information is so much more valuable than financial data. *The Independent.* [Online]. https://www.independent.co.uk/life-style/gadgets-and-tech/news/nhs-cyber-attack-medical-data-records-stolen-why-so-valuable-to-sell-financial-a7733171.html.
4. KPMG. (2015). *Health care and cyber security: Increasing threats require increased capabilities.* [Online]. https://assets.kpmg/content/dam/kpmg/pdf/2015/09/cyber-health-care-survey-kpmg-2015.pdf.
5. HIPAA. (2018). Healthcare data breach statistics. *HIPAA Journal.* [Online]. https://www.hipaajournal.com/healthcare-data-breach-statistics/.
6. Brad, E. (2018). 4 dead in Mercy Hospital shooting after gunman goes on rampage. *CBS Chicago.* [Online]. https://chicago.cbslocal.com/2018/11/19/mercy-hospital-gunman-officer-killed/.
7. Adelafa, L. (2018). *Healthcare experiences twice the number of cyber attacks as other industries.* [Online]. https://www.csoonline.com/article/3260191/healthcare-experiences-twice-the-number-of-cyber-attacks-as-other-industries.html.
8. British Standard Institute (BSI). (2014). *BS11200: Crisis Management – guidance and good practice.* s.l.: BSI.

9. Deloitte. (2016). Cyber crisis management: Readiness, response, and recovery. *Deloitte*. [Online]. https://www.google.com/url?sa=t&rct=j&q=&esrc=s &source=web&cd=16&cad=rja&uact=8&ved=2ahUKEwij0amRn_3lAhXISx UIHeu5AWAQFjAPegQICRAC&url=https%3A%2F%2Fwww2.deloitte. com%2Fcontent%2Fdam%2FDeloitte%2Fde%2FDocuments%2Frisk%2FDeloi tte-Cyber-crisis-management-Rea.

10. EU. (2008). *Council Directive 2008/114/EC*. [Online]. https://eur-lex. europa.eu/legal-content/EN/TXT/?uri=uriserv%3AOJ.L_ 2008.345.01.0075.01.ENG.

11. EU. (2016). *The Directive on security of network and information systems (NIS Directive)*. [Online]. https://ec.europa.eu/digital-single-market/en/ network-and-information-security-nis-directive.

12. EU. (2013). *Decision No 1082/2013/EU of the European Parliament and of the Council of 22 October 2013 on serious cross-border threats to health and repealing Decision No 2119/98/EC*. [Online]. https://ec.europa.eu/health/ sites/health/files/preparedness_response.

13. EU. (2017). *Cybersecurity Act*. [Online]. https://eur-lex.europa.eu/legal-content/EN/TXT/?uri=COM:2017:0477:FIN.

14. EU. (2016). *Regulation (EU) 2016/679 of the European Parliament and of The Council of 27 April 2016 on the protection of natural persons with regard to the processing of personal data and on the free movement of such data, and repealing Directive 95/46/EC (GDPR)*. [Online]. https://eur-lex.europa.eu/legal-content/EN/TXT/PDF/?uri=CELEX:32016R0679&from=EN.

15. EU. (2017). *Regulation (EU) 2017/746*. [Online]. https://eur-lex.europa. eu/legal-content/EN/TXT/?uri=CELEX:32017R0746.

16. EU. (2017). *Regulation (EU) 2017/745*. [Online]. https://eur-lex.europa. eu/legal-content/EN/TXT/?uri=CELEX:32017R0745.

17. ENISA. (2016). *Good practice guide on vulnerability disclosure. From challenges to recommendations*. [Online]. https://www.enisa.europa.eu/publications/ vulnerability-disclosure.

18. Mikušová, M., & Horváthová, P. (2019). Prepared for a crisis? Basic elements of crisis management in an organisation. *Economic Research-Ekonomska Istraživanja, 32*(1), 1844–1868.

19. Hellenic National Defence General Staff. (2019). *Hellenic National Defence General Staff*. [Online]. http://www.geetha.mil.gr/en/hndgs-en/history-en.html.

20. NIS. (2019). *NIS*. [Online]. http://www.nis.gr/portal/page/portal/NIS/.

21. ENISA. (2019). *Greek National Cyber Security Strategy*. [Online]. https://www.enisa.europa.eu/topics/national-cyber-security-strategies/ncss-map/national-cyber-security-strategies-interactive-map/strategies/national-cyber-security-strategy-greece/view.
22. Data Protection Authority. (2019). *Data protection authority*. [Online]. https://www.dpa.gr/portal/page?_pageid=33,40911&_dad=portal&_schema=PORTAL.
23. Gika, D. (2017). *Operational readiness plan at hospital unit level for Natural and Technological Hazards*. [Online]. https://pergamos.lib.uoa.gr/uoa/dl/frontend/file/lib/default/data/1332514/theFile/1332519.

CHAPTER 29

Multiple Drone Platform for Emergency Response Missions

Ilias Gkotsis, Paraskevi Petsioti, Georgios Eftychidis, Maria Terzi, and Panayiotis Kolios

29.1 Introduction

Several technologies have been used for natural disasters detection and monitoring including ground sensors, watchers, satellite imaging, etc. However, these methods are not yet able to offer a fast and reliable solution for a real-time detection and monitoring. Unmanned aerial vehicles (UAVs) have a significant raise of use. Drone-based monitoring can provide a low-cost, easy-to-operate, and rapid imaging solution [1] especially in

I. Gkotsis (✉) · P. Petsioti · G. Eftychidis
Center for Security Studies, Athens, Greece
e-mail: i.gkotsis@kemea-research.gr; p.petsioti@kemea-research.gr;
g.eftychidis@kemea-research.gr

M. Terzi · P. Kolios
KIOS Research and Innovation Center of Excellence, University of Cyprus,
Nicosia, Cyprus
e-mail: terzi.maria@ucy.ac.cy; kolios.panayiotis@ucy.ac.cy

public safety missions. Over the last years, UAVs are used in several emergency response missions, around the world, during a fire, in search and rescue, during or after a flood event, in delivering first aids, etc. They have also significant rise in different sectors, such as photogrammetry, infrastructure inspection, safety and security, etc. According to PwC research, more and more companies are interested in the drone industry, developing hardware tools for drones, such as for multispectral sensors, hyperspectral, RGB, goggles for visual observer, etc. The type of payloads (sensors that UAVs carry) plays an important role in the ability of UAV-based monitoring and detection of the points of interest [2]. As an example, thermal sensors can recognize the difference of the temperature between objects, and as a result detection of humans could occur even on no light conditions. In general, the industry sector analyzes and adjusts to the needs of the market, developing respective innovative technologies. This was also SWIFTERS' driving force that this chapter is analyzing, presenting an innovative solution that addresses end users' needs.

As depicted in Fig. 29.1, in 2015 the total value of drone solutions was $127,3b, with the highest amount referring to the sector of infrastructures (45.2% of the total amount), and security taking the fourth position, with 10.5%. Adding to this, the EU states that there are over 1700

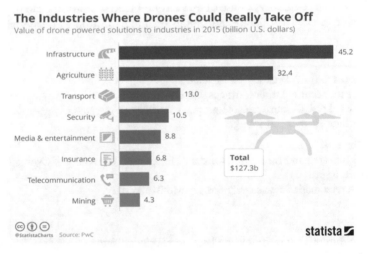

Fig. 29.1 The industries where drones could really take off, 2015. (Statista, Source PwC [10])

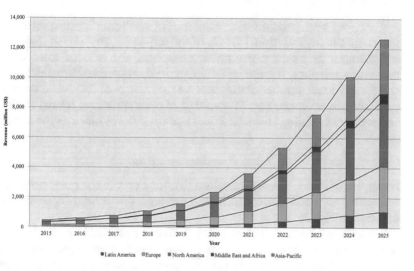

Fig. 29.2 Projected commercial drone revenue from 2015 to 2025 (in million US dollars). (Source: de Miguel Molina and Santamarina Campos [9])

different types of drones produced by official manufacturers (with approximately one-third made in the EU) [3, 4].

Following the above, based on Fig. 29.2, the UAVs' revenue for the next years and specifically, from 2016 to 2025, will have about 40% raise per year. This is significant revenue for the uses of UAVs, showing that they will be used more from day to day, as, for example, by first responders, and most probably not only one drone but a swarm of them.

29.2 END-USER REQUIREMENTS

29.2.1 Requirements Gathering Methodology

Taking advantage from the evolution in drone sector, SWIFTERS developed a platform that will support emergency response operations. In order to address as much as possible of the end users' needs and requirements, a questionnaire regarding the use of UAS and swarms during civil protection missions was shaped and distributed to both project end users and external stakeholders, such as civil protection authorities, emergency management services, and law enforcement agencies. In some cases, the questionnaire has been combined with open discussion and interviews.

Part I of the questionnaire consisted of multiple-choice questions, regarding the organizational profile of the responder, the type of disasters and missions that each organization deals with, the tasks of emergency evacuation missions, the use of UAVs, payloads related to their missions, and the expected number of the users for the control of the swarm during a mission. In addition, Part II consisted of open questions related to the end user needs during the most important operational missions for them.

29.2.2   Requirements Analysis

In respect of Part I, the organizations that took part in the questionnaire, the most common disasters addressed are fires, floods, and landsides, as depicted in Fig. 29.3.

With regard to the types of missions that the stakeholders are implementing during emergency response, those of high priority are related to real-time information, situation monitoring, and search and rescue (see Fig. 29.4).

In addition, 86% of the organizations they already have purchased a UAV, and they used them in the following missions: security and safety surveillance, mapping, monitoring of broken embankments, emergency evacuation procedures, exercises, fire prevention, etc. For those end users that do have hands on experience with the UAV usage but also for the rest, the most essential payloads consist of (a) optical camera and (b) video and data link. GPS and IR/thermal camera are the payloads following with slight difference of importance.

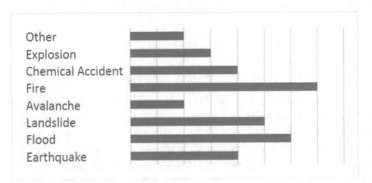

Fig. 29.3 Types of disasters that the responding organization deals with

Fig. 29.4   Types of disasters that the responding organization conducts during emergency response

Finally, most of the end users (43%) have declared the combination of one pilot and one payload controller as the most appropriate for swarm control. On the other hand, none of the end users find appropriate to use one pilot/controller for the whole swarm.

In respect to Part II, the responses were varied according the missions of each organization. The most common answer is to have a tool to detect people (also in the sea, wounded or in a gorge). Also, another requirement is the real-time information (thermal, RGB and video) for situational awareness and monitoring the affected area even at night, in an autonomous way, sending all this information back to a command center. With this tool they could find hospitality areas and evacuation routes for the victims and persons in danger, etc. Another important requirement is that of delivering packages such as first aids, water, life vest, etc. Finally, another requirement is to reduce the number of the resources and response time needed.

29.3   ROLES AND CAPABILITIES OF DRONES

The use of more than one drone in the emergency response and evacuation operations after a disaster could be helpful, faster, and less expensive. Having a swarm of drones means that the end user will have the capability of using multiple drones for the same mission/task or that each drone will have a specific role that can be applied simultaneously. Based on the aforementioned gathered requirements, which have also been combined with others coming from literature review, scientific research, lessons learned, and best practices, the following roles for UAVs with respective capabilities have been identified and implemented during SWIFTERS:

Real-time information in a disaster event, first responders need real timely delivery of accurate data to make correct decisions [5], provide evidence of the impact and depict the extent of damage, guide emergency responders, and have situational awareness. Valuable information is related to location of entrapped people, preliminary damage assessment, environmental monitoring, status of access or evacuation roads/paths, potential hazards, etc.

Reconnaissance and mapping (RAM) high-resolution images and 3D topographical mapping could help first responders by identify disaster prone areas or pinpointing critical infrastructures that need to be secured or restored immediately.

Primary humanitarian aid and supply delivery specific UAVs can deliver supplies whether the first responders could not have access, keeping victims alive for appropriate time to be rescued (e.g., first-aid kits, bottles of water, tools, mobile phone, etc.).

Search and rescue (SaR) drones could search for and rescue people who may be lost, dead, wounded, or in the sea, by detecting and locating them, providing shortest/easiest path to the rescuers to reach them or evacuated them, reducing response time, searching during low or no light conditions, and improving safety of first responders.

Restoring telecommunication during an event, drones can restore the damaged communication infrastructure or could serve as temporary airborne warning and control systems (AWACS) platforms, sending Wi-Fi and cell phone coverage.

Monitoring, forecasting, and early warnings drones could monitor and send information for further analysis, for example, for forecasts and become a sign for early warning [6, 7].

29.4   SWIFTERS Platform

29.4.1   Architecture

Based on the aforementioned requirements and roles, the SWIFTERS platform was designed to provide a modular, extendable, scalable, stable, and reliable infrastructure to enable end users' functional and nonfunctional requirements. It is focused on enabling the connection, communication, and control of a UAV swarm as well as on hosting multiple algorithms such as UAV swarm routing and computer vision. The system architecture, presented in Fig. 29.5, consists of five main layers with different components. Each layer is briefly presented below and thoroughly discussed in an individual section.

29.4.2   Graphical User Interfaces

The graphical user interfaces enable the platform to be accessible to the end users and be utilized to the best of its performance and to the maximum of the features offered. It must be designed to be light in terms of memory requirements and most importantly be user-friendly and provide a good experience to its targeted users. The SWIFTERS ground control platform offers two graphical user interfaces: (a) the web GUI and (b) the control GUI. Each of the interfaces is designed with different purposes.

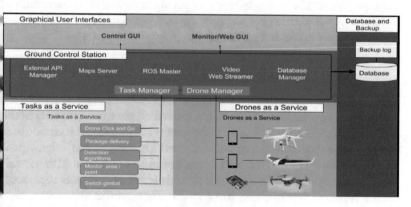

Fig. 29.5   System architecture of the SWIFTERS platform

The web GUI serves as a monitoring interface to the swarm platform, easily accessible by any web browser from any type of device. Through this interface, users will be able to view the operation taking place on a map, the position, and state of the UAVs. User will be able to watch the video feed from all UAVs with potentially enabled the detection of people and cars.

The control GUI is designed for the first responders who will control and instruct the swarm of drones to perform different tasks. In a simple and intuitive yet multifeatured graphical user interface, users will be able to view everything the web GUI offers and as an extension issue operational tasks to the drones. The control GUI is presented in Fig. 29.6.

The control GUI allows the operator to design a mission consisting of one or more tasks. The platform consists of different components: a top menu, a base map, info panels, command buttons, and the live feeds of the cameras. The top menu includes, among others, the available tasks such as the monitoring an area and monitoring a specific point. The main map shows the location of each drone and the current battery level of the drone. Additionally, the map is used to visualize the commands send to the drones such to monitor a particular area or specific points on the map. Info panels are provided to present important information regarding the state of the drones such as the current position (latitude, longitude, altitude) and state (battery level, any warnings, or obstacles). At the bottom of the user interface, quick access command buttons are provided to instruct drones to monitor an area, to synchronized take off, land, and return home. Additionally, buttons are provided to enable the detection of people and vehicles on the drone's camera footage. Any detections are simultaneously presented on the map.

29.4.3   Ground Control Station

External API Manager. The external API manager is responsible for handling all requests for resources not hosted within the boundaries of the SWIFTERS platforms. It has an important role since it is the component that can enhance our system with external knowledge and other opensource services. The external API manager can serve external APIs to multiple resources such as for maps, for altitude levels of buildings to an area, or for weather data such as wind levels. The external API manager is hosted on the ground control station and is able to communicate with both the ROS master and the database manager.

Fig. 29.6 Graphical user interface

Map Server. The map server is a fundamental part of the ground control station, and it is responsible for serving offline and online maps to the ground control application. Offline maps are very important for first responders who most of the times operate in locations with insufficient or no Internet coverage. The map server hosts satellite maps of the countries of interest and serves them so they can be accessed by any client within the same network as the map server.

ROS Master. The ROS master in the ground control station initiates and hosts the communication between all the UAVs, events, and clients connected to the platform. We note that although the ROS master is part of the ground control station, it can be hosted on either the same or a completely different machine of the rest of the components. This is important since the ROS master is a crucial part of our system that required perpetual execution. If the ROS master is unable to initiate the communication network and is unable to respond to requests or transfer data, it means that our system will not be able to operate. Therefore, the ROS master can be hosted on a second machine and have duplicated versions, and its reliability shall be evaluated before sending mission commands to UAVs.

Video Web Streamer. Providing real-time video feed from the UAVs connected to the platform is very important for the first responders. The aerial view provides them with information for situational assessment and allows them to better design the operation. Also, it helps them monitor events and locate people, buildings, and other points of interest. Video web streamer has the role of handling the video received from the UAVs and making it available to the graphical user interfaces that the end users will operate. Similarly, to the ROS master, this is part of the ground control station; however, it can still be operating from a different machine. Such an option will allow the expensive in terms of resources video web streamer operate effectively, without interfering with the performance of the other modules.

Database Manager. The database manager is responsible for serving an API to make the database accessible. Using the database manager, the UAVs and other software clients – such as the user interfaces – can perform asynchronous requests to add and get data from the database. In order for a request to be performed, the database manager checks that the request comes from within the network and no access to external users is provided to the database. For example, every UAV at every second reports its latitude, longitude, and altitude by making a request to the database manager. The database manager receives the requests and after assessing their origin proceeds with saving the data to the appropriate table. The database manager can be hosted on any machine; however, it shall be hosted near the database as well as the drone manager to avoid any latency in the communication between the UAVs and the database.

Drone Manager. The drone manager is based in the ground control station, and its role is focused on handling the connection to the UAVs as a service. It is responsible for identifying which UAVs are connected to the system – that is, which UAV is connected (identifiable by manufacture type and id), what equipment is attached to the UAV, and what is its battery level. Also, it is responsible for handling the operation of the drones and sending the user commands into UAV-understandable commands to the appropriate modules. It is one of the most important modules for enabling a UAV swarm, and it communicates with almost all the other components in the system.

Task Manager. The task manager, also located in the ground control station, is the module that enables all the developed UAV and UAV swarm cooperation algorithms to be accessible and usable by the graphical user interfaces. It can provide any developed algorithms as a service, and it can be further extended without interfering with the rest of the system. Any algorithm selected by the user via the task manager will receive data from the drone manager and any other component of the system required – such as the graphical user interface and the external API manager; it will be executed on the ground control station and in collaboration with the drone manager translated to a set of commands and send to the appropriate UAVs.

Database and Backup. The database is responsible for storing all the data available to the system. The data include all the characteristics of the drones, logs with their status (such as battery levels) at every second, any mission commands sent by the graphical user interface, all images received from drones, and any other data handled in the SWIFTERS platform. All data are encrypted and are stored only and as long as the data serve the requirement mission. The database is accessible only from the database manager – a decision made to protect its reliability and security from external threads.

29.4.4   Algorithms

SWIFTERS integrates multi-drone tasking algorithms to effectively enable users to execute parallel tasks using individual drones while also undertaking multi-drone tasks. The algorithm considers (a) the demand for the requested tasks and (b) the capacity of each drone to complete that task. SWIFTERS accepts as input the number of UAVs, the number of tasks, the maximum total mission time if there is one, the location of the depot, the speed of the UAV, and the recharging time required to replenish battery levels by some amount.

29.4.5   Functionalities

As discussed, the SWIFTERS platform enables the utilization of multiple drones in emergency repose missions. Using this platform, first responders can monitor the real-time state and position of each of the connected drones. Users can also instruct drones to visit a particular location in a click-and-go fashion and scan an area to detect people and vehicles (or other important objects) while viewing live feeds form the camera views of the drones.

Monitoring Point. The SWIFTERS platform enables users to instruct individual drones to visit specific points, through a dedicated command As shown below, the user can simply select the drone and the points on the map and click "GO" to instruct the drone to visit the given path (Fig. 29.7).

Monitoring Area. The SWIFTERS platform enables scanning an area using multiple drones. In this task, the underlying vehicle routing algorithms are used to compute the path of each drone based on the battery levels of the drone and the given altitude. This can be achieved by clicking the monitoring area button at the bottom of the screen, then selecting the drones, selecting the area of interest, clicking visualize paths, and then clicking "GO (Fig. 29.8)."

Detecting People and Vehicles. SWIFTERS enables the detection of people and vehicles and other important objects using the TensorFlow framework. TensorFlow is an open- source machine learning library and framework which enables the training, testing, and usage of machine learning

Fig. 29.7 User Interface during monitoring points task

models in a plethora of tasks including object detection on images. Using TensorFlow, we relied on a pretrained deep learning model to detect objects on the video received from the UAV [8].

The user can enable the detection by clicking on the "Enable Detections" button shown in the user interface. Upon detecting an object, the module sends the detected object frames to the ground control station of the SWIFTERS platform using ROS messages. The platform then sends the detected frames to the database and to the user interface which displays them on the map (as a pin) using the current location of the drone. This allows the user of the SWIFTERS platform to receive alters of the objects detected and their location and view the image with the detected object (Fig. 29.9).

29.5   Lessons Learned

SWIFTERS has participated in various search and rescue, maritime, fire, and flooding exercises, such as:

- "Lellapa" 2018 fire drill, in Cyprus. The drill saw the involvement and collaboration of multiple emergency response departments in

Fig. 29.8   User interface during monitoring area

Cyprus including the Fire Service, Forestry Department, the non-profit Emergency Response Unit, and the Ministry of Foreign Affairs Command Centre.

- "ARGONAUT" at the port of Larnaca, Cyprus. During the exercise, a swarm of drones was instructed via the SWIFTERS UAV swam platform to monitor the area near the Larnaca Port. The live video feeds of the drones were transmitted via the SWIFTERS platform, and object detection algorithms were used to detect boats. Additionally, the video captured via the SWIFTERS platform was live streamed over the Internet to allow interested parties to observe the evolution of the exercise.

- "ALKAIOS II" in the port of Mytilene, Greece. The purpose of the exercise was testing the coordination and cooperation of actors involved to deal with large-scale maritime accidents (including environmental impact of oil spills), as well as to enhance the sense of security of the island's residents and visitors, particularly in view of the tourist season. During the exercise, a swarm of drones was instructed via the SWIFTERS UAV swam platform (tested for the first time in a full-scale exercise) to monitor the accident. The live

Fig. 29.9 Drone video feed and detection module https://www.statista.com/chart/5729/the-industries-where-drones-could-really-take-off/

video feeds of the drones were transmitted via the SWIFTERS platform to the coordination center to provide situational awareness. Additionally, the SWIFTERS platform was used to search and detect people, boats, and vehicles.

During the exercises the software platform was presented with explicit focus on presenting the value it can offer in emergency response missions. During the exercises, the monitoring points and monitoring area with multiple UAVs as well as the people and vehicle detection features were demonstrated to the participants. In every demonstration, time was allocated after the exercise for gathering feedback and listening to the first responders' needs. Based on this feedback, but also from that coming from other demos and trainings, the following lessons have been learned, categorized under a set of key quality indicators, as depicted in the following Table 29.1.

Table 29.1 Lessons learned on swarm of UAVs

Key quality indicators	Using UAVs in swarm
Evacuation time	It will take much *less time for an evacuee* to be saved
	The total time of the evacuation will be decreased significantly
Response time	The *reaction time will be decreased*, due to the automatic procedures of the system
	There will be *no delay*
Human costs	The evacuees will stay *in danger less time*, which means that more people will be saved
	Delivering *supplies will save more peoples*
Capacity and source planning indicators	Preparation for and proceed with evacuation will be significantly *faster*, due to automatic deployment and support/replace of human resources
Cost-effectiveness	Mission total *operational cost will decrease*, due to operational time reduction
	Insurance costs will decrease, as less disruptive impacts will occur
	The macroeconomic impact of the disaster will point the *advantage of using UAVs*

29.6   CONCLUSIONS

A single UAV to simultaneously undertake all the roles needed in evacuation and other types of operations following a disaster is a for sure a great challenge. Planning and monitoring of the operation, as well as receiving and exploiting of data recorded by the sensors, can potentially be more useful/beneficial when using swarm of UAVs that act simultaneously, instead of using a single UAV. Furthermore, several UAVs, even from different agencies, can be coordinated and operated based on the interoperability of the platforms. Different payloads carried by unmanned vehicles can provide different type of information, which can be combined and assessed to provide situation awareness in a broad area.

In this direction, this chapter provides a summary of the work conducted during the 2 years and the respective results of the SWIFTERS project. The presented project results supported the development of a software package, featuring emergency operation planning capabilities enabled by UAV swarms, such as task allocation to individual UAVs of the swarm (e.g., monitoring emergency event progress, identifying stranded survivors, marking the evacuation path, etc.), and path planning for each UAV. In doing this, the SWIFTERS project supported the allocation of emergency response and evacuation operations to UAVs in an intelligent way with the ultimate goal of improving response efficiency and reducing evacuation times.

Acknowledgments The work has received funding from the European Union Civil Protection Call for proposals 2017 for prevention and preparedness projects in the field of civil protection and marine pollution under grant agreement – 783299 – SWIFTERS.

REFERENCES

1. Afghah, F., Razi, A., Chakareski, J., & Ashdown, J. (2019). Wildfire monitoring in remote areas using autonomous unmanned aerial vehicles. In *IEEE INFOCOM 2019 – IEEE Conference on Computer Communications Workshops (INFOCOM WKSHPS)*, Paris, France, pp. 835–840.
2. Barbedo, J. G. A. (2019). A review on the use of unmanned aerial vehicles and imaging sensors for monitoring and assessing plant stresses. *Drones, 3*(2), 40.
3. European Commission Memo 14/259, Remotely Piloted Aviation Systems (RPAS) – Frequently Asked Questions (2014).
4. Fox, S. J. (2017). The rise of the drones: Framework and governance – Why risk it! *Journal of Air Law and Commerce, 82*, 683–715.
5. Chen, D., et al. (2013). Natural disaster monitoring with wireless sensor networks: A case study of data-intensive applications upon low-cost scalable systems. *Mobile Networks and Applications, 18*(5), 651–663.
6. Erdelj, M., Natalizio, E., Chowdhury, K. R., & Akyildiz, I. F. (2017). Help from the sky: Leveraging UAVs for disaster management. *IEEE Pervasive Computing, 16*(1), 24–32. https://doi.org/10.1109/MPRV.
7. Remote Piloted Airborne Systems (RPAS) and the Emergency Services, EENA Operations Document – RPAS and the Emergency Service (2015).
8. Gąszczak, A., Breckon, T. P., & Han, J. (2011). Real-time people and vehicle detection from UAV imagery. In *Proceeding of SPIE: Intelligent Robots and Computer Vision XXVIII: Algorithms and Techniques*, San Francisco, California, USA, p. 78780B-1-13.
9. de Miguel Molina, M., & Santamarina Campos, V. (Eds.). (2018). *Ethics and civil drones* (Springer Briefs in Law). Cham: Springer.
10. https://www.statista.com/chart/5729/the-industries-where-drones-could-really-take-off/

CHAPTER 30

Towards to Integrate a Multilayer Machine Learning Data Fusion Approach into Crisis Classification and Risk Assessment of Extreme Natural Events

Gerasimos Antzoulatos, Ilias Koulalis,
Anastasios Karakostas, Stefanos Vrochidis,
and Ioannis Kompatsiaris

30.1 INTRODUCTION

Natural disasters can be defined as a combination of natural hazardous, extreme and unexpected threats, such as floods, wildfires, earthquakes, hurricanes, thunderstorms, etc. that may cause significant damages and losses of human lives, especially in highly vulnerable communities that are incapable of withstanding the adversities arising from them [11, 34, 50]. A recent report of Centre for Research on the Epidemiology of Disasters

G. Antzoulatos (✉) · I. Koulalis · A. Karakostas
S. Vrochidis · I. Kompatsiaris
Centre for Research and Technology Hellas, Thessaloniki, Greece
e-mail: gantzoulatos@iti.gr; iliask@iti.gr; akarakos@iti.gr;
stefanos@iti.gr; ikom@iti.gr

© The Author(s), under exclusive license to Springer Nature   513
Switzerland AG 2021
B. Akhgar et al. (eds.), *Technology Development for Security*
Practitioners, Security Informatics and Law Enforcement,
https://doi.org/10.1007/978-3-030-69460-9_30

(CRED) based on data from the Emergency Events Database (EM-DAT[1]) states that in 2018, there were 315 natural disaster events recorded with 11,804 deaths, over 68.5 million people affected and around US $132 billion in economic losses across the world. Natural disasters are localised and have a very severe impact on local economies and communities, and recovery usually takes a very long time [34].

It is worth to mention that this decade saw a significant decrease to the annual average of catastrophic events, of deaths, of people affected and of financial losses compared to the previous decade [19]. Contrary to the decrease that is seen and despite of many technological advances, the intensity, the frequency and extent of damage and devastation due to disasters are increasing year after year [19, 40]. Shock natural crisis events have short- and long-term devastating impact on the sustainable management and viability of natural, cultural and residential environments, the local and regional economies and societies. The climate change, the rapid and unplanned urbanisation, the population growth and the environmental degradation are factors that partially affect the escalation of the disastrous impacts from extreme natural events [40]. Especially, in low- and middle-income countries, disasters are a major contributor to entrench poverty and often lead to a downturn in the trajectory of socioeconomic development and exacerbate poverty (Sustainable Development Goal 1: End poverty in all its forms everywhere[2] [46, 47]. Thus, the utilisation of risk-based decision support systems that encapsulate the rise of technological achievements is increasingly important to deal with the complexity and uncertainty at extreme natural events [18, 34, 40, 43, 50].

In recent years, machine learning and deep learning algorithms have aided in solving domain-specific problems in the disaster risk management, including the prevention of new, the reduction of existing natural or human-made disaster risks as well as the monitoring of them [18, 24]. The rapid increase in the usage of machine learning techniques may be attributed partly to an unprecedented increase in the development and use of Internet of Things (IoT), the autonomous vehicle systems and unmanned aerial drones, which are equipped with a wide variety of sensors and cameras generating massive and diverse data that need to be processed in real-time to gain added value in the field of disaster risk assessment [24]. Despite advances in risk assessment techniques, the ability of real-time risk

---

[1] https://www.emdat.be/
[2] https://sustainabledevelopment.un.org/sdg1

assessment, which fostered with the knowledge extracted dynamically from the field of interest, is limited [7, 24, 39, 54]. As the needs for seamlessly and dynamic risk assessment increases, the development and utilisation of machine learning algorithms capable of addressing the emerging risk assessment challenges may also increase [24].

In this chapter, we attempt to briefly present the theoretical background behind disaster management, focusing on the risk assessment process. Moreover, we review state-of-the-art machine learning methodologies that have been started to penetrate into natural disaster management, aiming to tackle the challenges in that multisectoral and multidisciplinary field of disaster management and risk assessment process (Sect. 30.2). In Sect. 30.3, we describe the proposed methodological framework for the dynamic assessment of the severity of extreme crisis events and discuss the challenges involved and how to effectively address them (Sect. 30.4).

30.2 RELATED WORK

30.2.1 Risk Assessment

In literature, prominent studies have suggested various perspectives on risk, including the combination of expected consequences, probabilities and impacts [7, 39, 54]. In disaster management sciences, the risk is expressed as a function of hazard (probability) and vulnerability. The latter could be presented as the expected losses and the preparedness of our societies [20]. Figure 30.1 shows a two-dimensional risk matrix proposed by the European Commission in 2010 that relates the likelihood (probability) and impact (loss), for a graphical representation of multiple risks in a comparative way [40].

$$\text{Risk} = f\left(\text{Hazard,Vulnerability}\right) = f\left(\text{Hazard,Loss,Preparedness}\right) \quad (30.1)$$

In 2017, the UN Office for Disaster Risk Reduction (UNISDR), aiming to include the Sendai Framework for Disaster Risk Reduction 2015–2030 [47], has suggested a new definition concerning the *disaster risk*. Therefore, *disaster risk*[3] can be considered as the *potential loss of life, injury or destroyed or damaged assets which could occur to a system, society or a community in a specific period of time, determined probabilistically as a*

---

[3] https://www.undrr.org/terminology/disaster-risk

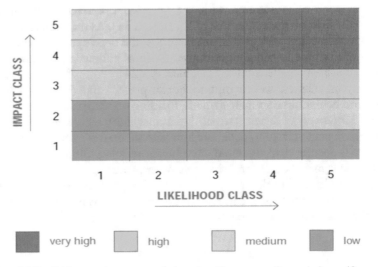

Fig. 30.1 Risk matrix proposed by the European Commission. (Source: European Commission [40])

function of hazard, exposure, vulnerability and capacity. Following the definition suggested in the Sendai Framework for Disaster Risk Reduction 2015–2030, the risk should compromise three elements [40]:

- *Hazard*: the adverse event causing the loss
- *Exposure*: the property, people, plant or environment that are threatened by the event
- *Vulnerability*: how the exposure at risk is vulnerable to an adverse event of that kind

The definition of disaster risk reflects the concept of hazardous events and disasters as the outcome of continuously present risk conditions. In the natural disaster management field, the notion of risk is difficult to clearly defined as there are serious constraints such as an inherent uncertainty, the availability of limited resources in impacted areas and dynamic changes in the environment [39, 40, 50, 54]. Hence, it is vital to properly prepare and protect society and the environment to understand the complicated notion of the risk. Risk assessment can be considered as the overall process of risk identification, analysis and evaluation leading to provide a framework to determine the effectiveness of disaster risk management, risk prevention and/or risk mitigation [40].

According to the UNISDR, *disaster risk assessment*[4] can be seen as *a qualitative or quantitative approach to determine the nature and extent of disaster risk by analysing potential hazards and evaluating existing conditions of exposure and vulnerability that together could harm people, property, services, livelihoods and the environment on which they depend.*

The above definitions incorporate the notion of knowledge implicitly. However, as knowledge is central to the risk assessment process, the definition of risk should be extended to make it explicit [8, 39, 54]. Namely, the above equation (Eq. 30.1) can be stated as follows:

$$\text{Risk} = f\left(\text{Hazard,Vulnerability,Knowledge}\right) \qquad (30.2)$$

Considering the knowledge dimension in the equation (Eq. 30.2), a three-dimensional risk matrix would be created, as depicted in Fig. 30.2. This formulation indicates that the outcomes of the risk assessment process are related explicitly with the underline background knowledge as

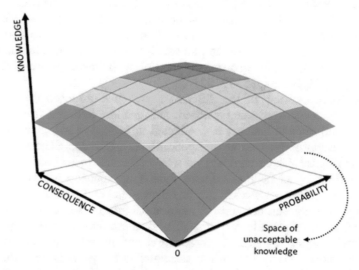

Fig. 30.2 Three-dimensional risk matrix integrate the dimension of the knowledge. (Source: Ref. [39])

---

[4] https://www.undrr.org/terminology/disaster-risk-assessment

well as the related assumptions made and parameter values assigned [54]. Extending the concept of risk assessment with the knowledge dimension requires the development of available knowledge models, which will represent and express the relevant uncertainties and use proper metrics (probabilities or others) to describe risks. Machine learning methodologies are suggested in this study as a possible option to tackle the challenges of the unpredictability and dynamically changing environment and deal with unknown and new events [39, 54].

30.2.2   Machine Learning in Disaster Management

The idea of using machine learning/deep Learning techniques in the field of disaster/crisis management is not new, as they have been used in image understanding, scene recognition and object detection [13, 14, 23, 25, 26, 28, 41, 45, 48, 51, 53]. Recently, machine/deep learning methodologies have been applied to monitor and detect hazardous natural disasters, such as floods, wildfires, etc. by analysing images and videos from UAVs, cameras or mobile devices as well as satellite images [2, 6, 12, 15, 17, 21, 22, 27, 29, 30, 33, 35, 44]. Furthermore, the analysis of visual and textual content receiving from social media posts has raised a lot of interest in the domain of disaster management as it allows crisis managers to discover useful, relevant information and knowledge about a catastrophic natural event, by analysing the growing volumes of public messages available [1, 9, 10, 16, 31, 32, 37, 38, 49, 52].

A detailed review of the different approaches proposed for natural disaster management using synthetic information retrieved from social media and satellite imagery is conducted in [43]. In [36], a systematic literature approach to report the application of artificial intelligence methods in disaster risk communication was conducted. In this review work, the authors conclude that research focuses on applying ML/DL techniques to achieve advances in two broad areas: (1) multi-parameter monitoring for early warning and (2) multimodal information extraction and classification to provide real-time situational awareness. State-of-the-art AI technologies applied in these areas are included in this study. Furthermore, in [50], the authors attempted to analyse the role of the big data in natural disaster risk management (DRM) and contact a review of the literature aiming to map the big data sources with the achievements in different disaster risk management phases and technological advances.

30.3 The Proposed Methodological Approach

In the framework of the EU Horizon 2020 funded project beAWARE,[5] a *crisis classification* component was designed and developed aiming to provide twofold functionalities in an innovative and beneficial manner for the authorities and crisis managers. This component consists of two main modules that cover the pre-emergency and emergency phases of the disaster management cycle, namely, the *Early Warning module* and the *Real-time Monitoring and Risk Assessment module*, described briefly in the following subsections.

30.3.1 High-Level Architecture of the Early Warning Module

The *Early Warning module* assesses the overall severity level of an impending crisis by fusing heterogeneous sensory data and forecasts from various resources. This goal performs in three main steps (Fig. 30.3), which are described briefly below:

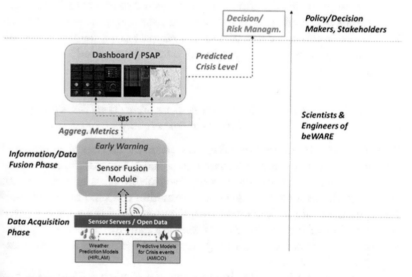

Fig. 30.3 High-Level Architecture of Early Warning module

[5] https://beaware-project.eu

- *Data acquisition*: Computational routines were implemented to allow understanding the database structures of the open data services of the Finnish Meteorological Institute (FMI), FROST-Server and geoServer, as well as of spreadsheets (XLSX file extension) containing monitoring data at daily or sub-daily intervals. Currently, the data acquisition tool fetches the sensory data from the OGC SensorThings API,[6] which follows the OGC standard.[7] However, it could be extended and customised to receive various types of sensing data following different protocols.
- *Sensor Fusion Module*: A rule-based approach has been developed to fuse the obtained data and estimate the overall crisis level in the region of interest. Data from multiple sensors are combined to extract the best estimate about the crisis level of smaller or larger regions of interest, based on the preset parameters. In [4] a detailed description of this approach has been exhibited.
- *Visualisation module*: Spatial data coming from different sources, in diverse formats, coordinate systems, and modalities are converted to a common representation and are projected to user friendly environments which allow users to investigate and assess the results interactively. In this context, topics for risk/impact maps, damage maps and early warning are produced and propagated for visual representation.

30.3.2   High-Level Architecture of the Real-Time Monitoring and Risk Assessment Module

The *Real-time Monitoring and Risk Assessment module* has been designed and developed to acquire and analyse data and information from heterogeneous sources to assess the severity level of the crisis in real time. In particular, this module leverages the information collected by citizens and first responders at the field along with data received from sensors located in the region of interest, to dynamically estimate the crisis severity level and positively impact on the emergency response at operational and strategic levels (Fig. 30.4).

Similarly, to the Early Warning module, the *Real-time Monitoring and Risk Assessment module* relies on a multilayer fusion approach that consists

---

[6] https://www.opengeospatial.org/standards/sensorthings
[7] https://www.opengeospatial.org/

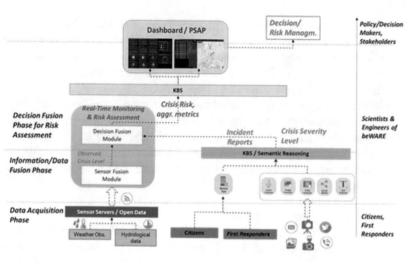

Fig. 30.4 High-Level Architecture of Real-Time Monitoring and Risk Assessment module

of a data/information fusion layer and a decision fusion layer. In the information/data fusion phase, the *Sensor Fusion Module* obtains the real-time sensory data and combines them by implementing a rule-based approach. The outcome is fed to the *Decision Fusion Module,* where the analysis results of the various beAWARE modules are combined to deliver an overall assessment of the severity of an on-going crisis. Specifically, the analysis of images and video data taken by drones and mobile or fixed cameras, voice messages and social media messages are forwarded to the *Decision Fusion Module* to be combined at a higher level and result in reliable re-estimates regarding the crisis level.

Recent work [3] has used a simplified approach to the above architecture based on the results of images analysed with techniques from the machine/deep learning field, in a fire scenario. The *Decision Fusion Module* was fuelled with spatial clustering capabilities to group the fire incidents and provide local-based estimations for the severity level of the on-going fire event.

A more complete version of the *Real-time Monitoring and Risk Assessment module* was demonstrated in the flood pilot in Vicenza, within the framework of the beAWARE[8] project. There, the *Sensor Fusion Module* obtained sensory data for the river water level as well as for the weather conditions in the region of interest. In the *Decision Fusion Module*, the aggregated sensory results were combined with the analysis of the incident reports, obtained from citizens and first responders, as well as with other multimedia analysis results. Historical information from flood risk maps was utilised in order to strengthen the decision-making process. The module dynamically assessed the on-going flood crisis level in the region of interest providing useful information to the crisis managers. The results of the aforementioned flood pilot use case are presented briefly in the following section.

30.3.2.1 Results from Flood Pilot
During the emergency phase, the *Real-Time Monitoring and Risk Assessment* component is activated, aiming to track and inform authorities and decision-makers regarding the evolution of the potential flood crisis event. Sensors have been installed in the specific locations at the region of Vicenza enabling to measure and convey in real-time weather (temperature, humidity, precipitation, etc.) and hydrological (e.g., river water level) observations (Fig. 30.5).

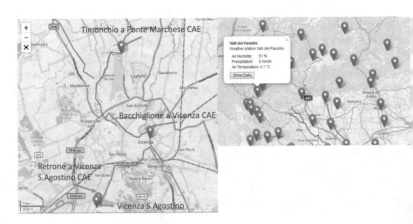

Fig. 30.5 Weather stations in Vicenza region in which sensors provide real-time observation of river water level and precipitation

[8] https://beaware-project.eu/

The analysis of acquired observations includes (a) the estimation of the flood crisis severity level employing a weighted average approach and (b) the creation and propagation of the appropriate messages to raise awareness of the civil protection authorities and crisis managers concerning the status of the river water level, precipitation level at specific weather stations and the ongoing flood severity level in the region of interest. The number of generated messages depends on the number of observations that exceeds predefined thresholds specified by authorities. Using various artificial datasets that can simulate different conditions, the *Real-Time Monitoring and Risk Assessment* component needs around 13.10 ± 2.08 seconds on average to generate approximately 10 messages for river water level and 12 messages for precipitation.

Another functionality of the *Real-Time Monitoring and Risk Assessment* component that it is worth to mention here is the process to assess dynamically the risk of ongoing flood crisis events that exploit information from citizens' mobile application. The following algorithm uses local information reported voluntarily by citizens to estimate the current flood crisis risk based on details given about people in danger, water level in adjacent buildings or other historical assets, etc. Briefly, the steps of the algorithm are the following:

Step 1. *Data Acquisition*: includes the processes to request the data related to the evolution of the flood crisis from citizens' mobile application via properly designed reports. Those data are retrieved from the beAWARE Knowledge Base Ontology.

Step 2. *Data Analysis*: estimates the risk assessment by the exploitation the receiving information from the citizens. It includes the calculations of hazard, exposure, vulnerability and, finally, the hydraulic risk and severity of each incident. The obtained information is enriched by linking with data extracted from risk maps provided by local authorities (AAWA) in ArcGIS files. Those files are stored in beAWARE's geoServer and relate to historical river water level observations, the exposure assets and their vulnerability in the Vicenza region as well as the severity level and risk estimations.

Step 3. *Store* the incident report and results of the analysis to the local database and *create* the appropriate messages to update the status of each incident displayed in the Public-safety answering point (PSAP).

Step 4. *Calculate* the accumulated *risk assessment* relying on the severity of all obtained incident reports.

Step 5. *Store* the results to the local database and *create* the appropriate messages to the dashboard in order to update the corresponding plots.

To evaluate the risk assessment algorithm, messages from various locations near the selected points of interest, such as hospitals (red polygon), other health care facilities (light brown polygons), public buildings (brown polygons) and places of relief (green polygons), are sent in the beAWARE system as shown in Fig. 30.6, other health care facilities (light brown polygons), public buildings (brown polygons) and places of relief (green polygons) are sent in the beAWARE system as shown in Fig. 30.6.

The distribution of the 50 incoming incident messages on the map is presented in Fig. 30.7. The majority of the incoming messages, 36% (18 out of 50), were categorised as severe by the risk assessment algorithm (Step 2), while the other 26% (13 out of 50) were rated as Extreme. Also, five messages (10%) were categorised as moderate by the algorithm and the rest (14, i.e. 28%) as of minor severity.

Risk analysis

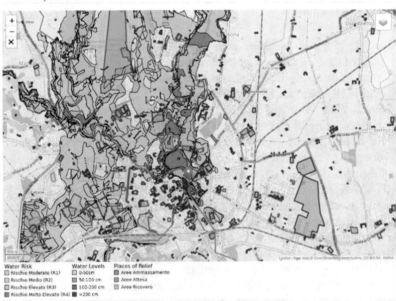

Fig. 30.6 Points of interest in Vicenza district. (Source: https://beaware.server. de/servlet/is/696/)

Fig. 30.7 Incoming incidents and their severity characterised by risk assessment algorithm

The most critical issue and significant challenge in the risk assessment algorithm is to handle incoming urgent messages from impacted citizens, directly and accurately. Thus, execution time is a critical metric that we want to evaluate in relation with the number of incoming messages. As expected, notable overhead in the runtime introduced when the risk assessment algorithm posts requests to the geoServer to link data from the risk maps (green bars in Fig. 30.8). It is worth noting that from those 30 cases, the 20 need to employ the geoServer services to estimate the hazard value and detect the exposure elements. However, the execution time does not exceed 3.2 seconds.

30.3.3 Enhance Crisis Classification Module with Machine Learning Techniques

The above proposed multi-level fusion approach for estimating the crisis severity level and risk assessment could be extended by incorporating state-of-the-art machine learning methodologies. This section will present some thoughts in this direction. The proposed integrated framework is illustrated in the Fig. 30.9 and consists of two main levels: the *Early*

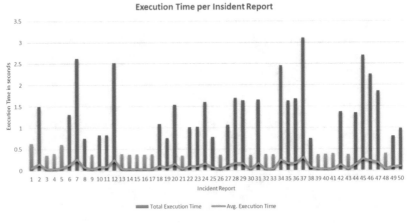

Fig. 30.8   Execution time (total and average) per incident report

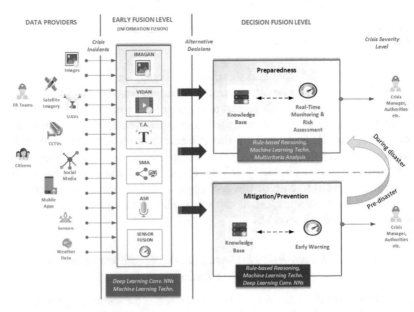

Fig. 30.9   Methodological schema for using ML/DL techniques in risk assessment process

Fusion Level and the *Decision Fusion Level*, aiming to synthesise data from different sources and adapt the latest machine learning methodologies, serving different phases of disaster risk management continuous process.

The heterogeneous data sources provide various kinds of data into the system, from sensors' measurements (real numbers) to videos from drones (UAVs) and satellite images, asynchronous and seamless. The obtained data indicate some specific crisis incidents captured from citizens or first responders or even from machines, such as sensors located in the field of interest, satellite imagery and UAVs during their flights. To extract useful and valuable information, the system should analyse and fuse the data from each modality and generate alternative decisions (modality decisions) regarding a crisis event, mostly in long-term mitigation/prevention (pre-crisis) and short-term preparation and prevention (i.e. "preparedness") phase. These alternatives should be further analysed, semantically enriched and combined in a higher level providing a set of potential decisions to crisis managers. The ultimate objective is to enable crisis managers and authorities to make actionable decisions, by effectively and efficiently handling an upcoming natural crisis event or during the early stages of a natural disaster.

In the beAWARE framework, the Early Fusion Level is considered as a collection of independent components whose purpose is to discover crisis events as emerged from raw data analysis. In general, such components can be integrated into the system to perform concept extraction from social media, audio, sensor data, images or videos. Depending on each component's design principles and capabilities, the set of extracted concepts can reveal critical insights regarding the extreme natural hazardous events, such as location, type of crisis event and number of people involved. At this level, the features extracted from the input data of various modalities and summarised as *visual*, *textual*, *audio* and *sensing* features usually have different representations. The analytical units, such as image analysis (IMAGAN), video analysis (VIDAN), text analysis (TA), social media analysis (SMA), automatic speech recognition (ASR) and Sensor Fusion Module, fuse heterogeneous features by employing machine learning methodologies (i.e. support vector machines, neural networks, Deep-Learning Convolution Neural Networks (CNNs), etc.) to create unified representations at the semantic level about the severity and the critical characteristics of a crisis event. Innovative machine learning techniques and deep learning neural networks could be included at the *Early Fusion Level*, to perform analysis task on the integrated multiple media and its associated features or the intermediate decisions and which is referred as *multimedia fusion* [5].

In the proposed schema, the *Decision Fusion Level* covers both the pre-disaster and the during disaster phases. In these phases, the development and deployment of innovative machine learning methodologies to fuse the analysis modules' outcomes intelligently play a significant role in the disaster risk management process.

Natural disasters are difficult to forecast because of the complexity of the physical phenomena and the variability of the parameters involved [50]. In the mitigation/prevention phase, machine learning methods can allow experts and crisis managers to identify timely geographical and infrastructure risks. Analysing satellite images and mapping the results to GIS maps assist in the long-term assessment and risk reduction phase. Additionally, the combination of the crowdsourcing and machine learning techniques to interpret social media and user-generated disaster data supports risk assessment processes. Likewise, exploiting predictive analytical results and combining them with other alternatives, using intelligent techniques, enable the crisis level estimation before the disaster occurs.

The enhancement of prediction capability becomes by using sensory data along with the integrating big earth observation data. Hence, the forecasting models better understand the underlying physical schemes providing valuable estimations regarding the weather and the characteristics of natural disasters, such as floods, wildfires, heatwaves, etc.

Effective detection and seamlessly monitoring are very helpful in improving the management of disasters. In the preparedness phase of a natural disaster, the analysis of remote sensing data, which is available in different spatial and temporal resolutions, along with real-time obtained data from sensors and user-generated data from social media posts, which contain multimedia and textual content, facilitates a better understanding of the context of location, timing, causes and impacts of natural disasters. Automatic extraction and discovery of novel, hidden knowledge from the acquired data relative to the disaster impacted areas facilitate the quick availability of the required information to the crisis managers and first responders to handle the ongoing crisis efficiently. Hence, the proposed schema encompasses machine learning techniques that enable the compound of the available alternatives generated from the analytical processes and result in the classification of the crisis event in terms of its severity level and estimation of its risk.

The general steps for the application of the above framework to the estimation of the crisis severity level and risk assessment employing machine learning techniques are the following [18]:

- *Define project's goals*: The targets of the machine learning techniques of each level of the above process (Early Fusion and Decision Fusion Level) in both phases of the disaster risk management process (pre-disaster and during a disaster) should be determined. The objective of the DRM project should be translated to the output variable that is targeted.
- *Data acquisition*: One of the significant steps in the implementation of ML techniques is the process of collecting the data. In the proposed framework, the data are generated from heterogeneous sources that have different specific characteristics. Thus, processes like the edit, modify, transform and harmonise of the obtained data assets should be adopted, before applying a machine learning technique. For instance, the analysis of satellite remote sensing imagery of many different kinds and resolutions has been already carried out by employing ML algorithms. Currently, work in the DRM sector often involves using high-resolution (sub-meter spatial accuracy to 10 m or so) panchromatic and multispectral imagery from satellites, drones, and airplanes, which produce a vast and varied volume of data streams. Thus, the appropriate resources for storing and preprocessing the obtained big data should be specified and implemented.
- *Training/validation data collection*: Labelled samples or reference data are required to train the model and validate the ML algorithm outputs. The collection of these labelled samples is often the most expensive part of ML projects.
- *Exploring datasets*: Exploratory data analysis is an important preprocessing step, as it assists in determining which machine learning algorithm and data that most suitable to apply. This analysis also clarifies which input variables are correlated with each other, what is most closely related to the output variable, the distributions of variables or whether a combination or transformation of the input variables should be carried out. Moreover, in this step, the outlier detection and removal processes should be adopted, as the existence of them

could skew the results dramatically by altering the variance in the data in disproportionate ways.

- *Choice of machine learning algorithm*: as the ML methodologies are data-driven, the choice of the appropriate algorithm depends on the size of the data set, the number of features and the available computational resources. The best way to decide is to analyse which algorithms have been used to tackle similar problems in the past. A common approach applies multiple models and selects the best performing one. However, it is important to understand and compare models that have been tuned optimally so that the comparison is actually assessing the model effectiveness and is not biased by the parametrisation behind it. In the proposed framework, the appropriate deep/machine learning methods should be chosen to serve the needs for analysis and fusion of the heterogeneous data in both levels of DRM process at precrisis and during the crisis phase.

- *Developing the code and training the ML algorithm*: the extensive application of ML techniques in many disciplines leads to the development of powerful algorithms in many programming languages and available in user interfaces. There are a number of readily available ML algorithms inside remote sensing and GIS software packages, some of which are free – like the GRASS GIS plug in for QGIS. In addition, any number of ML algorithms from open or proprietary libraries, such as TensorFlow[9] can be combined and customised to achieve any project's goals. On top of that, a number of customised ML services are available on cloud computing platforms like AWS, Azure and Google Cloud services. The training process of an ML model is very crucial and involves providing the learning algorithm (ML algorithm) along with the training data that it should be learned from. The term ML model refers to the model artefact that is created by the training process. The training data must contain a tuple of input attributes and the correct answer, known as a target or target attribute. The learning algorithm finds patterns in the training data that map the input data attributes to the target (class), and it outputs an ML model that captures these patterns. The training set is used to teach the model to distinguish the classes to be predicted by it. Each ML algorithm requires a number of model parameters to be determined. By evaluating trained model's accuracy

[9] https://www.tensorflow.org/

on the validation set, the different model parameter settings can be compared and finally would be chosen the best ones that suit the particular problem. The ultimate goal is to utilise the training model to get predictions on new data (testing set) for which you do not know the target.

- *Validation, reinforcement and re-running:* To attain the objectives of the training process, the ML model produces intermediate results that need to be validated in terms of accuracy. This is usually achieved by comparing the output data to a validation dataset that is considered the "truth", or accurate within a range acceptable for the project's goals.

- *Final data output:* Once training is complete, the generalisation ability of the ML model should be evaluated by the utilisation of the testing set. This set should not be employed during model development and is only used to check final model output accuracy. In some cases, the testing set is actually a new dataset, such as in a case where you want to apply a previously developed model to a new region. The final data output is achieved once the accuracy of the output dataset is deemed adequate for the goal of the project.

30.4   DISCUSSION

Despite the progress made in the utilisation of machine learning methodologies in risk assessment and general in all the phases of disaster management, there are still challenges that should be addressed. Machine learning techniques are data-driven approaches, meaning that they are heavily related to the data's availability. Especially in the phase of the training, relevant data need to be employed to build effective machine learning models that will predict the evolution of natural hazards successfully. Moreover, machine learning algorithms can combine various types of data. However, the reliability of the ML models adequately relates to the targeted output. Irrelevant data may incur additional costs without improving the model predictions and often lead to misleading results [18, 43, 50].

Additionally, the use of social media and satellite imagery in disaster analysis involves challenges of collecting, handling and processing a diversified set of multimedia content information obtained from heterogeneous social media platforms [43]. Hence, the relevance and authenticity of content shared via social media and its association with disaster detection can be considered one of the key research challenges nowadays, as researchers stated in [43]. Furthermore, the crisis managers should analyse the huge volume of received data and parameters in a short time to make decisions and actions for handling an evolving crisis event. Additionally, their work is hindered by the highly unstable and rapidly changing environment making it sometimes very difficult or even impossible to receive the correct decision.

In conclusion, in this study, we proposed a unified framework that fosters tailoring technological achievements in machine learning and enables the crisis managers and authorities to manage the pre-emergency and emergency phases of a hazardous natural event efficiently. Specifically, the federated platform compiles processes that are able to multi-level fuse data and information from heterogeneous sources. The detected events and predictions are collected and further analysed to assist crisis managers in efficiently making decisions. Furthermore, as pointed out in [42], uncertainties of multiple data sources are inherent in the data and propagate to disaster preparedness and response models, affecting the knowledge generation and decreasing the end-users' trust. Thus, sophisticated machine learning methods can be integrated into the proposed framework to deal with uncertainties, by enhancing human awareness concerning them, eliminating unrelated data, aiding in prediction analysis and identifying optimal response strategies towards to efficient decision-making. Additionally, the integration of machine learning methodologies in the aforementioned natural hazard risk reduction framework permits the dynamical assessment of vulnerability and exposure. Hence, it constitutes a response system that enables adaptation to environmental changes and different climate-induced hazards and risks. However, further work should be made to evaluate the proposed framework in real conditions of extreme natural events case as well as transboundary and multi-hazard crisis.

Acknowledgement

 This research is supported by the EC-funded research and innovation programs H2020 beAWARE: "Enhancing decision support and management services in extreme weather climate events" under the grant agreement No.700475 and aqua3S: "Enhancing standardisation strategies to integrate innovative technologies for Safety and Security in existing water networks" under the grant agreement No.832876.

BIBLIOGRAPHY

1. Ahmad, K., Pogorelov, K., Riegler, M., Conci, N., & Halvorsen, P. (2017). CNN and GAN based satellite and social media data fusion for disaster detection. In *MediaEval Workshop on Working Notes Proceedings*.
2. Ahmad, S., Ahmad, K., Ahmad, N., & Conci, N. (2017). Convolutional neural networks for disaster images retrieval. In *CEUR Workshop Proceedings*.
3. Antzoulatos, G., Giannakeris, P., Koulalis, I., Karakostas, A., Vrochidis, S., & Kompatsiaris, I. (2020). A multi-layer fusion approach for real-time fire severity assessment based on multimedia incidents. In A. L. Hughes, F. McNeill, & C. Zobel (Eds.), *17th international conference on Information Systems for Crisis Response and Management (ISCRAM'20) on proceedings* (pp. 75–89). Blacksburg, VA.
4. Antzoulatos, G., Karakostas, A., Vrochidis, S., & Kompatsiaris, I. (2019). The crisis classification component to strengthen the early warning, risk assessment and decision support in extreme climate events. In *4th International Conference on Dynamic of Disasters – DOD 2019 Proceedings*, Kalamata.
5. Atrey, P., Anwar, H., El Saddik, A., & Kankanhalli, S. (2010). Multimodal fusion for multimedia analysis: A survey. *Multimedia Systems, 16*(6), 345–379.
6. Avalhais, L., Rodrigues, J., & Traina, A. (2016). Fire detection on unconstrained videos using color-aware spatial modeling and motion ow. In *IEEE 28th International Conference on Tools with Artificial Intelligence (ICTAI)*, pp. 913–920.
7. Aven, T., & Cox, L. A., Jr. (2016). National and global risk studies: How can the field of risk analysis contribute? *Risk Analysis, 36*(2), 186–190. https://doi.org/10.1111/risa.12584.
8. Aven, T., & Renn, O. (2010). *Risk management* (pp. 121–158). Berlin, Heidelberg: Springer.
9. Avgerinakis, K., Moumtzidou, A., Andreadis, S., Michail, E., Gialampoukidis, I., Vrochidis, S., & Kompatsiaris, I. (2017). Visual and textual analysis of social media and satellite images for food detection@ multimedia satellite task mediaeval 2017. In *MediaEval Workshop 2017 on Multimedia Satellite Task*.
10. Bischke, B., Bhardwaj, P., Gautam, A., Helber, P., Borth, D., & Dengel, A. (2017). Detection of flooding events in social multimedia and satellite imagery using deep neural networks. In *MediaEval Workshop 2017 on Working Notes Proceedings*.
11. Blaikie, P., Cannon, T., Davis, I., & Wisner, B. (1994). *At risk: Natural hazards, people vulnerability and disasters* (1st ed.). London: Routledge.
12. Chaudhuri, N., & Bose, I. (2020). Exploring the role of deep neural networks for post-disaster decision support. *Decision Support Systems, 130*, 113234, Elsevier B.V.

13. Chen, L.-C., Papandreou, G., Kokkinos, I., Murphy, K., & Yuille, A. (2016). DeepLab: Semantic image segmentation with deep convolutional nets, atrous convolution, and fully connected CRFs. In *CoRR*, arXiv:1606.00915.

14. Chen, L.-C., Zhu, Y., Papandreou, G., Schroff, F., & Hartwig, A. (2018). Encoder-decoder with atrous separable convolution for semantic image segmentation. In *European Conference on Computer Vision (ECCV) on Proceedings*, pp. 801–818.

15. Chino, D., Avalhais, L., Rodrigues, J., & Traina, A. (2015). BoWFire: Detection of fire in still images by integrating pixel color and texture analysis. In *28th SIBGRAPI Conference on Graphics, Patterns and Images (SIBGRAPI)* (pp. 95–102). Piscataway: IEEE.

16. Dao, M. S., Minh, P. Q. N., Nguyen, D., & Tien, D. (2017). A domain-based late-fusion for disaster image retrieval from social media. In *MediaEval Workshop 2017 on Multimedia Satellite Task*.

17. Erdelj, M., Natalizio, E., Chowdhury, K. R., & Akyildiz, I. F. (2017). Help from the sky: Leveraging UAVs for disaster management. *IEEE Pervasive Computing, 16*(1), 24–32.

18. Global Facility for Disaster Reduction and Recovery (GFDRR). *Machine learning for disaster risk management*. https://www.gfdrr.org/sites/default/files/publication/181222_WorldBank_DisasterRiskManagement_Ebook_D6.pdf. Last accessed 17 Jan 2020.

19. Centre for Research on the Epidemiology of Disasters (CRED) – Emergency Events Database (EM-DAT). *Natural disasters 2018*. https://emdat.be/sites/default/files/adsr_2018.pdf. Last accessed 17 Jan 2020.

20. Prelipcean, G., & Mircea, B. (2011). Emerging applications of decision support systems (DSS) in crisis management. In J. Chiang (Ed.), *Efficient decision support systems – Practice and challenges in multidisciplinary domains*. IntechOpen.

21. Gebrehiwot, A., Hashemi-Beni, L., Thompson, G., Kordjamshidi, P., & Langan, T. (2019). Deep convolutional neural network for flood extent mapping using unmanned aerial vehicles data. *Sensors, 19*(7), 1486.

22. Giannakeris, P., Avgerinakis, K., Karakostas, A., Vrochidis, S., & Kompatsiaris, I. (2018). People and vehicles in danger – A fire and flood detection system in social media. In *2018 IEEE 13th Image, Video, and Multidimensional Signal Processing Workshop (IVMSP)* (pp. 1–5). Piscataway: IEEE.

23. He, K., Zhang, X., Ren, S., & Sun, J. (2016). Deep residual learning for image recognition. In *IEEE Conference on Computer Vision and Pattern Recognition on Proceedings*, pp. 770–778.

24. Hegde, J., & Rokseth, B. (2020). Applications of machine learning methods for engineering risk assessment – A review. *Safety Science, 122*, 104492. https://doi.org/10.1016/j.ssci.2019.09.015.

25. Huang, J., Rathod, V., Sun, C., Zhu, M., Korattikara, A., Fathi, A., Fischer, I., Wojna, Z., Song, Y., Guadarrama, S., & Murphy, K. (2017). Speed/accuracy trade-offs for modern convolutional object detectors. In *IEEE CVPR.*

26. Krizhevsky, A., Sutskever, I., & Hinton, G. E. (2012). ImageNet classification with deep convolutional neural networks. In *Advances in Neural Information Processing Systems*, pp. 1097–1105.

27. Kwak, Y. (2017). Nationwide flood monitoring for disaster risk reduction using multiple satellite data. In M. Konecny, & W. Kainz (Eds.). *ISPRS International Journal of Geo-Information*, *6*(7), 203.

28. Long, J., Shelhamer, E., & Darrell, T. (2015). Fully convolutional networks for semantic segmentation. In *IEEE Conference on Computer Vision and Pattern Recognition on Proceedings*, pp. 3431–3440.

29. Lopez-Fuentes, L., deWeijer, J., Bolanos, M., & Skinnemoen, H. (2017). Multi-modal deep learning approach for flood detection. In *MediaEval Workshop 2017 on Multimedia Satellite Task.*

30. Mettes, P., Tan, R., & Veltkamp, R. (2017). Water detection through spatio-temporal invariant descriptors. *Computer Vision and Image Understanding*, *154*, 182–191.

31. Mojaddadi, H., Pradhan, B., Nampak, H., Ahmad, N., & Halim bin Ghazali, A. (2017). Ensemble machine-learning-based geospatial approach for flood risk assessment using multi-sensor remote-sensing data and GIS. *Geomatics, Natural Hazards and Risk*, *8*(2), 1080–1102. https://doi.org/10.108 0/19475705.2017.1294113.

32. Moumtzidou, A., Andreadis, S., Gialampoukidis, I., Karakostas, A., Vrochidis, S., & Kompatsiaris, I. (2018). Flood relevance estimation from visual and textual content in social media streams. In *WWW '18 Companion Proceedings of the Web Conference 2018*, pp. 1621–1627. https://doi. org/10.1145/3184558.3191620.

33. Muhammad, K., Ahmad, J., & Wook Baik, S. (2017). Early fire detection using convolutional neural networks during surveillance for effective disaster management. *Neurocomputing, 288*, 30–42, Elsevier.

34. Newman, J., Maier, H., Riddell, G., Zecchin, A., Daniell, J., Schaefer, M., van Delden, H., Khazai, B., O'Flaherty, M., & Newland, C. (2017). Review of literature on decision support systems for natural hazard risk reduction: Current status and future research directions. *Environmental Modelling & Software, 96*, 378–409, Elsevier. https://doi.org/10.1016/j. envsoft.2017.06.042.

35. Nogueira, K., Fadel, S., Dourado, I., Werneck, R., Munoz, J., Penatti, O., Calumby, R., Li, L., Santos, J., & Torres, R. (2017). Data-driven flood detection using neural networks. In *MediaEval Workshop 2017 on Multimedia Satellite Task.*

36. Ogie, R., Rho, J., & Clarke, R. (2018). Artificial intelligence in disaster risk communication: A systematic literature review. In *5th International Conference on Information and Communication Technologies for Disaster Management (ICT-DM)*, pp. 1–8. https://doi.org/10.1109/ICT-DM.2018.8636380.

37. Ogie, R., Forehead, H., Clarke, R., & Perez, P. (2018). Participation patterns and reliability of human sensing in crowd-sourced disaster management. *Information Systems Frontiers, 20*(4), 713–728, Springer Nature. https://doi.org/10.1007/s10796-017-9790-y.

38. Opella, J., & Hernandez, A. (2019). Developing a flood risk assessment using support vector machine and convolutional neural network: A conceptual framework. In *IEEE 15th International Colloquium on Signal Processing Its Applications (CSPA)*, pp. 260–265.

39. Paltrinieri, N., Comfort, L., & Reniers, G. (2019). Learning about risk: Machine learning for risk assessment. *Safety Science, 118*, 475–486, Elsevier Ltd. https://doi.org/10.1016/j.ssci.2019.06.001.

40. Poljansek, K., Marin Ferrer, M., De Groeve, T., & Clark, I. (2017). *Science for Disaster Risk Management 2017: Knowing better and losing less.* Number EUR 28034. Publications Office of the European Union. https://doi.org/10.2788/842809.

41. Ren, S., He, K., Girshick, R., & Sun, J. (2015). Faster R-CNN: Towards real-time object detection with region proposal networks. In *Advances in Neural Information Processing Systems*, arXiv:1506.01497, pp. 91–99.

42. Sacha, D., Senaratne, H., Kwon, B. C., Ellis, G., & Keim, D. A. (2015). The role of uncertainty, awareness, and trust in visual analytics. *IEEE Transactions on Visualization and Computer Graphics, 22*, 240–249.

43. Said, N., Ahmad, K., Riegler, M., Pogorelov, K., Hassan, L., Ahmad, N., & Conci, N. (2019). Natural disasters detection in social media and satellite imagery: A survey. *Multimedia Tools and Applications, 78*(22), 31267–31302. https://doi.org/10.1007/s11042-019-07942-1.

44. Sharma, J., Granmo, O. C., Goodwin, M., Fidje, J. T. (2017). Deep Convolutional Neural Networks for Fire Detection in Images. In: Boracchi G., Iliadis L., Jayne C., Likas A. (eds) Engineering Applications of Neural Networks. EANN 2017. *Communications in Computer and Information Science, 744.* Springer, Cham. https://doi.org/10.1007/978-3-319-65172-9_16.

45. Simonyan, K., & Zisserman, A. (2014). Very deep convolutional networks for large-scale image recognition. In *CoRR*, arXiv:1409.1556.

46. UNISDR. (2015). *Making Development Sustainable: The Future of Disaster Risk Management. Global Assessment Report on Disaster Risk Reduction.* United Nations Office for Disaster Risk Reduction (UNISDR), Geneva, Switzerland.

47. UNISDR. (2015). *Sendai framework for disaster risk reduction 2015–2030.* https://www.undrr.org/publication/sendai-framework-disaster-risk-reduction-2015-2030.

48. Wen, L., Du, D., Cai, Z., Lei, Z., Chang, M.-C., Qi, H., Lim, J., Yang, M.-H., & Lyu, S. (2015). DETRAC: A new benchmark and protocol for multi-object tracking. In *CoRR*, abs/1511.04136.
49. Xiong, J., Li, J., Cheng, W., Wang, N., & Guo, L. (2019). A GIS-based support vector machine model for flash flood vulnerability assessment and mapping in China. *ISPRS International Journal of Geo-Information, 8*(7), 297.
50. Yu, M., Yang, C., & Li, Y. (2018). Big data in natural disaster management: A review. *Geosciences, 8*(5), 165. https://doi.org/10.3390/geosciences8050165.
51. Zhang, Z., Zhang, X., Peng, C., Xue, X., & Sun, J. (2018). ExFuse: Enhancing feature fusion for semantic segmentation. In *European Conference on Computer Vision (ECCV) on Proceedings*, pp. 269–284.
52. Zhao, Z., & Larson, M. (2017). Retrieving social flooding images based on multimodal information. In *MediaEval Workshop on Working Notes Proceedings*.
53. Zhou, B., Lapedriza, A., Xiao, J., Torralba, A., & Oliva, A. (2014). Learning deep features for scene recognition using places database. In *Advances in Neural Information Processing Systems*, pp. 487–495.
54. Zio, E. (2018). The future of risk assessment. *Reliability Engineering & System Safety, 177*, 176–190. https://doi.org/10.1016/j.ress.2018.04.020.

Correction to: FOLDOUT: A Through Foliage Surveillance System for Border Security

Christos Bolakis, Vasiliki Mantzana, Pantelis Michalis,
Aggelos Vassileiou, Roman Pflugfelder,
Martin Litzenberger, Michael Hubner, Gaetano Pastore,
Domenico Oricchio, Marie Desplas, Marie Ansart,
Maria Rosaria Santovito, Giulia Pica, Luis Patino,
James Ferryman, and Andreas Kriechbaum-Zabini

CORRECTION TO:
CHAPTER 16 IN: B. AKHGAR ET AL. (EDS.), *TECHNOLOGY DEVELOPMENT FOR SECURITY PRACTITIONERS*, SECURITY INFORMATICS AND LAW ENFORCEMENT,
HTTPS://DOI.ORG/10.1007/978-3-030-69460-9_16

The updated online version of the chapter can be found at
https://doi.org/10.1007/978-3-030-69460-9_16

C1

Chapter 16, "FOLDOUT: A Through Foliage Surveillance System for Border Security" was previously published non-open access. It has now been changed to open access under a CC BY 4.0 license and the copyright holder updated to 'The Author(s)'. The book has also been updated with this change.

Index

Printed in the United States
by Baker & Taylor Publisher Services